An Overview of M-theory

" A Unifying Model of Our Universe "

Edited by Paul F. Kisak

Contents

Chapter 1

Introduction to M-theory

In non-technical terms, **M-theory** presents an idea about the basic substance of the universe. So far no experimental evidence exists showing that M-theory is a description of the real world. Interest in this theory is mainly driven by mathematical elegance.

1.1 Background

In the early years of the 20th century, the atom – long believed to be the smallest building-block of matter – was proven to consist of even smaller components called protons, neutrons and electrons, which are known as subatomic particles. Starting in the 1960s, other subatomic particles began being discovered. In the 1970s, it was discovered that protons and neutrons (and other hadrons) are themselves made up of smaller particles called quarks. The Standard Model is the set of rules that describes the interactions of these particles.

In the 1980s, a new mathematical model of theoretical physics, called string theory, emerged. It showed how all the particles and all of the forms of energy in the universe could be constructed by hypothetical one-dimensional "strings", infinitesimal building-blocks that have only the dimension of length, but not height or width.

However, to make string theory mathematically consistent, the universe the strings exist in must have ten dimensions. This contradicts the experience that our real universe has four dimensions: three space dimensions (height, width, and length) and one time dimension. To "save" their theory, string theorists therefore added the explanation that the additional six dimensions exist but cannot be detected directly; this was explained by sophisticated mathematical objects called Calabi–Yau manifolds. The number of dimensions was later increased to 11 based on various interpretations of the 10-dimensional theory that led to five partial theo-ries, as described below. Supergravity theory also played a significant part in establishing the necessity of the 11th dimension.

These "strings" vibrate in multiple dimensions, and depending on how they vibrate, they might be seen in three-dimensional space as matter, light or gravity. It is the vibration of the string which determines whether it appears to be matter or energy, and every form of matter or energy is the result of the vibration of strings.

String theory, as mentioned above, ran into a problem: another version of the equations was discovered, then another, and then another. Eventually, five major string theories were developed. The main differences between the theories were principally the number of dimensions in which the strings developed, and their characteristics (some were open loops, some were closed loops, etc.). Furthermore, all these theories appeared to be workable. Scientists were not comfortable with five seemingly contradictory sets of equations to describe the same thing.

In 1994, Edward Witten of the Institute for Advanced Study and other researchers suggested that the five different versions of string theory might be describing the same thing seen from different perspectives. They proposed a unifying theory called "M-theory", in which the "M" is not specifically defined but is generally understood to stand for "membrane". The words "matrix", "master", "mother", "monster", "mystery" and "magic" have also been claimed. M-theory brought all of the string theories together. It did this by asserting that strings are really one-dimensional slices of a two-dimensional membrane vibrating in 11-dimensional space.

1.2 Status

M-theory is not complete, but the underlying structure of the mathematics has been established and is in agreement with all the string theories. Furthermore, it has passed many tests of internal mathematical consistency.

However, so far no experimental support of the M-theory exists. Some physicists are skeptical that this approach will ever lead to a physical theory describing our real world due to fundamental issues.[1]

Nevertheless, some cosmologists are drawn to M-theory because of its mathematical elegance and relative simplicity, triggering the hope that the simplicity is a reason why it may describe our world. Physicist and author Michio Kaku has remarked that M-theory may present us with a "Theory of Everything" which is so concise that its underlying formula would fit on a T-shirt.[2] Stephen Hawking originally believed that M-theory may be the ultimate theory but later suggested that the search for understanding of mathematics and physics will never be complete.[3] However, Hawking later changed his mind and stated, "M-theory is the *only* candidate for a complete theory of the universe."[4] Hawking and Leonard Mlodinow, in the popular-science book *The Grand Design*, take a philosophical position to support a view of the universe as a multiverse, and define it in the book as model-dependent realism which along with a sum-over-histories approach (see Path integral formulation of Quantum mechanics) to the universe as a whole, is used to claim that M-theory is the only candidate for a complete theory of the universe, mainly due to lack of viable alternatives.

1.3   See also

- Superstring theory

1.4   References

[1] Lee Smolin, April 2007: Response to review of The Trouble with Physics by Joe Polchinski

[2] Kaku, M. "M-Theory: The Mother of all SuperStrings". Retrieved 2013-08-06.

[3] Hawking, S. (3 October 2003). "Gödel and the end of physics". Retrieved 2013-08-06.

[4] Hawking, Stephen (2010). *The Grand Design*. Bantam Books. ISBN 055338466X.

1.5   Further reading

- Greene, B. (1999). *The Elegant Universe: Superstrings, Hidden Dimensions, and the Quest for the Ultimate Theory*. W.W. Norton. ISBN 0-375-70811-1.

- Greene, B. (2004). *The Fabric of the Cosmos: Space, Time, and the Texture of Reality*. Alfred A. Knopf. ISBN 0-375-41288-3.

- Miemic, A.; Schnakenburg, I. (2006). "Basics of M-theory". *Fortschritte der Physik* **54** (1): 5–72. arXiv:hep-th/0509137. Bibcode:2006ForPh..54....5M. doi:10.1002/prop.200510256.

- Musser, G. (2008). *The Complete Idiot's Guide to String Theory*. Alpha Books. ISBN 978-1-59257-702-6.

- Smolin, L. (2006). *The Trouble with Physics*. Houghton Mifflin. ISBN 978-0-618-55105-7.

- Woit, P. (2006). *Not Even Wrong: The Failure of String Theory and the Continuing Challenge to Unify the Laws of Physics*. Basic Books. ISBN 0-465-09275-6.

1.6   External links

- The Elegant Universe - A Three-Hour miniseries with Brian Greene by NOVA (original PBS Broadcast Dates: October 28, 8-10 p.m. and November 4, 8-9 p.m., 2003). Various images, texts, videos and animations explaining string theory and M-theory.

- Superstringtheory.com - The "Official String Theory Web Site", created by Patricia Schwarz. Excellent references on string theory and M-theory for the layperson and expert.

Chapter 2

History of string theory

The **history of string theory** spans several decades of intense research including two superstring revolutions. Through the combined efforts of many different researchers, string theory has developed into a broad and varied subject with connections to quantum gravity, particle and condensed matter physics, cosmology, and pure mathematics.

2.1 1943–1959: S-matrix

String theory is an outgrowth of a research program begun by Werner Heisenberg in 1943, picked up and advocated by many prominent theorists starting in the late 1950s and throughout the 1960s, which was discarded and marginalized in the 1970s to disappear by the 1980s. It was forgotten because a few of the ideas were deeply mistaken, because some of its mathematical methods were alien, and because quantum chromodynamics supplanted it as an approach to the strong interactions.

The program was called the S-matrix theory, and it was a radical rethinking of the foundation of physical law. By the 1940s it was clear that the proton and the neutron were not pointlike particles like the electron. Their magnetic moment differed greatly from that of a pointlike spin-1/2 charged particle, too much to attribute the difference to a small perturbation. Their interactions were so strong that they scattered like a small sphere, not like a point. Heisenberg proposed that the strongly interacting particles were in fact extended objects, and because there are difficulties of principle with extended relativistic particles, he proposed that the notion of a space-time point broke down at nuclear scales.

Without space and time, it is difficult to formulate a physical theory. Heisenberg believed that the solution to this problem is to focus on the observable quantities—those things measurable by experiments. An experiment only sees a microscopic quantity if it can be transferred by a series of events to the classical devices that surround the experimental chamber. The objects that fly to infinity are stable particles, in quantum superpositions of different momentum states.

Heisenberg proposed that even when space and time are unreliable, the notion of momentum state, which is defined far away from the experimental chamber, still works. The physical quantity he proposed as fundamental is the quantum mechanical amplitude for a group of incoming particles to turn into a group of outgoing particles, and he did not admit that there were any steps in between.

The S-matrix is the quantity that describes how a superposition of incoming particles turn into outgoing ones. Heisenberg proposed to study the S-matrix directly, without any assumptions about space-time structure. But when transitions from the far-past to the far-future occur in one step with no intermediate steps, it is difficult to calculate anything. In quantum field theory, the intermediate steps are the fluctuations of fields or equivalently the fluctuations of virtual particles. In this proposed S-matrix theory, there are no local quantities at all.

Heisenberg proposed to use unitarity to determine the S-matrix. In all conceivable situations, the sum of the squares of the amplitudes must be equal to 1. This property can determine the amplitude in a quantum field theory order by order in a perturbation series once the basic interactions are given, and in many quantum field theories the amplitudes grow too fast at high energies to make a unitary S-matrix. But without extra assumptions on the high-energy behavior unitarity is not enough to determine the scattering, and the proposal was ignored for many years.

Heisenberg's proposal was reinvigorated in the late 1950s when several theorists recognized that dispersion relations like those discovered by Hendrik Kramers and Ralph Kronig allow a notion of causality to be formulated, a notion that events in the future would not influence events in the past, even when the microscopic notion of past and future are not clearly defined. The dispersion relations were analytic properties of the S-matrix, and they were more stringent conditions than those that follow from unitarity alone.

Prominent advocates of this approach were Stanley Mandelstam and Geoffrey Chew. Mandelstam had discovered the double-dispersion relations, a new and powerful analytic form, in 1958, and believed that it would be the key to progress in the intractable strong interactions.

2.2 1959–1968: Regge theory and bootstrap models

By this time, many strongly interacting particles of ever higher spins had been discovered, and it became clear that they were not all fundamental. While Japanese physicist Shoichi Sakata proposed that the particles could be understood as bound states of just three of them—the proton, the neutron and the Lambda (see Sakata model), Geoffrey Chew believed that none of these particles are fundamental. Sakata's approach was reworked in the 1960s into the quark model by Murray Gell-Mann and George Zweig by making the charges of the hypothetical constituents fractional and rejecting the idea that they were observed particles. Chew's approach was then considered more mainstream because it did not introduce fractional charges and because it only focused on the experimentally measurable S-matrix elements, not on hypothetical pointlike constituents.

In 1959 Tullio Regge, a young theorist in Italy discovered that bound states in quantum mechanics can be organized into families with different angular momentum called Regge trajectories. This idea was generalized to relativistic quantum mechanics by Mandelstam, Vladimir Gribov and Marcel Froissart, using a mathematical method discovered decades earlier by Arnold Sommerfeld and Kenneth Marshall Watson.

In 1961 Geoffrey Chew and Steven Frautschi recognized that the mesons made Regge trajectories in straight lines, which implied, via Regge theory, that the scattering of these particles would have very strange behavior—it should fall off exponentially quickly at large angles. With this realization, theorists hoped to construct a theory of composite particles on Regge trajectories, whose scattering amplitudes had the asymptotic form demanded by Regge theory. Since the interactions fall off fast at large angles, the scattering theory would have to be somewhat holistic: Scattering off a pointlike constituent leads to large angular deviations at high energies.

2.3 1968–1974: dual resonance model

The first theory of this sort, the dual resonance model, was constructed by Gabriele Veneziano in 1968, who noted that the Euler Beta function could be used to describe 4-particle scattering amplitude data for particles on Regge trajectories. The Veneziano scattering amplitude was quickly generalized to an N-particle amplitude by Ziro Koba and Holger Bech Nielsen, and to what are now recognized as closed strings by Miguel Virasoro and Joel A. Shapiro. Dual resonance models for strong interactions were a popular subject of study 1968-1974.

2.4 1974–1984: superstring theory

In 1969–70, Yoichiro Nambu, Holger Bech Nielsen, and Leonard Susskind presented a physical interpretation of Euler's formula by representing nuclear forces as vibrating, one-dimensional strings. However, this string-based description of the strong force made many predictions that directly contradicted experimental findings. The scientific community lost interest in string theory as a theory of strong interactions in 1974 when quantum chromodynamics became the main focus of theoretical research.

In 1974 John H. Schwarz and Joel Scherk, and independently Tamiaki Yoneya, studied the boson-like patterns of string vibration and found that their properties exactly matched those of the graviton, the gravitational force's hypothetical "messenger" particle. Schwarz and Scherk argued that string theory had failed to catch on because physicists had underestimated its scope. This led to the development of bosonic string theory, which is still the version first taught to many students.

String theory is formulated in terms of the Polyakov action, which describes how strings move through space and time. Like springs, the strings want to contract to minimize their potential energy, but conservation of energy prevents them from disappearing, and instead they oscillate. By applying the ideas of quantum mechanics to strings it is possible to deduce the different vibrational modes of strings, and that each vibrational state appears to be a different particle. The mass of each particle, and the fashion with which it can interact, are determined by the way the string vibrates — in essence, by the "note" the string sounds. The scale of notes, each corresponding to a different kind of particle, is termed the "spectrum" of the theory.

Early models included both *open* strings, which have two distinct endpoints, and *closed* strings, where the endpoints are joined to make a complete loop. The two types of string behave in slightly different ways, yielding two spectra. Not all modern string theories use both types; some incorporate only the closed variety.

The earliest string model, which incorporated only bosons, has problems. Most importantly, the theory has a fundamental instability, believed to result in the decay of spacetime itself. Additionally, as the name implies, the spectrum of particles contains only bosons, particles like the photon that obey particular rules of behavior. While bosons are a critical ingredient of the Universe, they are not its only constituents. Investigating how a string theory may include fermions in its spectrum led to the invention of supersymmetry, a mathematical relation between bosons and fermions. String theories that include fermionic vibrations are now known as superstring theories; several different kinds have been described.

2.5 1984–1994: first superstring revolution

The first superstring revolution is a period of important discoveries roughly between 1984 and 1986. It was realised that string theory was capable of describing all elementary particles as well as the interactions between them. Hundreds of physicists started to work on string theory as the most promising idea to unify physical theories. The revolution was started by a discovery of anomaly cancellation in type I string theory via the Green–Schwarz mechanism in 1984. Several other ground-breaking discoveries, such as the heterotic string, were made in 1985. It was also realised in 1985 that to obtain $N = 1$ supersymmetry, the six small extra dimensions need to be compactified on a Calabi–Yau manifold.

Discover magazine in the November 1986 issue (vol 7, #11) featured a cover story written by Gary Taubes, "Everything's Now Tied to Strings", which explained string theory for a popular audience.

2.6 1994–2003: second superstring revolution

In the early 1990s, Edward Witten and others found strong evidence that the different superstring theories were different limits of a new 11-dimensional theory called M-theory.[1] These discoveries sparked the second superstring revolution that took place approximately between 1994 and 1997.

The different versions of superstring theory were unified, as long hoped, by new equivalences. These are known as S-duality, T-duality, U-duality, mirror symmetry, and conifold transitions. The different theories of strings were also connected to a new 11-dimensional theory called M-theory.

In the mid 1990s, Joseph Polchinski discovered that the theory requires the inclusion of higher-dimensional objects, called D-branes. These added an additional rich mathematical structure to the theory, and opened many possibilities for constructing realistic cosmological models in the theory.

In 1997 Juan Maldacena conjectured a relationship between string theory and a gauge theory called N = 4 supersymmetric Yang–Mills theory. This conjecture, called the AdS/CFT correspondence has generated a great deal of interest in the field and is now well accepted. It is a concrete realization of the holographic principle, which has far-reaching implications for black holes, locality and information in physics, and for the nature of the gravitational interaction.

2.7 2003–present

In 2003 the discovery of the string theory landscape, which suggests that string theory has a large number of inequivalent vacua, led to much discussion of what string theory might eventually be expected to predict, and how cosmology can be incorporated into the theory.[2]

2.8 Notes

[1] When Witten named it M-theory, he did not specify what the "M" stood for, presumably because he did not feel he had the right to name a theory he had not been able to fully describe. The "M" sometimes is said to stand for Mystery, or Magic, or Mother. More serious suggestions include Matrix or Membrane. Sheldon Glashow has noted that the "M" might be an upside down "W", standing for Witten. Others have suggested that the "M" in M-theory should stand for Missing, Monstrous or even Murky. According to Witten himself, as quoted in the PBS documentary based on Brian Greene's *The Elegant Universe*, the "M" in M-theory stands for "magic, mystery, or matrix according to taste."

[2] Rickles 2014, pp. 230–5 and 236 fn. 63.

2.9 References

- Dean Rickles (2014). *A Brief History of String Theory: From Dual Models to M-Theory*. Springer Science & Business Media. ISBN 978-3-642-45128-7.

2.10   Further reading

- Paul Frampton (1974).   *Dual Resonance Models*. Frontiers in Physics, W. A. Benjamin. ISBN 978-0-8053-2581-2.

- Shapiro, Joel A. (2007). "Reminiscence on the Birth of String Theory". arXiv:0711.3448.

- Andrea Cappelli; Elena Castellani; Filippo Colomo; Paolo Di Vecchia (2012). *The Birth of String Theory*. Cambridge University Press. ISBN 978-0-521-19790-8.

Chapter 3

String theory

For a more accessible and less technical introduction to this topic, see Introduction to M-theory.

In physics, **string theory** is a theoretical framework in which the point-like particles of particle physics are replaced by one-dimensional objects called strings. It describes how these strings propagate through space and interact with each other. On distance scales larger than the string scale, a string looks just like an ordinary particle, with its mass, charge, and other properties determined by the vibrational state of the string. In string theory, one of the many vibrational states of the string corresponds to the graviton, a quantum mechanical particle that carries gravitational force. Thus string theory is a theory of quantum gravity.

String theory is a broad and varied subject that attempts to address a number of deep questions of fundamental physics. String theory has been applied to a variety of problems in black hole physics, early universe cosmology, nuclear physics, and condensed matter physics, and it has stimulated a number of major developments in pure mathematics. Because string theory potentially provides a unified description of gravity and particle physics, it is a candidate for a theory of everything, a self-contained mathematical model that describes all fundamental forces and forms of matter. Despite much work on these problems, it is not known to what extent string theory describes the real world or how much freedom the theory allows to choose the details.

String theory was first studied in the late 1960s as a theory of the strong nuclear force, before being abandoned in favor of quantum chromodynamics. Subsequently, it was realized that the very properties that made string theory unsuitable as a theory of nuclear physics made it a promising candidate for a quantum theory of gravity. The earliest version of string theory, bosonic string theory, incorporated only the class of particles known as bosons. It later developed into superstring theory, which posits a connection called supersymmetry between bosons and the class of particles called fermions. Five consistent versions of super-

string theory were developed before it was conjectured in the mid-1990s that they were all different limiting cases of a single theory in eleven dimensions known as M-theory. In late 1997, theorists discovered an important relationship called the AdS/CFT correspondence, which relates string theory to another type of physical theory called a quantum field theory.

One of the challenges of string theory is that the full theory does not yet have a satisfactory definition in all circumstances. Another issue is that the theory is thought to describe an enormous landscape of possible universes, and this has complicated efforts to develop theories of particle physics based on string theory. These issues have led some in the community to criticize these approaches to physics and question the value of continued research on string theory unification.

3.1 Fundamentals

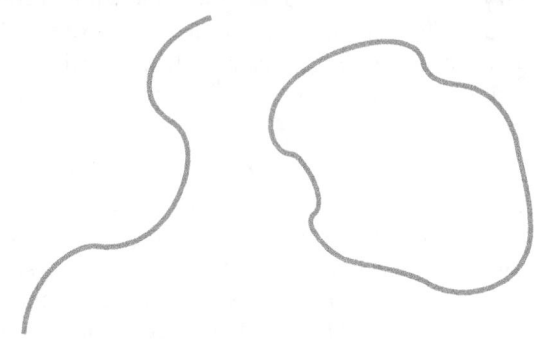

The fundamental objects of string theory are open and closed strings.

In the twentieth century, two theoretical frameworks emerged for formulating the laws of physics. One of these frameworks was Albert Einstein's general theory of relativity, a theory that explains the force of gravity and the structure of space and time. The other was quantum mechanics, a radically different formalism for describing physical phenomena using probability. By the late 1970s, these two frameworks had proven to be sufficient to explain most of the observed features of the universe, from elementary particles to atoms to the evolution of stars and the universe as a whole.[1]

In spite of these successes, there are still many problems that remain to be solved. One of the deepest problems in modern physics is the problem of quantum gravity.[1] The general theory of relativity is formulated within the framework of classical physics, whereas the other fundamental forces are described within the framework of quantum mechanics. A quantum theory of gravity is needed in order to reconcile general relativity with the principles of quantum mechanics, but difficulties arise when one attempts to apply the usual prescriptions of quantum theory to the force of gravity.[2] In addition to the problem of developing a consistent theory of quantum gravity, there are many other fundamental problems in the physics of atomic nuclei, black holes, and the early universe.[lower-alpha 1]

String theory is a theoretical framework that attempts to address these questions and many others. The starting point for string theory is the idea that the point-like particles of particle physics can also be modeled as one-dimensional objects called strings. String theory describes how strings propagate through space and interact with each other. In a given version of string theory, there is only one kind of string, which may look like a small loop or segment of ordinary string, and it can vibrate in different ways. On distance scales larger than the string scale, a string will look just like an ordinary particle, with its mass, charge, and other properties determined by the vibrational state of the string. In this way, all of the different elementary particles may be viewed as vibrating strings. In string theory, one of the vibrational states of the string gives rise to the graviton, a quantum mechanical particle that carries gravitational force. Thus string theory is a theory of quantum gravity.[3]

One of the main developments of the past several decades in string theory was the discovery of certain "dualities", mathematical transformations that identify one physical theory with another. Physicists studying string theory have discovered a number of these dualities between different versions of string theory, and this has led to the conjecture that all consistent versions of string theory are subsumed in a single framework known as M-theory.[4]

Studies of string theory have also yielded a number of results on the nature of black holes and the gravitational interaction. There are certain paradoxes that arise when one attempts to understand the quantum aspects of black holes, and work on string theory has attempted to clarify these issues. In late 1997 this line of work culminated in the discovery of the anti-de Sitter/conformal field theory correspondence or AdS/CFT.[5] This is a theoretical result which relates string theory to other physical theories which are better understood theoretically. The AdS/CFT correspondence has implications for the study of black holes and quantum gravity, and it has been applied to other subjects, including nuclear[6] and condensed matter physics.[7][8]

Since string theory incorporates all of the fundamental interactions, including gravity, many physicists hope that it fully describes our universe, making it a theory of everything. One of the goals of current research in string theory is to find a solution of the theory that reproduces the observed spectrum of elementary particles, with a small cosmological constant, containing dark matter and a plausible mechanism for cosmic inflation. While there has been progress toward these goals, it is not known to what extent string theory describes the real world or how much freedom the theory allows to choose the details.[9]

One of the challenges of string theory is that the full theory does not yet have a satisfactory definition in all circumstances. The scattering of strings is most straightforwardly defined using the techniques of perturbation theory, but it is not known in general how to define string theory nonperturbatively.[10] It is also not clear whether there is any principle by which string theory selects its vacuum state, the physical state that determines the properties of our universe.[11] These problems have led some in the community to criticize these approaches to the unification of physics and question the value of continued research on these problems.[12]

3.1.1 Strings

Main article: String (physics)
The application of quantum mechanics to physical objects

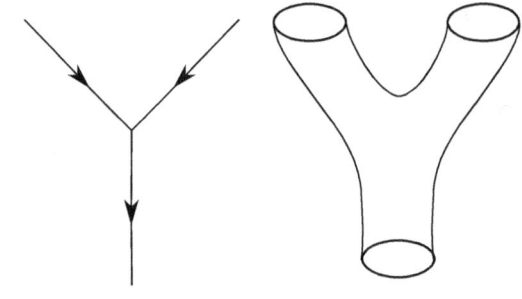

Interaction in the quantum world: worldlines of point-like particles or a worldsheet swept up by closed strings in string theory.

such as the electromagnetic field, which are extended in

space and time, is known as quantum field theory. In particle physics, quantum field theories form the basis for our understanding of elementary particles, which are modeled as excitations in the fundamental fields.[13]

In quantum field theory, one typically computes the probabilities of various physical events using the techniques of perturbation theory. Developed by Richard Feynman and others in the first half of the twentieth century, perturbative quantum field theory uses special diagrams called Feynman diagrams to organize computations. One imagines that these diagrams depict the paths of point-like particles and their interactions.[13]

The starting point for string theory is the idea that the point-like particles of quantum field theory can also be modeled as one-dimensional objects called strings.[14] The interaction of strings is most straightforwardly defined by generalizing the perturbation theory used in ordinary quantum field theory. At the level of Feynman diagrams, this means replacing the one-dimensional diagram representing the path of a point particle by a two-dimensional surface representing the motion of a string.[15] Unlike in quantum field theory, string theory does not yet have a full non-perturbative definition, so many of the theoretical questions that physicists would like to answer remain out of reach.[16]

In theories of particle physics based on string theory, the characteristic length scale of strings is assumed to be on the order of the Planck length, or 10^{-35} meters, the scale at which the effects of quantum gravity are believed to become significant.[15] On much larger length scales, such as the scales visible in physics laboratories, such objects would be indistinguishable from zero-dimensional point particles, and the vibrational state of the string would determine the type of particle. One of the vibrational states of a string corresponds to the graviton, a quantum mechanical particle that carries the gravitational force.[3]

The original version of string theory was bosonic string theory, but this version described only bosons, a class of particles which transmit forces between the matter particles, or fermions. Bosonic string theory was eventually superseded by theories called superstring theories. These theories describe both bosons and fermions, and they incorporate a theoretical idea called supersymmetry. This is a mathematical relation that exists in certain physical theories between the bosons and fermions. In theories with supersymmetry, each boson has a counterpart which is a fermion, and vice versa.[17]

There are several versions of superstring theory: type I, type IIA, type IIB, and two flavors of heterotic string theory ($SO(32)$ and $E_8 \times E_8$). The different theories allow different types of strings, and the particles that arise at low energies exhibit different symmetries. For example, the type I theory includes both open strings (which are segments with

endpoints) and closed strings (which form closed loops), while types IIA and IIB include only closed strings.[18]

3.1.2 Extra dimensions

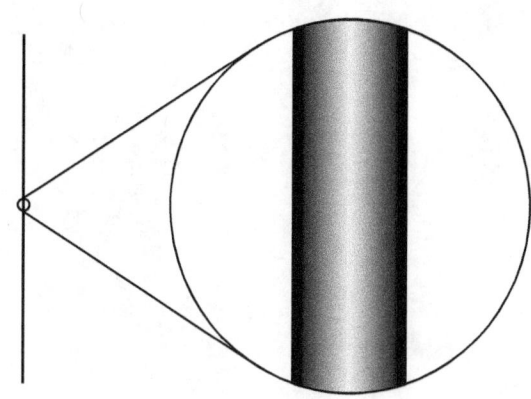

An example of compactification: At large distances, a two dimensional surface with one circular dimension looks one-dimensional.

In everyday life, there are three familiar dimensions of space: height, width and length. Einstein's general theory of relativity treats time as a dimension on par with the three spatial dimensions; in general relativity, space and time are not modeled as separate entities but are instead unified to a four-dimensional spacetime. In this framework, the phenomenon of gravity is viewed as a consequence of the geometry of spacetime.[19]

In spite of the fact that the universe is well described by four-dimensional spacetime, there are several reasons why physicists consider theories in other dimensions. In some cases, by modeling spacetime in a different number of dimensions, a theory becomes more mathematically tractable, and one can perform calculations and gain general insights more easily.[lower-alpha 2] There are also situations where theories in two or three spacetime dimensions are useful for describing phenomena in condensed matter physics.[20] Finally, there exist scenarios in which there could actually be more than four dimensions of spacetime which have nonetheless managed to escape detection.[21]

One notable feature of string theories is that these theories require extra dimensions of spacetime for their mathematical consistency. In bosonic string theory, spacetime is 26-dimensional, while in superstring theory it is ten-dimensional. In order to describe real physical phenomena using string theory, one must therefore imagine scenarios in which these extra dimensions would not be observed in experiments.[22]

Compactification is one way of modifying the number of dimensions in a physical theory. In compactification, some

A cross section of a quintic Calabi–Yau manifold

of the extra dimensions are assumed to "close up" on themselves to form circles.[23] In the limit where these curled up dimensions become very small, one obtains a theory in which spacetime has effectively a lower number of dimensions. A standard analogy for this is to consider a multidimensional object such as a garden hose. If the hose is viewed from a sufficient distance, it appears to have only one dimension, its length. However, as one approaches the hose, one discovers that it contains a second dimension, its circumference. Thus, an ant crawling on the surface of the hose would move in two dimensions.[24]

Compactification can be used to construct models in which spacetime is effectively four-dimensional. However, not every way of compactifying the extra dimensions produces a model with the right properties to describe nature. In a viable model of particle physics, the compact extra dimensions must be shaped like a Calabi–Yau manifold.[23] A Calabi–Yau manifold is a special space which is typically taken to be six-dimensional in applications to string theory. It is named after mathematicians Eugenio Calabi and Shing-Tung Yau.[25]

Another approach to reducing the number of dimensions is the so-called brane-world scenario. In this approach, physicists assume that the observable universe is a four-dimensional subspace of a higher dimensional space. In such models, the force-carrying bosons of particle physics arise from open strings with endpoints attached to the four-dimensional subspace, while gravity arises from closed strings propagating through the larger ambient space. This idea plays an important role in attempts to develop models

of real world physics based on string theory, and it provides a natural explanation for the weakness of gravity compared to the other fundamental forces.[26]

3.1.3 Dualities

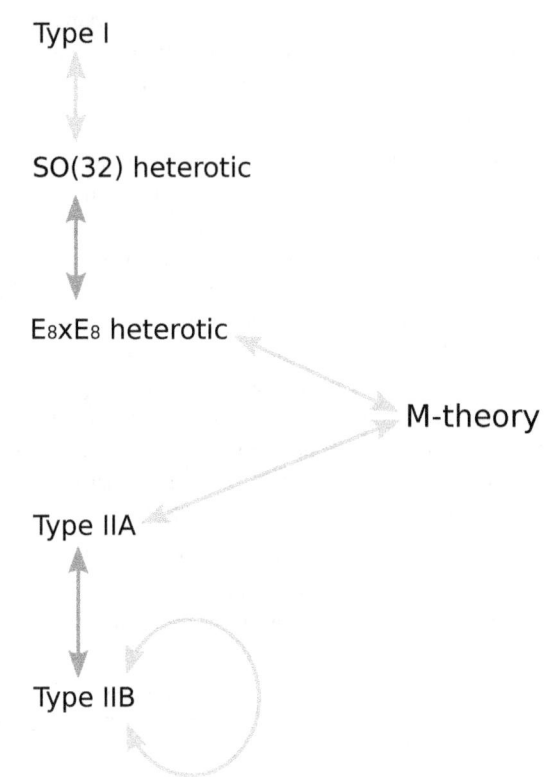

A diagram of string theory dualities. Yellow arrows indicate S-duality. Blue arrows indicate T-duality.

Main articles: S-duality and T-duality

One notable fact about string theory is that the different versions of the theory all turn out to be related in highly nontrivial ways. One of the relationships that can exist between different string theories is called S-duality. This is a relationship which says that a collection of strongly interacting particles in one theory can, in some cases, be viewed as a collection of weakly interacting particles in a completely different theory. Roughly speaking, a collection of particles is said to be strongly interacting if they combine and decay often and weakly interacting if they do so infrequently. Type I string theory turns out to be equivalent by S-duality to the *SO*(32) heterotic string theory. Similarly, type IIB string theory is related to itself in a nontrivial way by S-duality.[27]

Another relationship between different string theories is T-duality. Here one considers strings propagating around a circular extra dimension. T-duality states that a string propagating around a circle of radius R is equivalent to a string propagating around a circle of radius $1/R$ in the sense that all observable quantities in one description are identified with quantities in the dual description. For example, a string has momentum as it propagates around a circle, and it can also wind around the circle one or more times. The number of times the string winds around a circle is called the winding number. If a string has momentum p and winding number n in one description, it will have momentum n and winding number p in the dual description. For example, type IIA string theory is equivalent to type IIB string theory via T-duality, and the two versions of heterotic string theory are also related by T-duality.[27]

In general, the term *duality* refers to a situation where two seemingly different physical systems turn out to be equivalent in a nontrivial way. Two theories related by a duality need not be string theories. For example, Montonen–Olive duality is example of an S-duality relationship between quantum field theories. The AdS/CFT correspondence is example of a duality which relates string theory to a quantum field theory. If two theories are related by a duality, it means that one theory can be transformed in some way so that it ends up looking just like the other theory. The two theories are then said to be *dual* to one another under the transformation. Put differently, the two theories are mathematically different descriptions of the same phenomena.[28]

3.1.4 Branes

Main article: Brane
 In string theory and related theories, a brane is a physi-

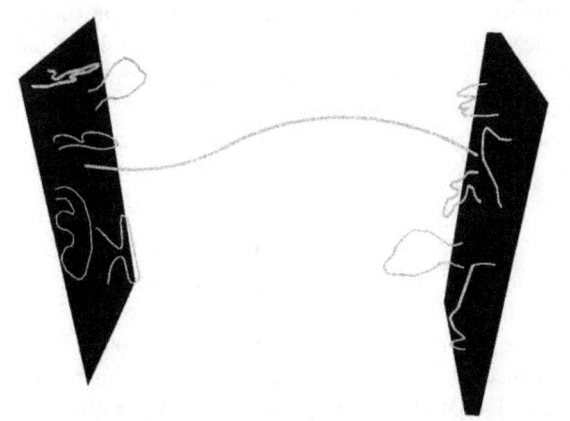

Open strings attached to a pair of D-branes

cal object that generalizes the notion of a point particle to higher dimensions. For example, a point particle can be

viewed as a brane of dimension zero, while a string can be viewed as a brane of dimension one. It is also possible to consider higher-dimensional branes. In dimension p, these are called p-branes. The word brane comes from the word "membrane" which refers to a two-dimensional brane.[29]

Branes are dynamical objects which can propagate through spacetime according to the rules of quantum mechanics. They have mass and can have other attributes such as charge. A p-brane sweeps out a $(p+1)$-dimensional volume in spacetime called its *worldvolume*. Physicists often study fields analogous to the electromagnetic field which live on the worldvolume of a brane.[29]

In string theory, D-branes are an important class of branes that arise when one considers open strings. As an open string propagates through spacetime, its endpoints are required to lie on a D-brane. The letter "D" in D-brane refers to a certain mathematical condition on the system known as the Dirichlet boundary condition. The study of D-branes in string theory has led to important results such as the AdS/CFT correspondence, which has shed light on many problems in quantum field theory.[30]

Branes are also frequently studied from a purely mathematical point of view. Mathematically, branes can be described as objects of certain categories, such as the derived category of coherent sheaves on a complex algebraic variety, or the Fukaya category of a symplectic manifold.[31] The connection between the physical notion of a brane and the mathematical notion of a category has led to important mathematical insights in the fields of algebraic and symplectic geometry[32] and representation theory.[33]

3.2 M-theory

Main article: M-theory

Prior to 1995, theorists believed that there were five consistent versions of superstring theory (type I, type IIA, type IIB, and two versions of heterotic string theory). This understanding changed in 1995 when Edward Witten suggested that the five theories were just special limiting cases of an eleven-dimensional theory called M-theory. Witten's conjecture was based on the work of a number of other physicists, including Ashoke Sen, Chris Hull, Paul Townsend, and Michael Duff. His announcement led to a flurry of research activity now known as the second superstring revolution.[34]

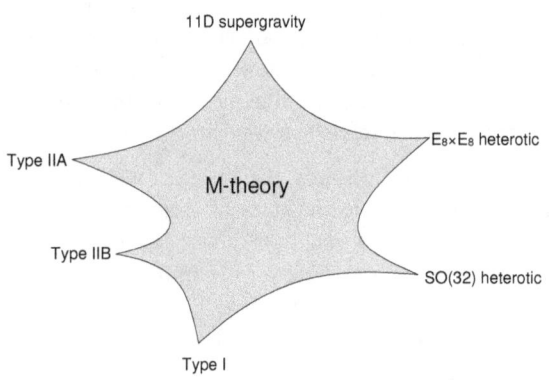

A schematic illustration of the relationship between M-theory, the five superstring theories, and eleven-dimensional supergravity. The shaded region represents a family of different physical scenarios that are possible in M-theory. In certain limiting cases corresponding to the cusps, it is natural to describe the physics using one of the six theories labeled there.

3.2.1 Unification of superstring theories

In the 1970s, many physicists became interested in supergravity theories, which combine general relativity with supersymmetry. Whereas general relativity makes sense in any number of dimensions, supergravity places an upper limit on the number of dimensions.[35] In 1978, work by Werner Nahm showed that the maximum spacetime dimension in which one can formulate a consistent supersymmetric theory is eleven.[36] In the same year, Eugene Cremmer, Bernard Julia, and Joel Scherk of the École Normale Supérieure showed that supergravity not only permits up to eleven dimensions but is in fact most elegant in this maximal number of dimensions.[37][38]

Initially, many physicists hoped that by compactifying eleven-dimensional supergravity, it might be possible to construct realistic models of our four-dimensional world. The hope was that such models would provide a unified description of the four fundamental forces of nature: electromagnetism, the strong and weak nuclear forces, and gravity. Interest in eleven-dimensional supergravity soon waned as various flaws in this scheme were discovered. One of the problems was that the laws of physics appear to distinguish between clockwise and counterclockwise, a phenomenon known as chirality. Edward Witten and others observed this chirality property cannot be readily derived by compactifying from eleven dimensions.[38]

In the first superstring revolution in 1984, many physicists turned to string theory as a unified theory of particle physics and quantum gravity. Unlike supergravity theory, string theory was able to accommodate the chirality of the standard model, and it provided a theory of gravity consistent with quantum effects.[38] Another feature of string theory that many physicists were drawn to in the 1980s and 1990s was its high degree of uniqueness. In ordinary particle theories, one can consider any collection of elementary particles whose classical behavior is described by an arbitrary Lagrangian. In string theory, the possibilities are much more constrained: by the 1990s, physicists had argued that there were only five consistent supersymmetric versions of the theory.[38]

Although there were only a handful of consistent superstring theories, it remained a mystery why there was not just one consistent formulation.[38] However, as physicists began to examine string theory more closely, they realized that these theories are related in intricate and nontrivial ways. They found that a system of strongly interacting strings can, in some cases, be viewed as a system of weakly interacting strings. This phenomenon is known as S-duality. It was studied by Ashoke Sen in the context of heterotic strings in four dimensions[39][40] and by Chris Hull and Paul Townsend in the context of the type IIB theory.[41] Theorists also found that different string theories may be related by T-duality. This duality implies that strings propagating on completely different spacetime geometries may be physically equivalent.[42]

At around the same time, as many physicists were studying the properties of strings, a small group of physicists was examining the possible applications of higher dimensional objects. In 1987, Eric Bergshoeff, Ergin Sezgin, and Paul Townsend showed that eleven-dimensional supergravity includes two-dimensional branes.[43] Intuitively, these objects look like sheets or membranes propagating through the eleven-dimensional spacetime. Shortly after this discovery, Michael Duff, Paul Howe, Takeo Inami, and Kellogg Stelle considered a particular compactification of eleven-dimensional supergravity with one of the dimensions curled up into a circle.[44] In this setting, one can imagine the membrane wrapping around the circular dimension. If the radius of the circle is sufficiently small, then this membrane looks just like a string in ten-dimensional spacetime. In fact, Duff and his collaborators showed that this construction reproduces exactly the strings appearing in type IIA superstring theory.[45]

Speaking at a string theory conference in 1995, Edward Witten made the surprising suggestion that all five superstring theories were in fact just different limiting cases of a single theory in eleven spacetime dimensions. Witten's announcement drew together all of the previous results on S- and T-duality and the appearance of higher dimensional branes in string theory.[46] In the months following Witten's announcement, hundreds of new papers appeared on the Internet confirming different parts of his proposal.[47] Today this flurry of work is known as the second superstring revolution.[48]

Initially, some physicists suggested that the new theory was a fundamental theory of membranes, but Witten was skeptical of the role of membranes in the theory. In a paper from 1996, Hořava and Witten wrote "As it has been proposed that the eleven-dimensional theory is a supermembrane theory but there are some reasons to doubt that interpretation, we will non-committally call it the M-theory, leaving to the future the relation of M to membranes."[49] In the absence of an understanding of the true meaning and structure of M-theory, Witten has suggested that the *M* should stand for "magic", "mystery", or "membrane" according to taste, and the true meaning of the title should be decided when a more fundamental formulation of the theory is known.[50]

3.2.2 Matrix theory

Main article: Matrix theory (physics)

In mathematics, a matrix is a rectangular array of numbers or other data. In physics, a matrix model is a particular kind of physical theory whose mathematical formulation involves the notion of a matrix in an important way. A matrix model describes the behavior of a set of matrices within the framework of quantum mechanics.[51]

One important example of a matrix model is the BFSS matrix model proposed by Tom Banks, Willy Fischler, Stephen Shenker, and Leonard Susskind in 1997. This theory describes the behavior of a set of nine large matrices. In their original paper, these authors showed, among other things, that the low energy limit of this matrix model is described by eleven-dimensional supergravity. These calculations led them to propose that the BFSS matrix model is exactly equivalent to M-theory. The BFSS matrix model can therefore be used as a prototype for a correct formulation of M-theory and a tool for investigating the properties of M-theory in a relatively simple setting.[51]

The development of the matrix model formulation of M-theory has led physicists to consider various connections between string theory and a branch of mathematics called noncommutative geometry. This subject is a generalization of ordinary geometry in which mathematicians define new geometric notions using tools from noncommutative algebra.[52] In a paper from 1998, Alain Connes, Michael R. Douglas, and Albert Schwarz showed that some aspects of matrix models and M-theory are described by a noncommutative quantum field theory, a special kind of physical theory in which spacetime is described mathematically using noncommutative geometry.[53] This established a link between matrix models and M-theory on the one hand, and noncommutative geometry on the other hand. It quickly led to the discovery of other important links between noncommutative geometry and various physical theories.[54][55]

3.3 Black holes

In general relativity, a black hole is defined as a region of spacetime in which the gravitational field is so strong that no particle or radiation can escape. In the currently accepted models of stellar evolution, black holes are thought to arise when massive stars undergo gravitational collapse, and many galaxies are thought to contain supermassive black holes at their centers. Black holes are also important for theoretical reasons, as they present profound challenges for theorists attempting to understand the quantum aspects of gravity. String theory has proved to be an important tool for investigating the theoretical properties of black holes because it provides a framework in which theorists can study their thermodynamics.[56]

3.3.1 Bekenstein–Hawking formula

In the branch of physics called statistical mechanics, entropy is a measure of the randomness or disorder of a physical system. This concept was studied in the 1870s by the Austrian physicist Ludwig Boltzmann, who showed that the thermodynamic properties of a gas could be derived from the combined properties of its many constituent molecules. Boltzmann argued that by averaging the behaviors of all the different molecules in a gas, one can understand macroscopic properties such as volume, temperature, and pressure. In addition, this perspective led him to give a precise definition of entropy as the natural logarithm of the number of different states of the molecules (also called *microstates*) that give rise to the same macroscopic features.[57]

In the twentieth century, physicists began to apply the same concepts to black holes. In most systems such as gases, the entropy scales with the volume. In the 1970s, the physicist Jacob Bekenstein suggested that the entropy of a black hole is instead proportional to the *surface area* of its event horizon, the boundary beyond which matter and radiation is lost to its gravitational attraction.[58] When combined with ideas of the physicist Stephen Hawking,[59] Bekenstein's work yielded a precise formula for the entropy of a black hole. The formula expresses the entropy S as

$$S = \frac{c^3 k A}{4 \hbar G}$$

where c is the speed of light, k is Boltzmann's constant, $\hbar$ is the reduced Planck constant, G is Newton's constant, and A is the surface area of the event horizon.[60]

Like any physical system, a black hole has an entropy defined in terms of the number of different microstates that lead to the same macroscopic features. The Bekenstein–Hawking entropy formula gives the expected value of the entropy of a black hole, but by the 1990s, physicists still lacked a derivation of this formula by counting microstates in a theory of quantum gravity. Finding such a derivation of this formula was considered an important test of the viability of any theory of quantum gravity such as string theory.[61]

3.3.2 Derivation within string theory

In a paper from 1996, Andrew Strominger and Cumrun Vafa showed how to derive the Beckenstein–Hawking formula for certain black holes in string theory.[62] Their calculation was based on the observation that D-branes—which look like fluctuating membranes when they are weakly interacting—become dense, massive objects with event horizons when the interactions are strong. In other words, a system of strongly interacting D-branes in string theory is indistinguishable from a black hole. Strominger and Vafa analyzed such D-brane systems and calculated the number of different ways of placing D-branes in spacetime so that their combined mass and charge is equal to a given mass and charge for the resulting black hole. Their calculation reproduced the Bekenstein–Hawking formula exactly, including the factor of 1/4.[63] Subsequent work by Strominger, Vafa, and others refined the original calculations and gave the precise values of the "quantum corrections" needed to describe very small black holes.[64][65]

The black holes that Strominger and Vafa considered in their original work were quite different from real astrophysical black holes. One difference was that Strominger and Vafa considered only extremal black holes in order to make the calculation tractable. These are defined as black holes with the lowest possible mass compatible with a given charge.[66] Strominger and Vafa also restricted attention to black holes in five-dimensional spacetime with unphysical supersymmetry.[67]

Although it was originally developed in this very particular and physically unrealistic context in string theory, the entropy calculation of Strominger and Vafa has led to a qualitative understanding of how black hole entropy can be accounted for in any theory of quantum gravity. Indeed, in 1998, Strominger argued that the original result could be generalized to an arbitrary consistent theory of quantum gravity without relying on strings or supersymmetry.[68] In collaboration with several other authors in 2010, he showed that some results on black hole entropy could be extended to non-extremal astrophysical black holes.[69][70]

3.4 AdS/CFT correspondence

Main article: AdS/CFT correspondence

One approach to formulating string theory and studying its properties is provided by the anti-de Sitter/conformal field theory (AdS/CFT) correspondence. This is a theoretical result which implies that string theory is in some cases equivalent to a quantum field theory. In addition to providing insights into the mathematical structure of string theory, the AdS/CFT correspondence has shed light on many aspects of quantum field theory in regimes where traditional calculational techniques are ineffective.[6] The AdS/CFT correspondence was first proposed by Juan Maldacena in late 1997.[71] Important aspects of the correspondence were elaborated in articles by Steven Gubser, Igor Klebanov, and Alexander Markovich Polyakov,[72] and by Edward Witten.[73] By 2010, Maldacena's article had over 7000 citations, becoming the most highly cited article in the field of high energy physics.[lower-alpha 3]

3.4.1 Overview of the correspondence

In the AdS/CFT correspondence, the geometry of spacetime is described in terms of a certain vacuum solution of Einstein's equation called anti-de Sitter space.[74] In very elementary terms, anti-de Sitter space is a mathematical model of spacetime in which the notion of distance between points (the metric) is different from the notion of distance in ordinary Euclidean geometry. It is closely related to hyperbolic space, which can be viewed as a disk as illustrated on the left.[75] This image shows a tessellation of a disk by triangles and squares. One can define the distance between points of this disk in such a way that all the triangles and squares are the same size and the circular outer boundary is infinitely far from any point in the interior.[76]

One can imagine a stack of hyperbolic disks where each disk represents the state of the universe at a given time. The resulting geometric object is three-dimensional anti-de Sitter space.[75] It looks like a solid cylinder in which any cross section is a copy of the hyperbolic disk. Time runs along the vertical direction in this picture. The surface of this cylinder plays an important role in the AdS/CFT correspondence. As with the hyperbolic plane, anti-de Sitter space is curved in such a way that any point in the interior is actually infinitely far from this boundary surface.[76]

This construction describes a hypothetical universe with only two space dimensions and one time dimension, but it can be generalized to any number of dimensions. Indeed, hyperbolic space can have more than two dimensions and one can "stack up" copies of hyperbolic space to get higher-dimensional models of anti-de Sitter space.[75]

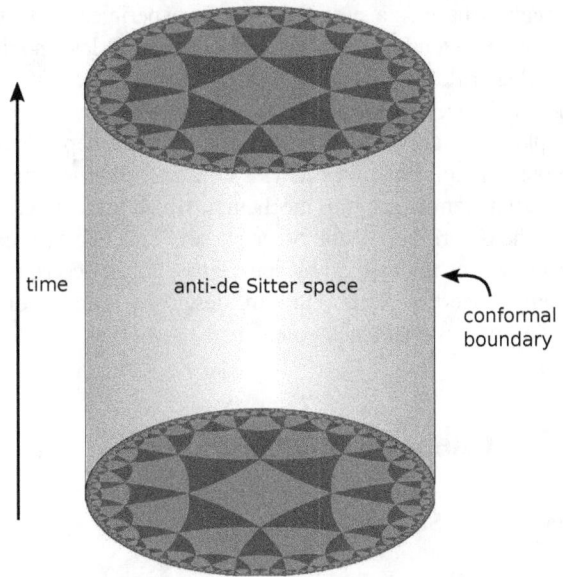

Three-dimensional anti-de Sitter space is like a stack of hyperbolic disks, each one representing the state of the universe at a given time. The resulting spacetime looks like a solid cylinder.

An important feature of anti-de Sitter space is its boundary (which looks like a cylinder in the case of three-dimensional anti-de Sitter space). One property of this boundary is that, within a small region on the surface around any given point, it looks just like Minkowski space, the model of spacetime used in nongravitational physics.[77] One can therefore consider an auxiliary theory in which "spacetime" is given by the boundary of anti-de Sitter space. This observation is the starting point for AdS/CFT correspondence, which states that the boundary of anti-de Sitter space can be regarded as the "spacetime" for a quantum field theory. The claim is that this quantum field theory is equivalent to a gravitational theory, such as string theory, in the bulk anti-de Sitter space in the sense that there is a "dictionary" for translating entities and calculations in one theory into their counterparts in the other theory. For example, a single particle in the gravitational theory might correspond to some collection of particles in the boundary theory. In addition, the predictions in the two theories are quantitatively identical so that if two particles have a 40 percent chance of colliding in the gravitational theory, then the corresponding collections in the boundary theory would also have a 40 percent chance of colliding.[78]

3.4.2 Applications to quantum gravity

The discovery of the AdS/CFT correspondence was a major advance in physicists' understanding of string theory and quantum gravity. One reason for this is that the correspon-dence provides a formulation of string theory in terms of quantum field theory, which is well understood by comparison. Another reason is that it provides a general framework in which physicists can study and attempt to resolve the paradoxes of black holes.[56]

In 1975, Stephen Hawking published a calculation which suggested that black holes are not completely black but emit a dim radiation due to quantum effects near the event horizon.[59] At first, Hawking's result posed a problem for theorists because it suggested that black holes destroy information. More precisely, Hawking's calculation seemed to conflict with one of the basic postulates of quantum mechanics, which states that physical systems evolve in time according to the Schrödinger equation. This property is usually referred to as unitarity of time evolution. The apparent contradiction between Hawking's calculation and the unitarity postulate of quantum mechanics came to be known as the black hole information paradox.[79]

The AdS/CFT correspondence resolves the black hole information paradox, at least to some extent, because it shows how a black hole can evolve in a manner consistent with quantum mechanics in some contexts. Indeed, one can consider black holes in the context of the AdS/CFT correspondence, and any such black hole corresponds to a configuration of particles on the boundary of anti-de Sitter space.[80] These particles obey the usual rules of quantum mechanics and in particular evolve in a unitary fashion, so the black hole must also evolve in a unitary fashion, respecting the principles of quantum mechanics.[81] In 2005, Hawking announced that the paradox had been settled in favor of information conservation by the AdS/CFT correspondence, and he suggested a concrete mechanism by which black holes might preserve information.[82]

3.4.3 Applications to quantum field theory

Main articles: AdS/QCD correspondence and AdS/CMT correspondence

In addition to its applications to theoretical problems in quantum gravity, the AdS/CFT correspondence has been applied to a variety of problems in quantum field theory. One physical system that has been studied using the AdS/CFT correspondence is the quark–gluon plasma, an exotic state of matter produced in particle accelerators. This state of matter arises for brief instants when heavy ions such as gold or lead nuclei are collided at high energies. Such collisions cause the quarks that make up atomic nuclei to deconfine at temperatures of approximately two trillion kelvins, conditions similar to those present at around 10^{-11} seconds after the Big Bang.[83]

The physics of the quark–gluon plasma is governed by a theory called quantum chromodynamics, but this the-

A magnet levitating above a high-temperature superconductor. Today some physicists are working to understand high-temperature superconductivity using the AdS/CFT correspondence.[7]

ory is mathematically intractable in problems involving the quark–gluon plasma.[lower-alpha 4] In an article appearing in 2005, Đàm Thanh Sơn and his collaborators showed that the AdS/CFT correspondence could be used to understand some aspects of the quark–gluon plasma by describing it in the language of string theory.[84] By applying the AdS/CFT correspondence, Sơn and his collaborators were able to describe the quark gluon plasma in terms of black holes in five-dimensional spacetime. The calculation showed that the ratio of two quantities associated with the quark–gluon plasma, the shear viscosity and volume density of entropy, should be approximately equal to a certain universal constant. In 2008, the predicted value of this ratio for the quark–gluon plasma was confirmed at the Relativistic Heavy Ion Collider at Brookhaven National Laboratory.[85][86]

The AdS/CFT correspondence has also been used to study aspects of condensed matter physics. Over the decades, experimental condensed matter physicists have discovered a number of exotic states of matter, including superconductors and superfluids. These states are described using the formalism of quantum field theory, but some phenomena are difficult to explain using standard field theoretic techniques. Some condensed matter theorists including Subir Sachdev hope that the AdS/CFT correspondence will make it possible to describe these systems in the language of string theory and learn more about their behavior.[85]

So far some success has been achieved in using string theory methods to describe the transition of a superfluid to an insulator. A superfluid is a system of electrically neutral atoms that flows without any friction. Such systems are often produced in the laboratory using liquid helium, but recently experimentalists have developed new ways of producing artificial superfluids by pouring trillions of cold atoms into a lattice of criss-crossing lasers. These atoms

initially behave as a superfluid, but as experimentalists increase the intensity of the lasers, they become less mobile and then suddenly transition to an insulating state. During the transition, the atoms behave in an unusual way. For example, the atoms slow to a halt at a rate that depends on the temperature and on Planck's constant, the fundamental parameter of quantum mechanics, which does not enter into the description of the other phases. This behavior has recently been understood by considering a dual description where properties of the fluid are described in terms of a higher dimensional black hole.[87]

3.5   Phenomenology

Main article: String phenomenology

In addition to being an idea of considerable theoretical interest, string theory provides a framework for constructing models of real world physics that combine general relativity and particle physics. Phenomenology is the branch of theoretical physics in which physicists construct realistic models of nature from more abstract theoretical ideas. String phenomenology is the part of string theory that attempts to construct realistic models based on string theory.

Partly because of theoretical and mathematical difficulties and partly because of the extremely high energies needed to test these theories experimentally, there is so far no experimental evidence that would unambiguously point to any of these models being a correct fundamental description of nature. This has led some in the community to criticize these approaches to unification and question the value of continued research on these problems.[12]

3.5.1   Particle physics

The currently accepted theory describing elementary particles and their interactions is known as the standard model of particle physics. This theory provides a unified description of three of the fundamental forces of nature: electromagnetism and the strong and weak nuclear forces. Despite its remarkable success in explaining a wide range of physical phenomena, the standard model cannot be a complete description of reality. This is because the standard model fails to incorporate the force of gravity and because of problems such as the hierarchy problem and the inability to explain the structure of fermion masses or dark matter.

String theory has been used to construct a variety of models of particle physics going beyond the standard model. Typically, such models are based on the idea of compactification. Starting with the ten- or eleven-dimensional spacetime of string or M-theory, physicists postulate a shape for the extra dimensions. By choosing this shape appropriately, they can construct models roughly similar to the standard model of particle physics, together with additional undiscovered particles.[88] One popular way of deriving realistic physics from string theory is to start with the heterotic theory in ten dimensions and assume that the six extra dimensions of spacetime are shaped like a six-dimensional Calabi–Yau manifold. Such compactifications offer many ways of extracting realistic physics from string theory. Other similar methods can be used to construct realistic models of our four-dimensional world based on M-theory.[89]

3.5.2 Cosmology

Main article: String cosmology

The Big Bang theory is the prevailing cosmological model

A map of the cosmic microwave background produced by the Wilkinson Microwave Anisotropy Probe

for the universe from the earliest known periods through its subsequent large-scale evolution. Despite its success in explaining many observed features of the universe including galactic redshifts, the relative abundance of light elements such as hydrogen and helium, and the existence of a cosmic microwave background, there are several questions that remain unanswered. For example, the standard Big Bang model does not explain why the universe appears to be same in all directions, why it appears flat on very large distance scales, or why certain hypothesized particles such as magnetic monopoles are not observed in experiments.[90]

Currently, the leading candidate for a theory going beyond the Big Bang is the theory of cosmic inflation. Developed by Alan Guth and others in the 1980s, inflation postulates a period of extremely rapid accelerated expansion of the universe prior to the expansion described by the standard Big Bang theory. The theory of cosmic inflation preserves the successes of the Big Bang while providing a natural explanation for some of the mysterious features of the universe.[91] The theory has also received striking support from observations of the cosmic microwave background,

the radiation that has filled the sky since around 380,000 years after the Big Bang.[92]

In the theory of inflation, the rapid initial expansion of the universe is caused by a hypothetical particle called the inflaton. The exact properties of this particle are not fixed by the theory but should ultimately be derived from a more fundamental theory such as string theory.[93] Indeed, there have been a number of attempts to identify an inflation within the spectrum of particles described by string theory and to study inflation using string theory. While these approaches might eventually find support in observational data such as measurements of the cosmic microwave background, the application of string theory to cosmology is still in its early stages.[94]

3.6 Connections to mathematics

In addition to influencing research in theoretical physics, string theory has stimulated a number of major developments in pure mathematics. Like many developing ideas in theoretical physics, string theory does not at present have a mathematically rigorous formulation in which all of its concepts can be defined precisely. As a result, physicists who study string theory are often guided by physical intuition to conjecture relationships between the seemingly different mathematical structures that are used to formalize different parts of the theory. These conjectures are later proved by mathematicians, and in this way, string theory serves as a source of new ideas in pure mathematics.[95]

3.6.1 Mirror symmetry

Main article: Mirror symmetry (string theory)

After Calabi–Yau manifolds had entered physics as a way to compactify extra dimensions in string theory, many physicists began studying these manifolds. In the late 1980s, several physicists noticed that given such a compactification of string theory, it is not possible to reconstruct uniquely a corresponding Calabi–Yau manifold.[96] Instead, two different versions of string theory, type IIA and type IIB, can be compactified on completely different Calabi–Yau manifolds giving rise to the same physics. In this situation, the manifolds are called mirror manifolds, and the relationship between the two physical theories is called mirror symmetry.[97]

Regardless of whether Calabi–Yau compactifications of string theory provide a correct description of nature, the existence of the mirror duality between different string theories has significant mathematical consequences. The Calabi–Yau manifolds used in string theory are of interest in pure mathematics, and mirror symmetry allows math-

The Clebsch cubic is an example of a kind of geometric object called an algebraic variety. A classical result of enumerative geometry states that there are exactly 27 straight lines that lie entirely on this surface.

ematicians to solve problems in enumerative geometry, a branch of mathematics concerned with counting the numbers of solutions to geometric questions.[31][98]

Enumerative geometry studies a class of geometric objects called algebraic varieties which are defined by the vanishing of polynomials. For example, the Clebsch cubic illustrated on the right is an algebraic variety defined using a certain polynomial of degree three in four variables. A celebrated result of nineteenth-century mathematicians Arthur Cayley and George Salmon states that there are exactly 27 straight lines that lie entirely on such a surface.[99]

Generalizing this problem, one can ask how many lines can be drawn on a quintic Calabi–Yau manifold, such as the one illustrated above, which is defined by a polynomial of degree five. This problem was solved by the nineteenth-century German mathematician Hermann Schubert, who found that there are exactly 2,875 such lines. In 1986, geometer Sheldon Katz proved that the number of curves, such as circles, that are defined by polynomials of degree two and lie entirely in the quintic is 609,250.[100]

By the year 1991, most of the classical problems of enumerative geometry had been solved and interest in enumerative geometry had begun to diminish.[101] The field was reinvigorated in May 1991 when physicists Philip Candelas, Xenia de la Ossa, Paul Green, and Linda Parks showed that mirror symmetry could be used to translate difficult mathematical questions about one Calabi–Yau manifold into easier questions about its mirror.[102] In particular, they used mirror symmetry to show that a six-dimensional Calabi–Yau

manifold can contain exactly 317,206,375 curves of degree three.[101] In addition to counting degree-three curves, Candelas and his collaborators obtained a number of more general results for counting rational curves which went far beyond the results obtained by mathematicians.[103]

Originally, these results of Candelas were justified on physical grounds. However, mathematicians generally prefer rigorous proofs that do not require an appeal to physical intuition. Inspired by physicists' work on mirror symmetry, mathematicians have therefore constructed their own arguments proving the enumerative predictions of mirror symmetry.[lower-alpha 5] Today mirror symmetry is an active area of research in mathematics, and mathematicians are working to develop a more complete mathematical understanding of mirror symmetry based on physicists' intuition.[104] Major approaches to mirror symmetry include the homological mirror symmetry program of Maxim Kontsevich[32] and the SYZ conjecture of Andrew Strominger, Shing-Tung Yau, and Eric Zaslow.[105]

3.6.2   Monstrous moonshine

Main article: Monstrous moonshine
Group theory is the branch of mathematics that studies the

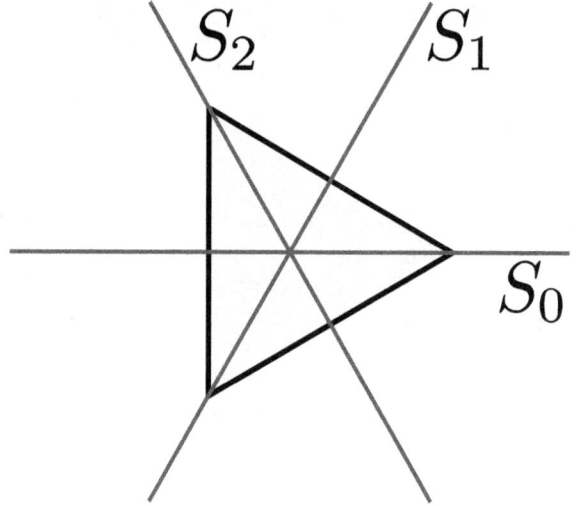

An equilateral triangle can be rotated through 120°, 240°, or 360°, or reflected in any of the three lines pictured without changing its shape.

concept of symmetry. For example, one can consider a geometric shape such as an equilateral triangle. There are various operations that one can perform on this triangle without changing its shape. One can rotate it through 120°, 240°, or 360°, or one can reflect in any of the lines labeled S_0, S_1, or S_2 in the picture. Each of these operations is called a *symmetry*, and the collection of these symmetries satisfies certain technical properties making it into what mathematicians call a group. In this particular example, the group is known as the dihedral group of order 6 because it has six elements. A general group may describe finitely many or infinitely many symmetries; if there are only finitely many symmetries, it is called a finite group.[106]

Mathematicians often strive for a classification (or list) of all mathematical objects of a given type. It is generally believed that finite groups are too diverse to admit a useful classification. A more modest but still challenging problem is to classify all finite *simple* groups. These are finite groups which may be used as building blocks for constructing arbitrary finite groups in the same way that prime numbers can be used to construct arbitrary whole numbers by taking products.[lower-alpha 6] One of the major achievements of contemporary group theory is the classification of finite simple groups, a mathematical theorem which provides a list of all possible finite simple groups.[107]

This classification theorem identifies several infinite families of groups as well as 26 additional groups which do not fit into any family. The latter groups are called the "sporadic" groups, and each one owes its existence to a remarkable combination of circumstances. The largest sporadic group, the so-called monster group, has over 10^{53} elements, more than a thousand times the number of atoms in the Earth.[108]

A graph of the j-function in the complex plane

A seemingly unrelated construction is the *j*-function of number theory. This object belongs to a special class of functions called modular functions, whose graphs form a certain kind of repeating pattern.[109] Although this function appears in a branch of mathematics which seems very different from the theory of finite groups, the two subjects turn out to be intimately related. In the late 1970s, mathematicians John McKay and John Thompson noticed that certain numbers arising in the analysis of the monster group (namely, the dimensions of its irreducible representations) are related to numbers that appear in a formula for the *j*-

function (namely, the coefficients of its Fourier series).[110] This relationship was further developed by John Horton Conway and Simon Norton[111] who called it monstrous moonshine because it seemed so far fetched.[112]

In 1992, Richard Borcherds constructed a bridge between the theory of modular functions and finite groups and, in the process, explained the observations of McKay and Thompson.[113][114] Borcherds' work used ideas from string theory in an essential way, extending earlier results of Igor Frenkel, James Lepowsky, and Arne Meurman, who had realized the monster group as the symmetries of a particular version of string theory.[115] In 1998, Borcherds was awarded the Fields medal for his work.[116]

Since the 1990s, the connection between string theory and moonshine has led to further results in mathematics and physics.[108] In 2010, physicists Tohru Eguchi, Hirosi Ooguri, and Yuji Tachikawa discovered connections between a different sporadic group, the Mathieu group M_{24}, and a certain version of string theory.[117] Miranda Cheng, John Duncan, and Jeffrey A. Harvey proposed a generalization of this moonshine phenomenon called umbral moonshine,[118] and their conjecture was proved mathematically by Duncan, Michael Griffin, and Ken Ono.[119] Witten has also speculated that the version of string theory appearing in monstrous moonshine might be related to a certain simplified model of gravity in three spacetime dimensions.[120]

3.7 History

Main article: History of string theory

3.7.1 Early results

Some of the structures reintroduced by string theory arose for the first time much earlier as part of the program of classical unification started by Albert Einstein. The first person to add a fifth dimension to a theory of gravity was Gunnar Nordström in 1914, who noted that gravity in five dimensions describes both gravity and electromagnetism in four. Nordström attempted to unify electromagnetism with his theory of gravitation, which was however superseded by Einstein's general relativity in 1919. Thereafter, German mathematician Theodor Kaluza combined the fifth dimension with general relativity, and only Kaluza is usually credited with the idea. In 1926, the Swedish physicist Oskar Klein gave a physical interpretation of the unobservable extra dimension—it is wrapped into a small circle. Einstein introduced a non-symmetric metric tensor, while much later Brans and Dicke added a scalar component to gravity. These ideas would be revived within string theory, where they are demanded by consistency conditions.

Leonard Susskind

String theory was originally developed during the late 1960s and early 1970s as a never completely successful theory of hadrons, the subatomic particles like the proton and neutron that feel the strong interaction. In the 1960s, Geoffrey Chew and Steven Frautschi discovered that the mesons make families called Regge trajectories with masses related to spins in a way that was later understood by Yoichiro Nambu, Holger Bech Nielsen and Leonard Susskind to be the relationship expected from rotating strings. Chew advocated making a theory for the interactions of these trajectories that did not presume that they were composed of any fundamental particles, but would construct their interactions from self-consistency conditions on the S-matrix. The S-matrix approach was started by Werner Heisenberg in the 1940s as a way of constructing a theory that did not rely on the local notions of space and time, which Heisenberg believed break down at the nuclear scale. While the scale was off by many orders of magnitude, the approach he advocated was ideally suited for a theory of quantum gravity.

Working with experimental data, R. Dolen, D. Horn and C. Schmid developed some sum rules for hadron exchange. When a particle and antiparticle scatter, virtual particles can

be exchanged in two qualitatively different ways. In the s-channel, the two particles annihilate to make temporary intermediate states that fall apart into the final state particles. In the t-channel, the particles exchange intermediate states by emission and absorption. In field theory, the two contributions add together, one giving a continuous background contribution, the other giving peaks at certain energies. In the data, it was clear that the peaks were stealing from the background—the authors interpreted this as saying that the t-channel contribution was dual to the s-channel one, meaning both described the whole amplitude and included the other.

Gabriele Veneziano

The result was widely advertised by Murray Gell-Mann, leading Gabriele Veneziano to construct a scattering amplitude that had the property of Dolen-Horn-Schmid duality, later renamed world-sheet duality. The amplitude needed poles where the particles appear, on straight line trajectories, and there is a special mathematical function whose poles are evenly spaced on half the real line— the Gamma function— which was widely used in Regge theory. By manipulating combinations of Gamma functions, Veneziano was able to find a consistent scattering amplitude with poles on straight lines, with mostly positive residues, which obeyed duality and had the appropriate Regge scaling at high energy. The amplitude could fit near-beam scattering data as well as other Regge type fits, and had a suggestive integral representation that could be used for generalization.

Over the next years, hundreds of physicists worked to complete the bootstrap program for this model, with many surprises. Veneziano himself discovered that for the scattering amplitude to describe the scattering of a particle that appears in the theory, an obvious self-consistency condition, the lightest particle must be a tachyon. Miguel Virasoro and Joel Shapiro found a different amplitude now understood to be that of closed strings, while Ziro Koba and Holger Nielsen generalized Veneziano's integral representation to multiparticle scattering. Veneziano and Sergio Fubini introduced an operator formalism for computing the scattering amplitudes that was a forerunner of world-sheet conformal theory, while Virasoro understood how to remove the poles with wrong-sign residues using a constraint on the states. Claud Lovelace calculated a loop amplitude, and noted that there is an inconsistency unless the dimension of the theory is 26. Charles Thorn, Peter Goddard and Richard Brower went on to prove that there are no wrong-sign propagating states in dimensions less than or equal to 26.

In 1969, Yoichiro Nambu, Holger Bech Nielsen, and Leonard Susskind recognized that the theory could be given a description in space and time in terms of strings. The scattering amplitudes were derived systematically from the action principle by Peter Goddard, Jeffrey Goldstone, Claudio Rebbi, and Charles Thorn, giving a space-time picture to the vertex operators introduced by Veneziano and Fubini and a geometrical interpretation to the Virasoro conditions.

In 1970, Pierre Ramond added fermions to the model, which led him to formulate a two-dimensional supersymmetry to cancel the wrong-sign states. John Schwarz and André Neveu added another sector to the fermi theory a short time later. In the fermion theories, the critical dimension was 10. Stanley Mandelstam formulated a world sheet conformal theory for both the bose and fermi case, giving a two-dimensional field theoretic path-integral to generate the operator formalism. Michio Kaku and Keiji Kikkawa gave a different formulation of the bosonic string, as a string field theory, with infinitely many particle types and with fields taking values not on points, but on loops and curves.

In 1974, Tamiaki Yoneya discovered that all the known string theories included a massless spin-two particle that obeyed the correct Ward identities to be a graviton. John Schwarz and Joel Scherk came to the same conclusion and made the bold leap to suggest that string theory was a theory of gravity, not a theory of hadrons. They reintroduced Kaluza–Klein theory as a way of making sense of the extra dimensions. At the same time, quantum chromodynamics was recognized as the correct theory of hadrons, shifting the attention of physicists and apparently leaving the bootstrap program in the dustbin of history.

String theory eventually made it out of the dustbin, but for the following decade all work on the theory was completely ignored. Still, the theory continued to develop at a steady pace thanks to the work of a handful of devotees. Ferdinando Gliozzi, Joel Scherk, and David Olive realized in 1976 that the original Ramond and Neveu Schwarz-strings were separately inconsistent and needed to be combined. The resulting theory did not have a tachyon, and was proven to have space-time supersymmetry by John Schwarz and Michael Green in 1981. The same year, Alexander Polyakov gave the theory a modern path integral formulation, and went on to develop conformal field theory extensively. In 1979, Daniel Friedan showed that the equations of motions of string theory, which are generalizations of the Einstein equations of General Relativity, emerge from the Renormalization group equations for the two-dimensional field theory. Schwarz and Green discovered T-duality, and constructed two superstring theories—IIA and IIB related by T-duality, and type I theories with open strings. The consistency conditions had been so strong, that the entire theory was nearly uniquely determined, with only a few discrete choices.

3.7.2 First superstring revolution

Edward Witten

In the early 1980s, Edward Witten discovered that most theories of quantum gravity could not accommodate chiral fermions like the neutrino. This led him, in collaboration with Luis Álvarez-Gaumé to study violations of the conservation laws in gravity theories with anomalies, concluding that type I string theories were inconsistent. Green and Schwarz discovered a contribution to the anomaly that Witten and Alvarez-Gaumé had missed, which restricted the gauge group of the type I string theory to be SO(32). In coming to understand this calculation, Edward Witten became convinced that string theory was truly a consistent theory of gravity, and he became a high-profile advocate. Following Witten's lead, between 1984 and 1986, hundreds of physicists started to work in this field, and this is sometimes called the first superstring revolution.

During this period, David Gross, Jeffrey Harvey, Emil Martinec, and Ryan Rohm discovered heterotic strings. The gauge group of these closed strings was two copies of E8, and either copy could easily and naturally include the standard model. Philip Candelas, Gary Horowitz, Andrew Strominger and Edward Witten found that the Calabi–Yau manifolds are the compactifications that preserve a realistic amount of supersymmetry, while Lance Dixon and others worked out the physical properties of orbifolds, distinctive geometrical singularities allowed in string theory. Cumrun Vafa generalized T-duality from circles to arbitrary manifolds, creating the mathematical field of mirror symmetry. Daniel Friedan, Emil Martinec and Stephen Shenker further developed the covariant quantization of the superstring using conformal field theory techniques. David Gross and Vipul Periwal discovered that string perturbation theory was divergent. Stephen Shenker showed it diverged much faster than in field theory suggesting that new nonperturbative objects were missing.

and identified these with the black-hole solutions of supergravity. These were understood to be the new objects suggested by the perturbative divergences, and they opened up a new field with rich mathematical structure. It quickly became clear that D-branes and other p-branes, not just strings, formed the matter content of the string theories, and the physical interpretation of the strings and branes was revealed—they are a type of black hole. Leonard Susskind had incorporated the holographic principle of Gerardus 't Hooft into string theory, identifying the long highly excited string states with ordinary thermal black hole states. As suggested by 't Hooft, the fluctuations of the black hole horizon, the world-sheet or world-volume theory, describes not only the degrees of freedom of the black hole, but all nearby objects too.

3.7.3 Second superstring revolution

In 1995, at the annual conference of string theorists at the University of Southern California (USC), Edward Witten gave a speech on string theory that in essence united the five string theories that existed at the time, and giving birth to a new 11-dimensional theory called M-theory. M-theory was also foreshadowed in the work of Paul Townsend at approximately the same time. The flurry of activity that began at this time is sometimes called the second superstring revolution.[34]

Joseph Polchinski

In the 1990s, Joseph Polchinski discovered that the theory requires higher-dimensional objects, called D-branes

Juan Maldacena

During this period, Tom Banks, Willy Fischler, Stephen Shenker and Leonard Susskind formulated matrix theory, a full holographic description of M-theory using IIA D0 branes.[51] This was the first definition of string theory that was fully non-perturbative and a concrete mathematical realization of the holographic principle. It is an example of a gauge-gravity duality and is now understood to be a special case of the AdS/CFT correspondence. Andrew Strominger and Cumrun Vafa calculated the entropy of certain configurations of D-branes and found agreement with the semiclassical answer for extreme charged black holes.[62] Petr Hořava and Witten found the eleven-dimensional formulation of the heterotic string theories, showing that orbifolds solve the chirality problem. Witten noted that the effective description of the physics of D-branes at low energies is by a supersymmetric gauge theory, and found geometrical interpretations of mathematical structures in gauge theory that he and Nathan Seiberg had earlier discovered in terms of the location of the branes.

In 1997, Juan Maldacena noted that the low energy excitations of a theory near a black hole consist of objects close to the horizon, which for extreme charged black holes looks like an anti-de Sitter space.[71] He noted that in this limit the gauge theory describes the string excitations near the branes. So he hypothesized that string theory on a near-horizon extreme-charged black-hole geometry, an anti-deSitter space times a sphere with flux, is equally well described by the low-energy limiting gauge theory, the $N = 4$ supersymmetric Yang–Mills theory. This hypothesis, which is called the AdS/CFT correspondence, was further developed by Steven Gubser, Igor Klebanov and Alexander Polyakov,[72] and by Edward Witten,[73] and it is now well-accepted. It is a concrete realization of the holographic principle, which has far-reaching implications for black holes, locality and information in physics, as well as the nature of the gravitational interaction.[56] Through this relationship, string theory has been shown to be related to gauge theories like quantum chromodynamics and this has led to more quantitative understanding of the behavior of hadrons, bringing string theory back to its roots.[84]

3.8 Criticism

3.8.1 Number of solutions

Main article: String theory landscape

To construct models of particle physics based on string theory, physicists typically begin by specifying a shape for the extra dimensions of spacetime. Each of these different shapes corresponds to a different possible universe, or "vacuum state", with a different collection of particles and forces. String theory as it is currently understood has an enormous number of vacuum states, typically estimated to be around 10^{500}, and these might be sufficiently diverse to accommodate almost any phenomena that might be observed at low energies.[121]

Many critics of string theory have expressed concerns about the large number of possible universes described by string theory. In his book *Not Even Wrong*, Peter Woit, a lecturer in the mathematics department at Columbia University, has argued that the large number of different physical scenarios renders string theory vacuous as a framework for constructing models of particle physics. According to Woit,

> The possible existence of, say, 10^{500} consistent different vacuum states for superstring theory probably destroys the hope of using the theory to predict anything. If one picks among this large set just those states whose properties agree with present experimental observations, it is likely there still will be such a large number of these that one can get just about whatever value one wants for the results of any new observation.[122]

Some physicists believe this large number of solutions is actually a virtue because it may allow a natural anthropic explanation of the observed values of physical constants, in particular the small value of the cosmological constant.[122] The anthropic principle is the idea that some of the numbers appearing in the laws of physics are not fixed by any fundamental principle but must be compatible with the evolution of intelligent life. In 1987, Steven Weinberg published an article in which he argued that the cosmological constant could not have been too large, or else galaxies and intelligent life would not have been able to develop.[123] Weinberg suggested that there might be a huge number of possible consistent universes, each with a different value of the cosmological constant, and observations indicate a small value of the cosmological constant only because humans happen to live in a universe that has allowed intelligent life, and hence observers, to exist.[124]

String theorist Leonard Susskind has argued that string theory provides a natural anthropic explanation of the small value of the cosmological constant.[125] According to Susskind, the different vacuum states of string theory might be realized as different universes within a larger multiverse. The fact that the observed universe has a small cosmological constant is just a tautological consequence of the fact that a small value is required for life to exist.[126] Many prominent theorists and critics have disagreed with Susskind's conclusions.[127] According to Woit, "in this case [anthropic reasoning] is nothing more than an excuse for failure. Speculative scientific ideas fail not just when they make incorrect predictions, but also when they turn out to be vacuous and incapable of predicting anything."[128]

3.8.2   Background independence

Main article: Background independence

One of the fundamental properties of Einstein's general theory of relativity is that it is background independent, meaning that the formulation of the theory does not in any way privilege a particular spacetime geometry.[129]

One of the main criticisms of string theory from early on is that it is not manifestly background independent. In string theory, one must typically specify a fixed reference geometry for spacetime, and all other possible geometries are described as perturbations of this fixed one. In his book *The Trouble With Physics*, physicist Lee Smolin of the Perimeter Institute for Theoretical Physics claims that this is the principal weakness of string theory as a theory of quantum gravity, saying that string theory has failed to incorporate this important insight from general relativity.[130]

Others have disagreed with Smolin's characterization of string theory. In a review of Smolin's book, string theorist Joseph Polchinski writes

> [Smolin] is mistaking an aspect of the mathematical language being used for one of the physics being described. New physical theories are often discovered using a mathematical language that is not the most suitable for them... In string theory it has always been clear that the physics is background-independent even if the language being used is not, and the search for more suitable language continues. Indeed, as Smolin belatedly notes, [AdS/CFT] provides a solution to this problem, one that is unexpected and powerful.[131]

Polchinski notes that an important open problem in quantum gravity is to develop holographic descriptions of gravity which do not require the gravitational field to be asymptotically anti-de Sitter.[131]

Smolin responded that the claims about background-independence, which Polchinski presents as "clear", are in fact only an unproven hope for future results, and Smolin is skeptical about them being true at all because of fundamental reasons: "If the strong form of the AdS/CFT conjecture is shown to be correct, then a very weak, and limited form of background will have been achieved. But ... this is still a big if". Smolin points out that current results about

the [AdS/CFT] conjecture rely on global super-symmetry as perturbative physics, "but the whole point of general relativity and quantum gravity is that the generic solutions are governed by no global symmetries because the geometry of spacetime is completely dynamical", which "makes it very non-trivial to show the strong form of the [AdS/CFT] conjecture, because it must extend to solutions of supergravity arbitrarily far from those with global symmetries in the bulk".[132] Smolin summarizes:

> It would be more accurate to say, "Some string theorists believe that the formulations of perturbative string theories and dualities between them that they study concretely are approximations to a deeper, background independent formulation. This missing background independent formulation is not just a different language for the theory, it is hoped to be the statement of the principles and laws that define the theory, from which everything studied so far would be derived as an approximation."[132]

3.8.3   Sociological issues

Since the superstring revolutions of the 1980s and 1990s, string theory has become the dominant paradigm of high energy theoretical physics.[133] Some string theorists have expressed the view that there does not exist an equally successful alternative theory addressing the deep questions of fundamental physics. In an interview from 1987, Nobel laureate David Gross made the following controversial comments about the reasons for the popularity of string theory:

> The most important [reason] is that there are no other good ideas around. That's what gets most people into it. When people started to get interested in string theory they didn't know anything about it. In fact, the first reaction of most people is that the theory is extremely ugly and unpleasant, at least that was the case a few years ago when the understanding of string theory was much less developed. It was difficult for people to learn about it and to be turned on. So I think the real reason why people have got attracted by it is because there is no other game in town. All other approaches of constructing grand unified theories, which were more conservative to begin with, and only gradually became more and more radical, have failed, and this game hasn't failed yet.[134]

Several other high profile theorists and commentators have expressed similar views, suggesting that there are no viable alternatives to string theory.[135]

Many critics of string theory have commented on this state of affairs. In his book criticizing string theory, Peter Woit views the status of string theory research as unhealthy and detrimental to the future of fundamental physics. He argues that the extreme popularity of string theory among theoretical physicists is partly a consequence of the financial structure of academia and the fierce competition for scarce resources.[136] In his book *The Road to Reality*, mathematical physicist Roger Penrose expresses similar views, stating "The often frantic competitiveness that this ease of communication engenders leads to 'bandwagon' effects, where researchers fear to be left behind if they do not join in."[137] Penrose also claims that the technical difficulty of modern physics forces young scientists to rely on the preferences of established researchers, rather than forging new paths of their own.[138] Lee Smolin expresses a slightly different position in his critique, claiming that string theory grew out of a tradition of particle physics which discourages speculation about the foundations of physics, while his preferred approach, loop quantum gravity, encourages more radical thinking. According to Smolin,

> String theory is a powerful, well-motivated idea and deserves much of the work that has been devoted to it. If it has so far failed, the principal reason is that its intrinsic flaws are closely tied to its strengths—and, of course, the story is unfinished, since string theory may well turn out to be part of the truth. The real question is not why we have expended so much energy on string theory but why we haven't expended nearly enough on alternative approaches.[139]

Smolin goes on to offer a number of prescriptions for how scientists might encourage a greater diversity of approaches to quantum gravity research.[140]

3.9 References

3.9.1 Notes

[1] For example, physicists are still working to understand the phenomenon of quark confinement, the paradoxes of black holes, and the origin of dark energy.

[2] For example, in the context of the AdS/CFT correspondence, theorists often formulate and study theories of gravity in unphysical numbers of spacetime dimensions.

[3] "Top Cited Articles during 2010 in hep-th". Retrieved 25 July 2013.

[4] More precisely, one cannot apply the methods of perturbative quantum field theory.

[5] Two independent mathematical proofs of mirror symmetry were given by Givental 1996, 1998 and Lian, Liu, Yau 1997, 1999, 2000.

[6] More precisely, a nontrivial group is called *simple* if its only normal subgroups are the trivial group and the group itself. The Jordan–Hölder theorem exhibits finite simple groups as the building blocks for all finite groups.

3.9.2 Citations

[1] Becker, Becker, and Schwarz 2007, p. 1

[2] Zwiebach 2009, p. 6

[3] Becker, Becker, and Schwarz 2007, pp. 2–3

[4] Becker, Becker, and Schwarz 2007, pp. 9–12

[5] Becker, Becker, and Schwarz 2007, pp. 14–15

[6] Klebanov and Maldacena 2009

[7] Merali 2011

[8] Sachdev 2013

[9] Becker, Becker, and Schwarz 2007, pp. 3, 15–16

[10] Becker, Becker, and Schwarz 2007, p. 8

[11] Becker, Becker, and Schwarz 13–14

[12] Woit 2006

[13] Zee 2010

[14] Becker, Becker, and Schwarz 2007, p. 2

[15] Becker, Becker, and Schwarz 2007, p. 6

[16] Zwiebach 2009, p. 12

[17] Becker, Becker, and Schwarz 2007, p. 4

[18] Zwiebach 2009, p. 324

[19] Wald 1984, p. 4

[20] Zee 2010, Parts V and VI

[21] Zwiebach 2009, p. 9

[22] Zwiebach 2009, p. 8

[23] Yau and Nadis 2010, Ch. 6

[24] Greene 2000, p. 186

[25] Yau and Nadis 2010, p. ix

[26] Randall and Sundrum 1999

[27] Becker, Becker, and Schwarz 2007

[28] Zwiebach 2009, p. 376

[29] Moore 2005, p. 214

[30] Moore 2005, p. 215

[31] Aspinwall et al. 2009

[32] Kontsevich 1995

[33] Kapustin and Witten 2007

[34] Duff 1998

[35] Duff 1998, p. 64

[36] Nahm 1978

[37] Cremmer, Julia, and Scherk 1978

[38] Duff 1998, p. 65

[39] Sen 1994a

[40] Sen 1994b

[41] Hull and Townsend 1995

[42] Duff 1998, p. 67

[43] Bergshoeff, Sezgin, and Townsend 1987

[44] Duff et al. 1987

[45] Duff 1998, p. 66

[46] Witten 1995

[47] Duff 1998, pp. 67–68

[48] Becker, Becker, and Schwarz 2007, p. 296

[49] Hořava and Witten 1996

[50] Duff 1996, sec. 1

[51] Banks et al. 1997

[52] Connes 1994

[53] Connes, Douglas, and Schwarz 1998

[54] Nekrasov and Schwarz 1998

[55] Seiberg and Witten 1999

[56] de Haro et al. 2013, p. 2

[57] Yau and Nadis 2010, p. 187–188

[58] Bekenstein 1973

[59] Hawking 1975

[60] Wald 1984, p. 417

[61] Yau and Nadis 2010, p. 189

[62] Strominger and Vafa 1996

[63] Yau and Nadis 2010, pp. 190–192

[64] Maldacena, Strominger, and Witten 1997

[65] Ooguri, Strominger, and Vafa 2004

[66] Yau and Nadis 2010, pp. 192–193

[67] Yau and Nadis 2010, pp. 194–195

[68] Strominger 1998

[69] Guica et al. 2009

[70] Castro, Maloney, and Strominger 2010

[71] Maldacena 1998

[72] Gubser, Klebanov, and Polyakov 1998

[73] Witten 1998

[74] Klebanov and Maldacena 2009, p. 28

[75] Maldacena 2005, p. 60

[76] Maldacena 2005, p. 61

[77] Zwiebach 2009, p. 552

[78] Maldacena 2005, pp. 61–62

[79] Susskind 2008

[80] Zwiebach 2009, p. 554

[81] Maldacena 2005, p. 63

[82] Hawking 2005

[83] Zwiebach 2009, p. 559

[84] Kovtun, Son, and Starinets 2001

[85] Merali 2011, p. 303

[86] Luzum and Romatschke 2008

[87] Sachdev 2013, p. 51

[88] Candelas et al. 1985

[89] Yau and Nadis 2010, pp. 147–150

[90] Becker, Becker, and Schwarz 2007, pp. 530–531

[91] Becker, Becker, and Schwarz 2007, p. 531

[92] Becker, Becker, and Schwarz 2007, p. 538

[93] Becker, Becker, and Schwarz 2007, p. 533

[94] Becker, Becker, and Schwarz 2007, pp. 539–543

[95] Deligne et al. 1999, p. 1

[96] Hori et al. 2003, p. xvii

[97] Aspinwall et al. 2009, p. 13

[98] Hori et al. 2003

[99] Yau and Nadis 2010, p. 167

[100] Yau and Nadis 2010, p. 166

[101] Yau and Nadis 2010, p. 169

[102] Candelas et al. 1991

[103] Yau and Nadis 2010, p. 171

[104] Hori et al. 2003, p. xix

[105] Strominger, Yau, and Zaslow 1996

[106] Dummit and Foote 2004

[107] Dummit and Foote 2004, pp. 102–103

[108] Klarreich 2015

[109] Gannon 2006, p. 2

[110] Gannon 2006, p. 4

[111] Conway and Norton 1979

[112] Gannon 2006, p. 5

[113] Gannon 2006, p. 8

[114] Borcherds 1992

[115] Frenkel, Lepowsky, and Meurman 1988

[116] Gannon 2006, p. 11

[117] Eguchi, Ooguri, and Tachikawa 2010

[118] Cheng, Duncan, and Harvey 2013

[119] Duncan, Griffin, and Ono 2015

[120] Witten 2007

[121] Woit 2006, pp. 240–242

[122] Woit 2006, p. 242

[123] Weinberg 1987

[124] Woit 2006, p. 243

[125] Susskind 2005

[126] Woit 2006, pp. 242–243

[127] Woit 2006, p. 240

[128] Woit 2006, p. 249

[129] Smolin 2006, p. 81

[130] Smolin 2006, p. 184

[131] Polchinski 2007

[132] Lee Smolin, April 2007:"Archived copy". Archived from the original on November 5, 2015. Retrieved December 31, 2015. Response to review of The Trouble with Physics by Joe Polchinski

[133] Penrose 2004, p. 1017

[134] Woit 2006, pp. 224–225

[135] Woit 2006, Ch. 16

[136] Woit 2006, p. 239

[137] Penrose 2004, p. 1018

[138] Penrose 2004, pp. 1019–1020

[139] Smolin 2006, p. 349

[140] Smolin 2006, Ch. 20

3.9.3 Bibliography

- Aspinwall, Paul; Bridgeland, Tom; Craw, Alastair; Douglas, Michael; Gross, Mark; Kapustin, Anton; Moore, Gregory; Segal, Graeme; Szendröi, Balázs; Wilson, P.M.H., eds. (2009). *Dirichlet Branes and Mirror Symmetry*. American Mathematical Society. ISBN 978-0-8218-3848-8.

- Banks, Tom; Fischler, Willy; Schenker, Stephen; Susskind, Leonard (1997). "M theory as a matrix model: A conjecture". *Physical Review D* **55** (8): 5112–5128. arXiv:hep-th/9610043. Bibcode:1997PhRvD..55.5112B. doi:10.1103/physrevd.55.5112.

- Becker, Katrin; Becker, Melanie; Schwarz, John (2007). *String theory and M-theory: A modern introduction*. Cambridge University Press. ISBN 978-0-521-86069-7.

- Bekenstein, Jacob (1973). "Black holes and entropy". *Physical Review D* **7** (8): 2333–2346. Bibcode:1973PhRvD...7.2333B. doi:10.1103/PhysRevD.7.2333.

- Bergshoeff, Eric; Sezgin, Ergin; Townsend, Paul (1987). "Supermembranes and eleven-dimensional supergravity". *Physics Letters B* **189** (1): 75–78. Bibcode:1987PhLB..189...75B. doi:10.1016/0370-2693(87)91272-X.

- Borcherds, Richard (1992). "Monstrous moonshine and Lie superalgebras". *Inventiones Mathematicae* **109** (1): 405–444. Bibcode:1992InMat.109..405B. doi:10.1007/BF01232032.

- Candelas, Philip; de la Ossa, Xenia; Green, Paul; Parks, Linda (1991). "A pair of Calabi–Yau manifolds as an exactly soluble superconformal field theory". *Nuclear Physics B* **359** (1): 21–74. Bibcode:1991NuPhB.359...21C. doi:10.1016/0550-3213(91)90292-6.

- Candelas, Philip; Horowitz, Gary; Strominger, Andrew; Witten, Edward (1985). "Vacuum configurations for superstrings". *Nuclear Physics B* **258**: 46–74. Bibcode:1985NuPhB.258...46C. doi:10.1016/0550-3213(85)90602-9.

- Castro, Alejandra; Maloney, Alexander; Strominger, Andrew (2010). "Hidden conformal symmetry of the Kerr black hole". *Physical Review D* **82** (2). arXiv:1004.0996. Bibcode:2010PhRvD..82b4008C. doi:10.1103/PhysRevD.82.024008.

- Cheng, Miranda; Duncan, John; Harvey, Jeffrey (2013). "Umbral Moonshine". arXiv:1204.2779.

- Connes, Alain (1994). *Noncommutative Geometry*. Academic Press. ISBN 978-0-12-185860-5.

- Connes, Alain; Douglas, Michael; Schwarz, Albert (1998). "Noncommutative geometry and matrix theory". *Journal of High Energy Physics*. 19981 (2): 003. arXiv:hep-th/9711162. Bibcode:1998JHEP...02..003C. doi:10.1088/1126-6708/1998/02/003.

- Conway, John; Norton, Simon (1979). "Monstrous moonshine". *Bull. London Math. Soc.* **11** (3): 308–339. doi:10.1112/blms/11.3.308.

- Cremmer, Eugene; Julia, Bernard; Scherk, Joel (1978). "Supergravity theory in eleven dimensions". *Physics Letters B* **76** (4): 409–412. Bibcode:1978PhLB...76..409C. doi:10.1016/0370-2693(78)90894-8.

- de Haro, Sebastian; Dieks, Dennis; 't Hooft, Gerard; Verlinde, Erik (2013). "Forty Years of String Theory Reflecting on the Foundations". *Foundations of Physics* **43** (1): 1–7. Bibcode:2013FoPh...43....1D. doi:10.1007/s10701-012-9691-3.

- Deligne, Pierre; Etingof, Pavel; Freed, Daniel; Jeffery, Lisa; Kazhdan, David; Morgan, John; Morrison, David; Witten, Edward, eds. (1999). *Quantum Fields and Strings: A Course for Mathematicians* **1**. American Mathematical Society. ISBN 978-0821820124.

- Duff, Michael (1996). "M-theory (the theory formerly known as strings)". *International Journal of Modern Physics A* **11** (32): 6523–41. arXiv:hep-th/9608117. Bibcode:1996IJMPA..11.5623D. doi:10.1142/S0217751X96002583.

- Duff, Michael (1998). "The theory formerly known as strings". *Scientific American* **278** (2): 64–9. doi:10.1038/scientificamerican0298-64.

- Duff, Michael; Howe, Paul; Inami, Takeo; Stelle, Kellogg (1987). "Superstrings in *D*=10 from supermembranes in *D*=11". *Nuclear Physics B* **191** (1): 70–74. Bibcode:1987PhLB..191...70D. doi:10.1016/0370-2693(87)91323-2.

- Dummit, David; Foote, Richard (2004). *Abstract Algebra*. Wiley. ISBN 978-0-471-43334-7.

- Duncan, John; Griffin, Michael; Ono, Ken (2015). "Proof of the Umbral Moonshine Conjecture". arXiv:1503.01472.

- Eguchi, Tohru; Ooguri, Hirosi; Tachikawa, Yuji (2011). "Notes on the K3 surface and the Mathieu group M_{24}". *Experimental Mathematics* **20** (1): 91–96. doi:10.1080/10586458.2011.544585.

- Frenkel, Igor; Lepowsky, James; Meurman, Arne (1988). *Vertex Operator Algebras and the Monster*. Pure and Applied Mathematics **134**. Academic Press. ISBN 0-12-267065-5.

- Gannon, Terry. *Moonshine Beyond the Monster: The Bridge Connecting Algebra, Modular Forms, and Physics*. Cambridge University Press.

- Givental, Alexander (1996). "Equivariant Gromov-Witten invariants". *International Mathematics Research Notices* **1996** (13): 613–663. doi:10.1155/S1073792896000414.

- Givental, Alexander (1998). "A mirror theorem for toric complete intersections". *Topological field theory, primitive forms and related topics*: 141–175. doi:10.1007/978-1-4612-0705-4_5. ISBN 978-1-4612-6874-1.

- Gubser, Steven; Klebanov, Igor; Polyakov, Alexander (1998). "Gauge theory correlators from non-critical string theory". *Physics Letters B* **428**: 105–114. arXiv:hep-th/9802109. Bibcode:1998PhLB..428..105G. doi:10.1016/S0370-2693(98)00377-3.

- Guica, Monica; Hartman, Thomas; Song, Wei; Strominger, Andrew (2009). "The Kerr/CFT Correspondence". *Physical Review D* **80** (12). arXiv:0809.4266. Bibcode:2009PhRvD..80l4008G. doi:10.1103/PhysRevD.80.124008.

- Hawking, Stephen (1975). "Particle creation by black holes". *Communications in Mathematical Physics* **43** (3): 199–220. Bibcode:1975CMaPh..43..199H. doi:10.1007/BF02345020.

• Hawking, Stephen (2005). "Information loss in black holes". *Physical Review D* **72** (8). arXiv:hep-th/0507171. Bibcode:2005PhRvD..72h4013H. doi:10.1103/PhysRevD.72.084013.

• Hořava, Petr; Witten, Edward (1996). "Heterotic and Type I string dynamics from eleven dimensions". *Nuclear Physics B* **460** (3): 506–524. arXiv:hep-th/9510209. Bibcode:1996NuPhB.460..506H. doi:10.1016/0550-3213(95)00621-4.

• Hori, Kentaro; Katz, Sheldon; Klemm, Albrecht; Pandharipande, Rahul; Thomas, Richard; Vafa, Cumrun; Vakil, Ravi; Zaslow, Eric, eds. (2003). *Mirror Symmetry* (PDF). American Mathematical Society. ISBN 0-8218-2955-6.

• Hull, Chris; Townsend, Paul (1995). "Unity of superstring dualities". *Nuclear Physics B* **4381** (1): 109–137. arXiv:hep-th/9410167. Bibcode:1995NuPhB.438..109H. doi:10.1016/0550-3213(94)00559-W.

• Kapustin, Anton; Witten, Edward (2007). "Electric-magnetic duality and the geometric Langlands program". *Communications in Number Theory and Physics* **1** (1): 1–236. arXiv:hep-th/0604151. Bibcode:2007CNTP....1....1K. doi:10.4310/cntp.2007.v1.n1.a1.

• Klarreich, Erica. "Mathematicians chase moonshine's shadow". *Quanta Magazine*. Retrieved March 2015.

• Klebanov, Igor; Maldacena, Juan (2009). "Solving Quantum Field Theories via Curved Spacetimes" (PDF). *Physics Today* **62**: 28–33. Bibcode:2009PhT....62a..28K. doi:10.1063/1.3074260. Archived from the original (PDF) on July 2, 2013. Retrieved May 2013.

• Kontsevich, Maxim (1995). "Homological algebra of mirror symmetry". *Proceedings of the International Congress of Mathematicians*: 120–139. arXiv:alg-geom/9411018. Bibcode:1994alg.geom.11018K.

• Kovtun, P. K.; Son, Dam T.; Starinets, A. O. (2001). "Viscosity in strongly interacting quantum field theories from black hole physics". *Physical Review Letters* **94** (11): 111601. arXiv:hep-th/0405231. Bibcode:2005PhRvL..94k1601K. doi:10.1103/PhysRevLett.94.111601. PMID 15903845.

• Lian, Bong; Liu, Kefeng; Yau, Shing-Tung (1997). "Mirror principle, I". *Asian Journal of Mathematics* **1**: 729–763. arXiv:alg-geom/9712011. Bibcode:1997alg.geom.12011L.

• Lian, Bong; Liu, Kefeng; Yau, Shing-Tung (1999a). "Mirror principle, II". *Asian Journal of Mathematics* **3**: 109–146. arXiv:math/9905006. Bibcode:1999math......5006L.

• Lian, Bong; Liu, Kefeng; Yau, Shing-Tung (1999b). "Mirror principle, III". *Asian Journal of Mathematics* **3**: 771–800. arXiv:math/9912038. Bibcode:1999math.....12038L.

• Lian, Bong; Liu, Kefeng; Yau, Shing-Tung (2000). "Mirror principle, IV". *Surveys in Differential Geometry* **7**: 475–496. arXiv:math/0007104. Bibcode:2000math......7104L. doi:10.4310/sdg.2002.v7.n1.a15.

• Luzum, Matthew; Romatschke, Paul (2008). "Conformal relativistic viscous hydrodynamics: Applications to RHIC results at $\sqrt{s_{NN}}=200$ GeV". *Physical Review C* **78** (3). arXiv:0804.4015. doi:10.1103/PhysRevC.78.034915.

• Maldacena, Juan (1998). "The Large N limit of superconformal field theories and supergravity". *Advances in Theoretical and Mathematical Physics* **2**: 231–252. arXiv:hep-th/9711200. Bibcode:1998AdTMP...2..231M. doi:10.1063/1.59653.

• Maldacena, Juan (2005). "The Illusion of Gravity" (PDF). *Scientific American* **293** (5): 56–63. Bibcode:2005SciAm.293e..56M. doi:10.1038/scientificamerican1105-56. PMID 16318027. Archived from the original (PDF) on November 1, 2014. Retrieved July 2013.

• Maldacena, Juan; Strominger, Andrew; Witten, Edward (1997). "Black hole entropy in M-theory". *Journal of High Energy Physics* **1997** (12). doi:10.1088/1126-6708/1997/12/002.

• Merali, Zeeya (2011). "Collaborative physics: string theory finds a bench mate". *Nature* **478** (7369): 302–304. Bibcode:2011Natur.478..302M. doi:10.1038/478302a. PMID 22012369.

• Moore, Gregory (2005). "What is ... a Brane?" (PDF). *Notices of the AMS* **52**: 214. Retrieved June 2013.

• Nahm, Walter (1978). "Supersymmetries and their representations". *Nuclear Physics B* **135** (1): 149–166. Bibcode:1978NuPhB.135..149N. doi:10.1016/0550-3213(78)90218-3.

• Nekrasov, Nikita; Schwarz, Albert (1998). "Instantons on noncommutative $\mathbf{R}^4$ and (2,0) superconformal six dimensional theory". *Communications in*

Mathematical Physics **198** (3): 689–703. arXiv:hep-th/9802068.   Bibcode:1998CMaPh.198..689N. doi:10.1007/s002200050490.

- Ooguri, Hirosi; Strominger, Andrew; Vafa, Cumrun (2004). "Black hole attractors and the topological string". *Physical Review D* **70** (10). doi:10.1103/physrevd.70.106007.

- Polchinski, Joseph (2007). "All Strung Out?". *American Scientist*. Retrieved April 2015.

- Penrose, Roger (2005). *The Road to Reality: A Complete Guide to the Laws of the Universe*. Knopf. ISBN 0-679-45443-8.

- Randall, Lisa; Sundrum, Raman (1999). "An alternative to compactification". *Physical Review Letters* **83** (23): 4690–4693. arXiv:hep-th/9906064.   Bibcode:1999PhRvL..83.4690R. doi:10.1103/PhysRevLett.83.4690.

- Sachdev, Subir (2013). "Strange and stringy". *Scientific American* **308** (44): 44–51.   Bibcode:2012SciAm.308a..44S. doi:10.1038/scientificamerican0113-44.

- Seiberg, Nathan; Witten, Edward (1999). "String Theory and Noncommutative Geometry". *Journal of High Energy Physics* **1999** (9): 032. arXiv:hep-th/9908142.   Bibcode:1999JHEP...09..032S. doi:10.1088/1126-6708/1999/09/032.

- Sen, Ashoke (1994a). "Strong-weak coupling duality in four-dimensional string theory". *International Journal of Modern Physics A* **9** (21): 3707–3750. arXiv:hep-th/9402002.   Bibcode:1994IJMPA...9.3707S. doi:10.1142/S0217751X94001497.

- Sen, Ashoke (1994b). "Dyon-monopole bound states, self-dual harmonic forms on the multi-monopole moduli space, and $SL(2,\mathbf{Z})$ invariance in string theory". *Physics Letters B* **329** (2): 217–221. arXiv:hep-th/9402032.   Bibcode:1994PhLB..329..217S. doi:10.1016/0370-2693(94)90763-3.

- Smolin, Lee (2006). *The Trouble with Physics: The Rise of String Theory, the Fall of a Science, and What Comes Next*. New York: Houghton Mifflin Co. ISBN 0-618-55105-0.

- Strominger, Andrew (1998). "Black hole entropy from near-horizon microstates". *Journal of High Energy Physics* **1998** (2): 009. arXiv:hep-th/9712251. Bibcode:1998JHEP...02..009S.   doi:10.1088/1126-6708/1998/02/009.

- Strominger, Andrew; Vafa, Cumrun (1996). "Microscopic origin of the Bekenstein–Hawking entropy". *Physics Letters B* **379** (1): 99–104. arXiv:hep-th/9601029.   Bibcode:1996PhLB..379...99S. doi:10.1016/0370-2693(96)00345-0.

- Strominger, Andrew; Yau, Shing-Tung; Zaslow, Eric (1996). "Mirror symmetry is T-duality". *Nuclear Physics B* **479** (1): 243–259. arXiv:hep-th/9606040. Bibcode:1996NuPhB.479..243S. doi:10.1016/0550-3213(96)00434-8.

- Susskind, Leonard (2005). *The Cosmic Landscape: String Theory and the Illusion of Intelligent Design*. Back Bay Books. ISBN 978-0316013338.

- Susskind, Leonard (2008). *The Black Hole War: My Battle with Stephen Hawking to Make the World Safe for Quantum Mechanics*. Little, Brown and Company. ISBN 978-0-316-01641-4.

- Wald, Robert (1984). *General Relativity*. University of Chicago Press. ISBN 978-0-226-87033-5.

- Weinberg, Steven (1987). *Anthropic bound on the cosmological constant* **59**. Physical Review Letters. p. 2607.

- Witten, Edward (1995). "String theory dynamics in various dimensions". *Nuclear Physics B* **443** (1): 85–126. arXiv:hep-th/9503124. Bibcode:1995NuPhB.443...85W. doi:10.1016/0550-3213(95)00158-O.

- Witten, Edward (1998). "Anti-de Sitter space and holography". *Advances in Theoretical and Mathematical Physics* **2**: 253–291. arXiv:hep-th/9802150. Bibcode:1998AdTMP...2..253W.

- Witten, Edward (2007). "Three-dimensional gravity revisited". arXiv:0706.3359 [hep-th].

- Woit, Peter (2006). *Not Even Wrong: The Failure of String Theory and the Search for Unity in Physical Law*. Basic Books. p. 105. ISBN 0-465-09275-6.

- Yau, Shing-Tung; Nadis, Steve (2010). *The Shape of Inner Space: String Theory and the Geometry of the Universe's Hidden Dimensions*. Basic Books. ISBN 978-0-465-02023-2.

- Zee, Anthony (2010). *Quantum Field Theory in a Nutshell* (2nd ed.). Princeton University Press. ISBN 978-0-691-14034-6.

- Zwiebach, Barton (2009). *A First Course in String Theory*. Cambridge University Press. ISBN 978-0-521-88032-9.

3.10 Further reading

3.10.1 Popularizations

General

- Greene, Brian (2003). *The Elegant Universe: Superstrings, Hidden Dimensions, and the Quest for the Ultimate Theory*. New York: W.W. Norton & Company. ISBN 0-393-05858-1.

- Greene, Brian (2004). *The Fabric of the Cosmos: Space, Time, and the Texture of Reality*. New York: Alfred A. Knopf. ISBN 0-375-41288-3.

Critical

- Penrose, Roger (2005). *The Road to Reality: A Complete Guide to the Laws of the Universe*. Knopf. ISBN 0-679-45443-8.

- Smolin, Lee (2006). *The Trouble with Physics: The Rise of String Theory, the Fall of a Science, and What Comes Next*. New York: Houghton Mifflin Co. ISBN 0-618-55105-0.

- Woit, Peter (2006). *Not Even Wrong: The Failure of String Theory And the Search for Unity in Physical Law*. London: Jonathan Cape &: New York: Basic Books. ISBN 978-0-465-09275-8.

3.10.2 Textbooks

For physicists

- Becker, Katrin; Becker, Melanie; Schwarz, John (2007). *String Theory and M-theory: A Modern Introduction*. Cambridge University Press. ISBN 978-0-521-86069-7.

- Green, Michael; Schwarz, John; Witten, Edward (2012). *Superstring theory. Vol. 1: Introduction*. Cambridge University Press. ISBN 978-1107029118.

- Green, Michael; Schwarz, John; Witten, Edward (2012). *Superstring theory. Vol. 2: Loop amplitudes, anomalies and phenomenology*. Cambridge University Press. ISBN 978-1107029132.

- Polchinski, Joseph (1998). *String Theory Vol. 1: An Introduction to the Bosonic String*. Cambridge University Press. ISBN 0-521-63303-6.

- Polchinski, Joseph (1998). *String Theory Vol. 2: Superstring Theory and Beyond*. Cambridge University Press. ISBN 0-521-63304-4.

- Zwiebach, Barton (2009). *A First Course in String Theory*. Cambridge University Press. ISBN 978-0-521-88032-9.

For mathematicians

- Deligne, Pierre; Etingof, Pavel; Freed, Daniel; Jeffery, Lisa; Kazhdan, David; Morgan, John; Morrison, David; Witten, Edward, eds. (1999). *Quantum Fields and Strings: A Course for Mathematicians, Vol. 2*. American Mathematical Society. ISBN 978-0821819883.

3.11 External links

- *The Elegant Universe*—A three-hour miniseries with Brian Greene by *NOVA* (original PBS Broadcast Dates: October 28, 8–10 p.m. and November 4, 8–9 p.m., 2003). Various images, texts, videos and animations explaining string theory.

- Not Even Wrong—A blog critical of string theory

- The Official String Theory Web Site

- Why String Theory—An introduction to string theory.

Chapter 4

Superstring theory

"Superstring" redirects here. For the converse relation of "substring", see Superstring (formal languages). For the bundle of firecrackers, see Superstring (fireworks).

Superstring theory is an attempt to explain all of the particles and fundamental forces of nature in one theory by modelling them as vibrations of tiny supersymmetric strings.

'Superstring theory' is a shorthand for **supersymmetric string theory** because unlike bosonic string theory, it is the version of string theory that incorporates fermions and supersymmetry.

Since the second superstring revolution, the five superstring theories are regarded as different limits of a single theory tentatively called M-theory, or simply string theory.

4.1 Background

The deepest problem in theoretical physics is harmonizing the theory of general relativity, which describes gravitation and applies to large-scale structures (stars, galaxies, super clusters), with quantum mechanics, which describes the other three fundamental forces acting on the atomic scale.

The development of a quantum field theory of a force invariably results in infinite possibilities. Physicists have developed mathematical techniques (renormalization) to eliminate these infinities that work for three of the four fundamental forces—electromagnetic, strong nuclear and weak nuclear forces—but not for gravity. The development of a quantum theory of gravity must therefore come about by different means than those used for the other forces.[1]

According to the theory, the fundamental constituents of reality are strings of the Planck length (about 10^{-33} cm) that vibrate at resonant frequencies. Every string, in theory, has a unique resonance, or harmonic. Different harmonics determine different fundamental particles. The tension in a string is on the order of the Planck force (10^{44} newtons). The graviton (the proposed messenger particle of the gravitational force), for example, is predicted by the theory to be a string with wave amplitude zero.

4.2 History

Main article History of string theory

Since its beginnings in late sixties, the theory was developed through several decades of intense research and combined effort of numerous scientists. It has developed into a broad and varied subject with connections to quantum gravity, particle and condensed matter physics, cosmology, and pure mathematics.

4.3 Lack of experimental evidence

Superstring theory is based on supersymmetry. No supersymmetric particles have been discovered and recent research at LHC and Tevatron has excluded some of the ranges.[2][3][4][5] For instance, the mass constraint of the Minimal Supersymmetric Standard Model squarks has been up to 1.1 TeV, and gluinos up to 500 GeV.[6] No report on suggesting large extra dimensions has been delivered from LHC. There have been no principles so far to limit the number of vacua in the concept of a landscape of vacua.[7]

Some particle physicists became disappointed[8] by the lack of experimental verification of supersymmetry, and some have already discarded it; Jon Butterworth at the University College London said that we had no sign of supersymmetry, even in higher energy region, excluding the superpartners of the top quark up to a few TeV. Ben Allanach at the University of Cambridge states that if we do not discover any new particles in the next trial at the LHC, then we can say it is unlikely to discover supersymmetry at CERN in the foreseeable future.[8]

4.4 Extra dimensions

See also: Why does consistency require 10 dimensions?

Our physical space is observed to have three large spatial dimensions and, along with time, is a boundless four-dimensional continuum known as spacetime. However, nothing prevents a theory from including more than 4 dimensions. In the case of string theory, consistency requires spacetime to have 10 (3+1+6) dimensions. The fact that we see only 3 dimensions of space can be explained by one of two mechanisms: either the extra dimensions are compactified on a very small scale, or else our world may live on a 3-dimensional submanifold corresponding to a brane, on which all known particles besides gravity would be restricted.

If the extra dimensions are compactified, then the extra six dimensions must be in the form of a Calabi–Yau manifold. Within the more complete framework of M-theory, they would have to take form of a G2 manifold. Calabi-Yaus are interesting mathematical spaces in their own right. A particular exact symmetry of string/M-theory called T-duality (which exchanges momentum modes for winding number and sends compact dimensions of radius R to radius 1/R),[9] has led to the discovery of equivalences between different Calabi-Yaus called Mirror Symmetry.

Superstring theory is not the first theory to propose extra spatial dimensions. It can be seen as building upon the Kaluza–Klein theory, which proposed a 4+1-dimensional theory of gravity. When compactified on a circle, the gravity in the extra dimension precisely describes electromagnetism from the perspective of the 3 remaining large space dimensions. Thus the original Kaluza–Klein theory is a prototype for the unification of gauge and gravity interactions, at least at the classical level, however it is known to be insufficient to describe nature for a variety of reasons (missing weak and strong forces, lack of parity violation, etc.) A more complex compact geometry is needed to reproduce the known gauge forces. Also, to obtain a consistent, fundamental, quantum theory requires the upgrade to string theory—not just the extra dimensions.

4.5 Number of superstring theories

Theoretical physicists were troubled by the existence of five separate string theories. A possible solution for this dilemma was suggested at the beginning of what is called the second superstring revolution in the 1990s, which suggests that the five string theories might be different limits of a single underlying theory, called M-theory. This remains a conjecture.[10]

The five consistent superstring theories are:

- The type I string has one supersymmetry in the ten-dimensional sense (16 supercharges). This theory is special in the sense that it is based on unoriented open and closed strings, while the rest are based on oriented closed strings.

- The type II string theories have two supersymmetries in the ten-dimensional sense (32 supercharges). There are actually two kinds of type II strings called type IIA and type IIB. They differ mainly in the fact that the IIA theory is non-chiral (parity conserving) while the IIB theory is chiral (parity violating).

- The heterotic string theories are based on a peculiar hybrid of a type I superstring and a bosonic string. There are two kinds of heterotic strings differing in their ten-dimensional gauge groups: the heterotic $E_8 \times E_8$ string and the heterotic SO(32) string. (The name heterotic SO(32) is slightly inaccurate since among the SO(32) Lie groups, string theory singles out a quotient Spin(32)/Z_2 that is not equivalent to SO(32).)

Chiral gauge theories can be inconsistent due to anomalies. This happens when certain one-loop Feynman diagrams cause a quantum mechanical breakdown of the gauge symmetry. The anomalies were canceled out via the Green–Schwarz mechanism.

Even though there are only five superstring theories, making detailed predictions for real experiments requires information about exactly what physical configuration the theory is in. This considerably complicates efforts to test string theory because there is an astronomically high number – 10^{500} or more – of configurations that meet some of the basic requirements to be consistent with our world. Along with the extreme remoteness of the Planck scale, this is the other major reason it is hard to test superstring theory.

Another approach to the number of superstring theories refers to the mathematical structure called composition algebra. In the findings of abstract algebra there are just seven composition algebras over the field of real numbers. In 1990 physicists R. Foot and G.C. Joshi in Australia stated that "the seven classical superstring theories are in one-to-one correspondence to the seven composition algebras."[11]

4.6 Integrating general relativity and quantum mechanics

General relativity typically deals with situations involving large mass objects in fairly large regions of spacetime whereas quantum mechanics is generally reserved for scenarios at the atomic scale (small spacetime regions). The two are very rarely used together, and the most common case that combines them is in the study of black holes. Having *peak density*, or the maximum amount of matter possible in a space, and very small area, the two must be used in synchrony to predict conditions in such places. Yet, when used together, the equations fall apart, spitting out impossible answers, such as imaginary distances and less than one dimension.

The major problem with their congruence is that, at Planck scale (a fundamental small unit of length) lengths, general relativity predicts a smooth, flowing surface, while quantum mechanics predicts a random, warped surface, neither of which are anywhere near compatible. Superstring theory resolves this issue, replacing the classical idea of point particles with strings. These strings have an average diameter of the Planck length, with extremely small variances, which completely ignores the quantum mechanical predictions of Planck-scale length dimensional warping. Also, these surfaces can be mapped as branes. These branes can be viewed as objects with a morphism between them. In this case, the morphism will be state of a string that stretches between brane A and brane B.

Singularities are avoided because the observed consequences of "Big Crunches" never reach zero size. In fact, should the universe begin a "big crunch" sort of process, string theory dictates that the universe could never be smaller than the size of one string, at which point it would actually begin expanding.

4.7   Mathematics

4.7.1   D-branes

D-branes are membrane-like objects in 10D string theory. They can be thought of as occurring as a result of a Kaluza–Klein compactification of 11D M-theory that contains membranes. Because compactification of a geometric theory produces extra vector fields the D-branes can be included in the action by adding an extra U(1) vector field to the string action.

$$\partial_z \to \partial_z + iA_z(z, \overline{z})$$

In **type I** open string theory, the ends of open strings are always attached to D-brane surfaces. A string theory with more gauge fields such as SU(2) gauge fields would then correspond to the compactification of some higher-dimensional theory above 11 dimensions, which is not thought to be possible to date. Furthemore, the tachyons attached to the D-branes, show, the instability of those d-branes with respect to the annihilation. We will consider that tachyon total energy is (or reflects) the total energy of the D-branes.

4.7.2   Why five superstring theories?

For a 10 dimensional supersymmetric theory we are allowed a 32-component Majorana spinor. This can be decomposed into a pair of 16-component Majorana-Weyl (chiral) spinors. There are then various ways to construct an invariant depending on whether these two spinors have the same or opposite chiralities:

The heterotic superstrings come in two types SO(32) and $E_8{\times}E_8$ as indicated above and the type I superstrings include open strings.

4.8   Beyond superstring theory

It is conceivable that the five superstring theories are approximated to a theory in higher dimensions possibly involving membranes. Because the action for this involves quartic terms and higher so is not Gaussian, the functional integrals are very difficult to solve and so this has confounded the top theoretical physicists. Edward Witten has popularised the concept of a theory in 11 dimensions M-theory involving membranes interpolating from the known symmetries of superstring theory. It may turn out that there exist membrane models or other non-membrane models in higher dimensions—which may become acceptable when we find new unknown symmetries of nature, such as non-commutative geometry. It is thought, however, that 16 is probably the maximum since O(16) is a maximal subgroup of E8 the largest exceptional lie group and also is more than large enough to contain the Standard Model. Quartic integrals of the non-functional kind are easier to solve so there is hope for the future. This is the series solution, which is always convergent when a is non-zero and negative:

$$\int_{-\infty}^{\infty} \exp(ax^4 + bx^3 + cx^2 + dx + f)\,dx =$$

$$e^f \sum_{n,m,p=0}^{\infty} \frac{b^{4n}}{(4n)!} \frac{c^{2m}}{(2m)!} \frac{d^{4p}}{(4p)!} \frac{\Gamma(3n+m+p+\frac{1}{4})}{a^{3n+m+p+\frac{1}{4}}}$$

In the case of membranes the series would correspond to sums of various membrane interactions that are not seen in string theory.

4.8.1   Compactification

Investigating theories of higher dimensions often involves looking at the 10 dimensional superstring theory and interpreting some of the more obscure results in terms of compactified dimensions. For example, D-branes are seen as compactified membranes from 11D M-theory. Theories of higher dimensions such as 12D F-theory and beyond produce other effects, such as gauge terms higher than $U(1)$. The components of the extra vector fields (A) in the D-brane actions can be thought of as extra coordinates (X) in disguise. However, the *known* symmetries including supersymmetry currently restrict the spinors to 32-components—which limits the number of dimensions to 11 (or 12 if you include two time dimensions.) Some commentators (e.g., John Baez et al.) have speculated that the exceptional lie groups E_6, E_7 and E_8 having maximum orthogonal subgroups $O(10)$, $O(12)$ and $O(16)$ may be related to theories in 10, 12 and 16 dimensions; 10 dimensions corresponding to string theory and the 12 and 16 dimensional theories being yet undiscovered but would be theories based on 3-branes and 7-branes respectively. However this is a minority view within the string community. Since E_7 is in some sense F_4 quaternified and E_8 is F_4 octonified, then the 12 and 16 dimensional theories, if they did exist, may involve the noncommutative geometry based on the quaternions and octonions respectively. From the above discussion, it can be seen that physicists have many ideas for extending superstring theory beyond the current 10 dimensional theory, but so far none have been successful.

4.8.2 Kac–Moody algebras

Since strings can have an infinite number of modes, the symmetry used to describe string theory is based on infinite dimensional Lie algebras. Some Kac–Moody algebras that have been considered as symmetries for M-theory have been E_{10} and E_{11} and their supersymmetric extensions.

4.9 See also

- AdS/CFT correspondence

- dS/CFT correspondence

- Grand unification theory

- Large Hadron Collider

- List of string theory topics

- Quantum gravity

- String field theory

4.10 Notes

[1] Polchinski, Joseph. *String Theory: Volume I*. Cambridge University Press, p. 4.

[2] Woit, Peter (February 22, 2011). "Implications of Initial LHC Searches for Supersymmetry".

[3] Cassel, S.; Ghilencea, D. M.; Kraml, S.; Lessa, A.; Ross, G. G. (2011). "Fine-tuning implications for complementary dark matter and LHC SUSY searches". *Journal of High Energy Physics* **2011** (5): 120. arXiv:1101.4664. Bibcode:2011JHEP...05..120C. doi:10.1007/JHEP05(2011)120.

[4] Falkowski, Adam (Jester) (February 16, 2011). "What LHC tells about SUSY". *resonaances.blogspot.com*. Archived from the original on March 22, 2014. Retrieved March 22, 2014.

[5] Tapper, Alex (24 March 2010). "Early SUSY searches at the LHC" (PDF). Imperial College London.

[6] CMS Collaboration (2011). "Search for Supersymmetry at the LHC in Events with Jets and Missing Transverse Energy". *Physical Review Letters* **107** (22): 221804. arXiv:1109.2352. Bibcode:2011PhRvL.107v1804C. doi:10.1103/PhysRevLett.107.221804. PMID 22182023.

[7] Shifman, M. (2012). "Frontiers Beyond the Standard Model: Reflections and Impressionistic Portrait of the Conference". *Modern Physics Letters A* **27** (40): 1230043. Bibcode:2012MPLA...2730043S. doi:10.1142/S0217732312300431.

[8] Jha, Alok (August 6, 2013). "One year on from the Higgs boson find, has physics hit the buffers?". *The Guardian*. photograph: Harold Cunningham/Getty Images (London: GMG). ISSN 0261-3077. OCLC 60623878. Archived from the original on March 22, 2014. Retrieved March 22, 2014.

[9] Polchinski, Joseph. *String Theory: Volume I*. Cambridge University Press, p. 247.

[10] Polchinski, Joseph. *String Theory: Volume II*. Cambridge University Press, p. 198.

[11] Foot, R.; Joshi, G. C. (1990). "Nonstandard signature of spacetime, superstrings, and the split composition algebras". *Letters in Mathematical Physics* **19**: 65–71. Bibcode:1990LMaPh..19...65F. doi:10.1007/BF00402262.

4.11 References

- Kaku, Michio (1999). *Introduction to Superstring and M-Theory* (2nd ed.). New York, USA: Springer-Verlag.

- Shen, Sinyan (1982). *Introduction to Superfluidity* (2nd ed.). Beijing, China: Science Press.

- Greene, Brian (2000). *The Elegant Universe: Superstrings, Hidden Dimensions, and the Quest for the Ultimate Theory.* Random House Inc.

4.12   External links

- Wellcome Collection video on superstring theory

- The Official Superstring theory website: http://superstringtheory.com/index.html

Chapter 5

String duality

This article is about string duality. For other forms of duality, see Duality (disambiguation).

String duality is a class of symmetries in physics that link different string theories, theories which assume that the fundamental building blocks of the universe are strings instead of point particles.

5.1 Overview

Before the so-called "duality revolution" there were believed to be five distinct versions of string theory, plus the (unstable) bosonic and gluonic theories.

Note that in the type IIA and type IIB string theories closed strings are allowed to move everywhere throughout the ten-dimensional space-time (called the *bulk*), while open strings have their ends attached to D-branes, which are membranes of lower dimensionality (their dimension is odd - 1,3,5,7 or 9 - in type IIA and even - 0,2,4,6 or 8 - in type IIB, including the time direction).

Before the 1990s, string theorists believed there were five distinct superstring theories: type I, types IIA and IIB, and the two heterotic string theories (SO(32) and $E_8 \times E_8$). The thinking was that out of these five candidate theories, only one was the actual theory of everything, and that theory was the theory whose low energy limit, with ten dimensions spacetime compactified down to four, matched the physics observed in our world today. It is now known that the five superstring theories are not fundamental, but are instead different limits of a more fundamental theory, dubbed M-theory. These theories are related by transformations called dualities. If two theories are related by a duality transformation, each observable of the first theory can be mapped in some way to the second theory to yield equivalent predictions. The two theories are then said to be dual to one another under that transformation. Put differently, the two theories are two mathematically different descriptions of the same phenomena. A simple example of a duality is the equivalence of particle physics upon replacing matter with antimatter; describing our universe in terms of anti-particles would yield identical predictions for any possible experiment.

String dualities often link quantities that appear to be separate: Large and small distance scales, strong and weak coupling strengths. These quantities have always marked very distinct limits of behavior of a physical system, in both classical field theory and quantum particle physics. But strings can obscure the difference between large and small, strong and weak, and this is how these five very different theories end up being related.

5.2 T-duality

Main article: T-duality

Suppose we are in ten spacetime dimensions, which means we have nine space dimensions and one time. Take one of those nine space dimensions and make it a circle of radius R, so that traveling in that direction for a distance L = 2πR takes you around the circle and brings you back to where you started. A particle traveling around this circle will have a quantized momentum around the circle, because its momentum is linked to its wavelength (see Wave-particle duality), and 2πR must be a multiple of that. In fact, the particle momentum around the circle - and the contribution to its energy - is of the form n/R (in standard units, for an integer n), so that at large R there will be many more states compared to small R (for a given maximum energy). A string, in addition to traveling around the circle, may also wrap around it. The number of times the string winds around the circle is called the winding number, and that is also quantized (as it must be an integer). Winding around the circle requires energy, because the string must be stretched against its tension, so it contributes an amount of energy of the form wR/L_{st}^2 , where L_{st} is a constant called the *string length* and w is the winding number (an integer). Now (for a given maximum energy) there will be many different states (with different momenta) at large R, but there will also be many different states (with different windings) at small R. In fact, a theory with large R and a theory with small R are equivalent, where the role of momentum in the first is played by the winding in the second, and vice versa. Mathematically, taking R to L_{st}^2/R and switching n and w will yield the same equations. So exchanging momentum and winding modes of the string exchanges a large distance scale with a small distance scale.

This type of duality is called T-duality. T-duality relates type IIA superstring theory to type IIB superstring theory. That means if we take type IIA and Type IIB theory and compactify them both on a circle (one with a large radius and the other with a small radius) then switching the momentum and winding modes, and switching the distance scale, changes one theory into the other. The same is also true for the two heterotic theories. T-duality also relates type I superstring theory to both type IIA and type IIB superstring theories with certain boundary conditions (termed orientifold).

Formally, the location of the string on the circle is described by two fields living on it, one which is left-moving and another which is right-moving. The movement of the string center (and hence its momentum) is related to the sum of the fields, while the string stretch (and hence its winding number) is related to their difference. T-duality can be formally described by taking the left-moving field to minus itself, so that the sum and the difference are interchanged, leading to switching of momentum and winding.

5.3   S-duality

Main articles: S-duality and M-theory

Every force has a coupling constant, which is a measure of its strength, and determines the chances of one particle to emit or absorb another particle. For electromagnetism, the coupling constant is proportional to the square of the electric charge. When physicists study the quantum behavior of electromagnetism, they can't solve the whole theory exactly, because every particle may emit and absorb many other particles, which may also do the same, endlessly. So events of emission and absorption are considered as perturbations and are dealt with by a series of approximations, first assuming there is only one such event, then correcting the result for allowing two such events, etc. (this method is called Perturbation theory). This is a reasonable approximation only if the coupling constant is small, which is the case for electromagnetism. But if the coupling constant gets large, that method of calculation breaks down, and the lit-

tle pieces become worthless as an approximation to the real physics.

This also can happen in string theory. String theories have a coupling constant. But unlike in particle theories, the string coupling constant is not just a number, but depends on one of the oscillation modes of the string, called the dilaton. Exchanging the dilaton field with minus itself exchanges a very large coupling constant with a very small one. This symmetry is called S-duality. If two string theories are related by S-duality, then one theory with a strong coupling constant is the same as the other theory with weak coupling constant. The theory with strong coupling cannot be understood by means of perturbation theory, but the theory with weak coupling can. So if the two theories are related by S-duality, then we just need to understand the weak theory, and that is equivalent to understanding the strong theory.

Superstring theories related by S-duality are: type I superstring theory with heterotic SO(32) superstring theory, and type IIB theory with itself.

Furthermore, type IIA theory in strong coupling behaves like an 11-dimensional theory, with the dilaton field playing the role of an eleventh dimension. This 11-dimensional theory is known as M-theory.

Unlike the T-duality, however, S-duality has not been proven to even a physics level of rigor for any of the aforementioned cases. It remains, strictly speaking, a conjecture, although most string theorists believe in its validity.

5.4   See also

- U-duality

- Mirror Symmetry

Chapter 6

S-duality

This article is about S-duality (strong–weak duality) in physics. For the mathematical S-duality (Spanier–Whitehead duality), see S-duality (homotopy theory).

In theoretical physics, **S-duality** is an equivalence of two physical theories, which may be either quantum field theories or string theories. S-duality is useful for doing calculations in theoretical physics because it relates a theory in which calculations are difficult to a theory in which they are easier.[1]

In quantum field theory, S-duality generalizes a well known fact from classical electrodynamics, namely the invariance of Maxwell's equations under the interchange of electric and magnetic fields. One of the earliest known examples of S-duality in quantum field theory is Montonen–Olive duality which relates two versions of a quantum field theory called N = 4 supersymmetric Yang–Mills theory. Recent work of Anton Kapustin and Edward Witten suggests that Montonen–Olive duality is closely related to a research program in mathematics called the geometric Langlands program. Another realization of S-duality in quantum field theory is Seiberg duality, which relates two versions of a theory called N=1 supersymmetric Yang–Mills theory.

There are also many examples of S-duality in string theory. The existence of these string dualities implies that seemingly different formulations of string theory are actually physically equivalent. This led to the realization, in the mid-1990s, that all of the five consistent superstring theories are just different limiting cases of a single eleven-dimensional theory called M-theory.[2]

6.1   Overview

In quantum field theory and string theory, a coupling constant is a number that controls the strength of interactions in the theory. For example, the strength of gravity is described by a number called Newton's constant, which appears in Newton's law of gravity and also in the equations of Albert Einstein's general theory of relativity. Similarly, the strength of the electromagnetic force is described by a coupling constant, which is related to the charge carried by a single proton.

To compute observable quantities in quantum field theory or string theory, physicists typically apply the methods of perturbation theory. In perturbation theory, quantities called probability amplitudes, which determine the probability for various physical processes to occur, are expressed as sums of infinitely many terms, where each term is proportional to a power of the coupling constant g :

$$A = A_0 + A_1 g + A_2 g^2 + A_3 g^3 + \ldots$$

In order for such an expression to make sense, the coupling constant must be less than 1 so that the higher powers of g become negligibly small and the sum is finite. If the coupling constant is not less than 1, then the terms of this sum will grow larger and larger, and the expression gives a meaningless infinite answer. In this case the theory is said to be *strongly coupled*, and one cannot use perturbation theory to make predictions.

For certain theories, S-duality provides a way of doing computations at strong coupling by translating these computations into different computations in a weakly coupled theory. S-duality is a particular example of a general notion of duality in physics. The term *duality* refers to a situation where two seemingly different physical systems turn out to be equivalent in a nontrivial way. If two theories are related by a duality, it means that one theory can be transformed in some way so that it ends up looking just like the other theory. The two theories are then said to be *dual* to one another under the transformation. Put differently, the two theories are mathematically different descriptions of the same phenomena.

S-duality is useful because it relates a theory with coupling constant g to an equivalent theory with coupling constant $1/g$. Thus it relates a strongly coupled theory (where the coupling constant g is much greater than 1) to a weakly coupled theory (where the coupling constant $1/g$ is much less than 1 and computations are possible). For this reason, S-duality is called a **strong-weak duality**.

6.2 S-duality in quantum field theory

6.2.1 A symmetry of Maxwell's equations

In classical physics, the behavior of the electric and magnetic field is described by a system of equations known as Maxwell's equations. Working in the language of vector calculus and assuming that no electric charges or currents are present, these equations can be written[3]

$$\nabla \cdot \mathbf{E} = 0,$$
$$\nabla \cdot \mathbf{B} = 0,$$
$$\nabla \times \mathbf{E} = -\frac{\partial \mathbf{B}}{\partial t},$$
$$\nabla \times \mathbf{B} = \frac{1}{c^2}\frac{\partial \mathbf{E}}{\partial t}.$$

Here $\mathbf{E}$ is a vector (or more precisely a *vector field* whose magnitude and direction may vary from point to point in space) representing the electric field, $\mathbf{B}$ is a vector representing the magnetic field, t is time, and c is the speed of light. The other symbols in these equations refer to the divergence and curl, which are concepts from vector calculus.

An important property of these equations[4] is their invariance under the transformation that simultaneously replaces the electric field $\mathbf{E}$ by the magnetic field $\mathbf{B}$ and replaces $\mathbf{B}$ by $-1/c^2 \mathbf{E}$:

$$\mathbf{E} \to \mathbf{B}$$
$$\mathbf{B} \to -\frac{1}{c^2}\mathbf{E}.$$

In other words, given a pair of electric and magnetic fields that solve Maxwell's equations, it is possible to describe a new physical setup in which these electric and magnetic fields are essentially interchanged, and the new fields will again give a solution of Maxwell's equations. This situation is the most basic manifestation of S-duality in quantum field theory.

6.2.2 Montonen–Olive duality

Main article: Montonen–Olive duality

In quantum field theory, the electric and magnetic fields are unified into a single entity called the electromagnetic field, and this field is described by a special type of quantum field theory called a gauge theory or Yang–Mills theory. In a gauge theory, the physical fields have a high degree of symmetry which can be understood mathematically using the notion of a Lie group. This Lie group is known as the gauge group. The electromagnetic field is described by a very simple gauge theory corresponding to the abelian gauge group U(1), but there are other gauge theories with more complicated non-abelian gauge groups.[5]

It is natural to ask whether there is an analog in gauge theory of the symmetry interchanging the electric and magnetic fields in Maxwell's equations. The answer was given in the late 1970s by Claus Montonen and David Olive,[6] building on earlier work of Peter Goddard, Jean Nuyts, and Olive.[7] Their work provides an example of S-duality now known as Montonen–Olive duality. Montonen–Olive duality applies to a very special type of gauge theory called N = 4 supersymmetric Yang–Mills theory, and it says that two such theories may be equivalent in a certain precise sense.[1] If one of the theories has a gauge group G, then the dual theory has gauge group ^{L}G where ^{L}G denotes the Langlands dual group which is in general different from G.[8]

An important quantity in quantum field theory is complexified coupling constant. This is a complex number defined by the formula[9]

$$\tau = \frac{\theta}{2\pi} + \frac{4\pi i}{g^2}$$

where θ is the theta angle, a quantity appearing in the Lagrangian that defines the theory,[9] and g is the coupling constant. For example, in the Yang–Mills theory that describes the electromagnetic field, this number g is simply the elementary charge e carried by a single proton.[1] In addition to exchanging the gauge groups of the two theories, Montonen–Olive duality transforms a theory with complexified coupling coupling constant τ to a theory with complexified constant $-1/\tau$.[9]

6.2.3 Relation to the Langlands program

Main article: Langlands program

In mathematics, the classical Langlands correspondence is a collection of results and conjectures relating number theory to the branch of mathematics known as representation theory.[10] Formulated by Robert Langlands in the late 1960s, the Langlands correspondence is related to important conjectures in number theory such as the Taniyama-Shimura

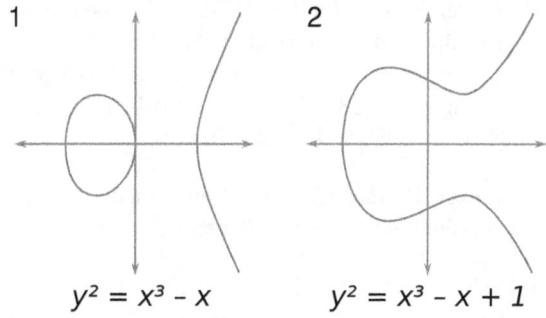

$$y^2 = x^3 - x \qquad y^2 = x^3 - x + 1$$

The geometric Langlands correspondence is a relationship between abstract geometric objects associated to an algebraic curve such as the elliptic curves illustrated above.

conjecture, which includes Fermat's last theorem as a special case.[10]

In spite of its importance in number theory, establishing the Langlands correspondence in the number theoretic context has proved extremely difficult.[10] As a result, some mathematicians have worked on a related conjecture known as the geometric Langlands correspondence. This is a geometric reformulation of the classical Langlands correspondence which is obtained by replacing the number fields appearing in the original version by function fields and applying techniques from algebraic geometry.[10]

In a paper from 2007, Anton Kapustin and Edward Witten suggested that the geometric Langlands correspondence can be viewed as a mathematical statement of Montonen–Olive duality.[11] Starting with two Yang–Mills theories related by S-duality, Kapustin and Witten showed that one can construct a pair of quantum field theories in two-dimensional spacetime. By analyzing what this dimensional reduction does to certain physical objects called D-branes, they showed that one can recover the mathematical ingredients of the geometric Langlands correspondence.[12] Their work shows that the Langlands correspondence is closely related to S-duality in quantum field theory, with possible applications in both subjects.[10]

6.2.4   Seiberg duality

Main article: Seiberg duality

Another realization of S-duality in quantum field theory is Seiberg duality, first introduced by Nathan Seiberg around 1995.[13] Unlike Montonen–Olive duality, which relates two versions of the maximally supersymmetric gauge theory in four-dimensional spacetime, Seiberg duality relates less symmetric theories called N=1 supersymmetric gauge theories. The two N=1 theories appearing in Seiberg dual-

ity are not identical, but they give rise to the same physics at large distances. Like Montonen–Olive duality, Seiberg duality generalizes the symmetry of Maxwell's equations that interchanges electric and magnetic fields.

6.3   S-duality in string theory

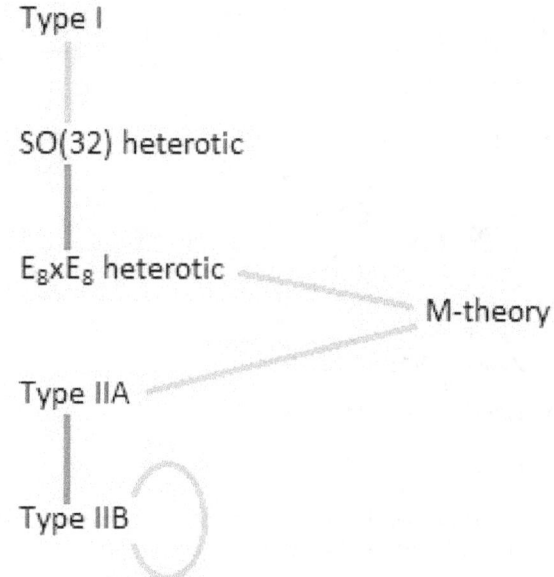

A diagram of string theory dualities. Yellow lines indicate S-duality. Blue lines indicate T-duality.

Up until the mid 1990s, physicists working on string theory believed there were five distinct versions of the theory: type I, type IIA, type IIB, and the two flavors of heterotic string theory (SO(32) and $E_8 \times E_8$). The different theories allow different types of strings, and the particles that arise at low energies exhibit different symmetries.

In the mid 1990s, physicists noticed that these five string theories are actually related by highly nontrivial dualities. One of these dualities is S-duality. The existence of S-duality in string theory was first proposed by Ashoke Sen in 1994.[14] It was shown that type IIB string theory with the coupling constant g is equivalent via S-duality to the same string theory with the coupling constant $1/g$. Similarly, type I string theory with the coupling g is equivalent to the SO(32) heterotic string theory with the coupling constant $1/g$.

The existence of these dualities showed that the five string theories were in fact not all distinct theories. In 1995, at the string theory conference at University of Southern California, Edward Witten made the surprising suggestion that all five of these theories were just different limits of a single theory now known as M-theory.[15] Witten's proposal was based on the observation that type IIA and $E_8 \times E_8$ heterotic string theories are closely related to a gravitational theory called eleven-dimensional supergravity. His announcement led to a flurry of work now known as the second superstring revolution.

6.4 See also

- T-duality

- Mirror symmetry

- AdS/CFT correspondence

6.5 Notes

[1] Frenkel 2009, p.2

[2] Zwiebach 2009, p.325

[3] Griffiths 1999, p.326

[4] Griffiths 1999, p.327

[5] For an introduction to quantum field theory in general including the basics of gauge theory, see Zee 2010.

[6] Montonen and Olive 1977

[7] Goddard, Nuyts, and Olive 1977

[8] Frenkel 2009, p.5

[9] Frenkel 2009, p.12

[10] Frenkel 2007

[11] Kapustin and Witten 2007

[12] Aspinwall et al. 2009, p.415

[13] Seiberg 1995

[14] Sen 1994

[15] Witten 1995

6.6 References

- Aspinwall, Paul; Bridgeland, Tom; Craw, Alastair; Douglas, Michael; Gross, Mark; Kapustin, Anton; Moore, Gregory; Segal, Graeme; Szendrői, Balázs; Wilson, P.M.H., eds. (2009). *Dirichlet Branes and Mirror Symmetry*. American Mathematical Society. ISBN 978-0-8218-3848-8.

- Frenkel, Edward (2007). "Lectures on the Langlands program and conformal field theory". *Frontiers in number theory, physics, and geometry II* (Springer): 387–533. arXiv:hep-th/0512172. Bibcode:2005hep.th...12172F.

- Frenkel, Edward (2009). "Gauge theory and Langlands duality". *Seminaire Bourbaki*.

- Goddard, Peter; Nuyts, Jean; Olive, David (1977). "Gauge theories and magnetic charge". *Nuclear Physics B* **125** (1): 1–28. Bibcode:1977NuPhB.125....1G. doi:10.1016/0550-3213(77)90221-8.

- Griffiths, David (1999). *Introduction to Electrodynamics*. New Jersey: Prentice-Hall.

- Kapustin, Anton; Witten, Edward (2007). "Electric-magnetic duality and the geometric Langlands program". *Communications in Number Theory and Physics* **1** (1): 1–236. arXiv:hep-th/0604151. Bibcode:2007CNTP....1....1K. doi:10.4310/cntp.2007.v1.n1.a1.

- Montonen, Claus; Olive, David (1977). "Magnetic monopoles as gauge particles?". *Physics Letters B* **72** (1): 117–120. Bibcode:1977PhLB...72..117M. doi:10.1016/0370-2693(77)90076-4.

- Seiberg, Nathan (1995). "Electric-magnetic duality in supersymmetric non-Abelian gauge theories". *Nuclear Physics B* **435** (1): 129–146. arXiv:hep-th/9411149. Bibcode:1995NuPhB.435..129S. doi:10.1016/0550-3213(94)00023-8.

- Sen, Ashoke (1994). "Strong-weak coupling duality in four-dimensional string theory". *International Journal of Modern Physics A* **9** (21): 3707–3750. arXiv:hep-th/9402002. Bibcode:1994IJMPA...9.3707S. doi:10.1142/S0217751X94001497.

- Witten, Edward (March 13–18, 1995). "Some problems of strong and weak coupling". *Proceedings of Strings '95: Future Perspectives in String Theory*. World Scientific.

- Witten, Edward (1995). "String theory dynamics in various dimensions". *Nuclear Physics B* **443** (1): 85–126. arXiv:hep-th/9503124. Bibcode:1995NuPhB.443...85W. doi:10.1016/0550-3213(95)00158-O.

- Zee, Anthony (2010). *Quantum Field Theory in a Nutshell* (2nd ed.). Princeton University Press. ISBN 978-0-691-14034-6.

- Zwiebach, Barton (2009). *A First Course in String Theory*. Cambridge University Press. ISBN 978-0-521-88032-9.

Chapter 7

T-duality

In theoretical physics, **T-duality** is an equivalence of two physical theories, which may be either quantum field theories or string theories. In the simplest example of this relationship, one of the theories describes strings propagating in an imaginary spacetime shaped like a circle of some radius R, while the other theory describes strings propagating on a spacetime shaped like a circle of radius $1/R$. The two theories are equivalent in the sense that all observable quantities in one description are identified with quantities in the dual description. For example, momentum in one description takes discrete values and is equal to the number of times the string winds around the circle in the dual description.

The idea of T-duality can be extended to more complicated theories, including superstring theories. The existence of these dualities implies that seemingly different superstring theories are actually physically equivalent. This led to the realization, in the mid-1990s, that all of the five consistent superstring theories are just different limiting cases of a single eleven-dimensional theory called M-theory.

In general, T-duality relates two theories with different spacetime geometries. In this way, T-duality suggests a possible scenario in which the classical notions of geometry break down in a theory of Planck scale physics.[1] The geometric relationships suggested by T-duality are also important in pure mathematics. Indeed, according to the SYZ conjecture of Andrew Strominger, Shing-Tung Yau, and Eric Zaslow, T-duality is closely related to another duality called mirror symmetry, which has important applications in a branch of mathematics called enumerative algebraic geometry.

7.1 Overview

7.1.1 Strings and duality

T-duality is a particular example of a general notion of duality in physics. The term *duality* refers to a situation where two seemingly different physical systems turn out to be equivalent in a nontrivial way. If two theories are related by a duality, it means that one theory can be transformed in some way so that it ends up looking just like the other theory. The two theories are then said to be *dual* to one another under the transformation. Put differently, the two theories are mathematically different descriptions of the same phenomena.

Like many of the dualities studied in theoretical physics, T-duality was discovered in the context of string theory.[2] In string theory, particles are modeled not as zero-dimensional points but as one-dimensional extended objects called strings. The physics of strings can be studied in various numbers of dimensions. In addition to three familiar dimensions from everyday experience (up/down, left/right, forward/backward), string theories may include one or more compact dimensions which are curled up into circles.

A standard analogy for this is to consider multidimensional object such as a garden hose.[3] If the hose is viewed from a sufficient distance, it appears to have only one dimension, its length. However, as one approaches the hose, one discovers that it contains a second dimension, its circumference. Thus, an ant crawling inside it would move in two dimensions. Such extra dimensions are important in T-duality, which relates a theory in which strings propagate on a circle of some radius R to a theory in which strings propagate on a circle of radius $1/R$.

7.1.2 Winding numbers

Main article: Winding number

In mathematics, the winding number of a curve in the plane around a given point is an integer representing the total number of times that curve travels counterclockwise around the point. The notion of winding number is important in the mathematical description of T-duality where it is used to measure the winding of strings around compact extra dimensions.

For example, the image below shows several examples of curves in the plane, illustrated in red. Each curve is assumed to be closed, meaning it has no endpoints, and is allowed to intersect itself. Each curve has an orientation given by the arrows in the picture. In each situation, there is a distinguished point in the plane, illustrated in black. The *winding number* of the curve around this distinguished point is equal to the total number of counterclockwise turns that the curve makes around this point.

When counting the total number of turns, counterclockwise turns count as positive, while clockwise turns counts as negative. For example, if the curve first circles the origin four times counterclockwise, and then circles the origin once clockwise, then the total winding number of the curve is three. According to this scheme, a curve that does not travel around the distinguished point at all has winding number zero, while a curve that travels clockwise around the point has negative winding number. Therefore, the winding number of a curve may be any integer. The pictures above show curves with winding numbers between −2 and 3:

7.1.3 Quantized momenta

The simplest theories in which T-duality arises are two-dimensional sigma models with circular target spaces. These are simple quantum field theories that describe propagation of strings in an imaginary spacetime shaped like a circle. The strings can thus be modeled as curves in the plane that are confined to lie in a circle, say of radius R, about the origin. In what follows, the strings are assumed to be closed (that is, without endpoints).

Denote this circle by S_R^1. One can think of this circle as a copy of the real line with two points identified if they differ by a multiple of the circle's circumference $2\pi R$. It follows that the state of a string at any given time can be represented as a function $\varphi(\theta)$ of a single real parameter θ. Such a function can be expanded in a Fourier series as

$$\varphi(\theta) = mR\theta + x + \sum_{n\neq 0} c_n e^{in\theta}$$

Here m denotes the winding number of the string around the circle, and the constant mode $x = c_0$ of the Fourier series has been singled out. Since this expression represents the configuration of a string at a fixed time, all coefficients (x and the c_n) are also functions of time.

Let $\dot{x}$ denote the time derivative of the constant mode x. This represents a type of momentum in the theory. One can show, using the fact that the strings considered here are closed, that this momentum can only take on discrete values of the form $\dot{x} = n/R$ for some integer n. In more

physical language, one says that the momentum spectrum is *quantized*.

7.1.4 An equivalence of theories

In the situation described above, the total energy, or Hamiltonian, of the string is given by the expression

$$H = (mR)^2 + \dot{x}^2 + \sum_n |\dot{c}_n|^2 + n^2|c_n|^2$$

Since the momenta of the theory are quantized, the first two terms in this formula are $(mR)^2 + (n/R)^2$, and this expression is unchanged when one simultaneously replaces the radius R by $1/R$ and exchanges the winding number m and the integer n. The summation in the expression for H is similarly unaffected by these changes, so the total energy is unchanged. In fact, this equivalence of Hamiltonians descends to an equivalence of two quantum mechanical theories: One of these theories describes strings propagating on a circle of radius R, while the other describes string propagating in a circle of radius $1/R$ with momentum and winding numbers interchanged. This equivalence of theories is the simplest manifestation of T-duality.

7.2 Superstrings

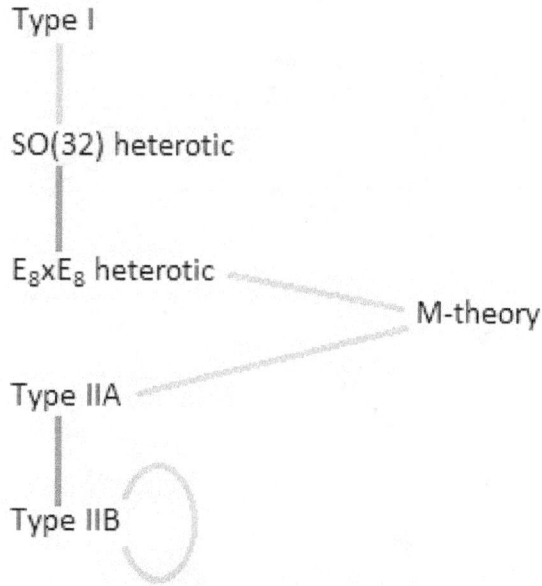

A diagram of string theory dualities. Yellow lines indicate S-duality. Blue lines indicate T-duality.

Up until the mid 1990s, physicists working on string theory believed there were five distinct versions of the theory: type

I, type IIA, type IIB, and the two flavors of heterotic string theory (SO(32) and $E_8 \times E_8$). The different theories allow different types of strings, and the particles that arise at low energies exhibit different symmetries.

In the mid 1990s, physicists noticed that these five string theories are actually related by highly nontrivial dualities. One of these dualities is T-duality. For example, it was shown that type IIA string theory is equivalent to type IIB string theory via T-duality and also that the two versions of heterotic string theory are related by T-duality.

The existence of these dualities showed that the five string theories were in fact not all distinct theories. In 1995, at the string theory conference at University of Southern California, Edward Witten made the surprising suggestion that all five of these theories were just different limits of a single theory now known as M-theory.[4] Witten's proposal was based on the observation that different superstring theories are linked by dualities and the fact that type IIA and $E_8 \times E_8$ heterotic string theories are closely related to a gravitational theory called eleven-dimensional supergravity. His announcement led to a flurry of work now known as the second superstring revolution.

7.3 Mirror symmetry

Main article: Mirror symmetry (string theory)

In string theory and algebraic geometry, the term "mirror

A hypersurface of a six-dimensional Calabi–Yau manifold.

symmetry" refers to a phenomenon involving complicated shapes called Calabi-Yau manifolds. These manifolds pro-

vide an interesting geometry on which strings can propagate, and the resulting theories may have applications in particle physics.[5] In the late 1980s, it was noticed that such a Calabi-Yau manifold does not uniquely determine the physics of the theory. Instead, one finds that there are *two* Calabi-Yau manifolds that give rise to the same physics.[6] These manifolds are said to be "mirror" to one another. This mirror duality is an important computational tool in string theory, and it has allowed mathematicians to solve difficult problems in enumerative geometry.[7]

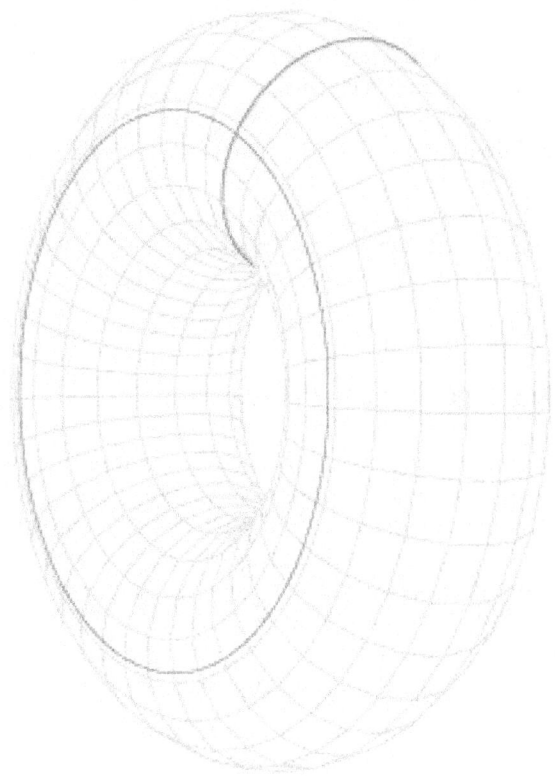

A torus is the cartesian product of two circles.

One approach to understanding mirror symmetry is the SYZ conjecture, which was suggested by Andrew Strominger, Shing-Tung Yau, and Eric Zaslow in 1996.[8] According to the SYZ conjecture, mirror symmetry can be understood by dividing a complicated Calabi-Yau manifold into simpler pieces and considering the effects of T-duality on these pieces.[9]

The simplest example of a Calabi-Yau manifold is a torus (a surface shaped like a donut). Such a surface can be viewed as the product of two circles. This means that the torus can be viewed as the union of a collection of longitudinal circles (such as the red circle in the image). There is an auxiliary space which says how these circles are organized, and this space is itself a circle (the pink circle). This space is said to *parametrize* the longitudinal circles on the torus. In this case, mirror symmetry is equivalent to T-duality acting on the longitudinal circles, changing their radii from R to $1/R$.

The SYZ conjecture generalizes this idea to the more complicated case of six-dimensional Calabi-Yau manifolds like the one illustrated above. As in the case of a torus, one can divide a six-dimensional Calabi-Yau manifold into simpler pieces, which in this case are 3-tori (three-dimensional objects which generalize the notion of a torus) parametrized by a 3-sphere (a three-dimensional generalization of a sphere).[10] T-duality can be extended from circles to the three-dimensional tori appearing in this decomposition, and the SYZ conjecture states that mirror symmetry is equivalent to the simultaneous application of T-duality to these three-dimensional tori.[11] In this way, the SYZ conjecture provides a geometric picture of how mirror symmetry acts on a Calabi-Yau manifold.

7.4   See also

- S-duality

- Mirror symmetry

- AdS/CFT correspondence

7.5   Notes

[1] Seiberg 2006

[2] Other dualities that arise in string theory are S-duality, U-duality, mirror symmetry, and the AdS/CFT correspondence.

[3] This analogy is used for example in Greene 2000, p.186

[4] Witten 1995

[5] Candelas et al. 1985

[6] Dixon 1988; Lerche, Vafa, and Warner 1989

[7] Zaslow 2008

[8] Strominger, Yau, and Zaslow 1996

[9] Yau and Nadis 2010, p.174

[10] More precisely, there is a 3-torus associated to every point on the three-sphere except at certain bad points, which correspond to singular tori. See Yau and Nadis 2010, pp.176–7.

[11] Yau and Nadis 2010, p.178

7.6   References

- Candelas, Philip; Horowitz, Gary; Strominger, Andrew; Witten, Edward (1985). "Vacuum configurations for superstrings". *Nuclear Physics B* **258**: 46–74. Bibcode:1985NuPhB.258...46C. doi:10.1016/0550-3213(85)90602-9.

- Dixon, Lance (1988). "Some world-sheet properties of superstring compactifications, on orbifolds and otherwise". *ICTP Ser. Theoret. Phys.* **4**: 67–126.

- Greene, Brian (2000). *The Elegant Universe: Superstrings, Hidden Dimensions, and the Quest for the Ultimate Theory*. Random House. ISBN 978-0-9650888-0-0.

- Lerche, Wolfgang; Vafa, Cumrun; Warner, Nicholas (1989). "Chiral rings in $N = 2$ superconformal theories". *Nuclear Physics B* **324** (2): 427–474. Bibcode:1989NuPhB.324..427L. doi:10.1016/0550-3213(89)90474-4.

- Seiberg, Nathan (2006). "Emergent Spacetime". arXiv:hep-th/0601234.

- Strominger, Andrew; Yau, Shing-Tung; Zaslow, Eric (1996). "Mirror symmetry is T-duality". *Nuclear Physics B* **479** (1): 243–259. arXiv:hep-th/9606040. Bibcode:1996NuPhB.479..243S. doi:10.1016/0550-3213(96)00434-8.

- Witten, Edward (March 13–18, 1995). "Some problems of strong and weak coupling". *Proceedings of Strings '95: Future Perspectives in String Theory*. World Scientific.

- Witten, Edward (1995). "String theory dynamics in various dimensions". *Nuclear Physics B* **443** (1): 85–126. arXiv:hep-th/9503124. Bibcode:1995NuPhB.443...85W. doi:10.1016/0550-3213(95)00158-O.

- Yau, Shing-Tung; Nadis, Steve (2010). *The Shape of Inner Space: String Theory and the Geometry of the Universe's Hidden Dimensions*. Basic Books. ISBN 978-0-465-02023-2.

- Zaslow, Eric (2008). "Mirror Symmetry". In Gowers, Timothy. *The Princeton Companion to Mathematics*. ISBN 978-0-691-11880-2.

Chapter 8

D-brane

In string theory, **D-branes** are a class of extended objects upon which open strings can end with Dirichlet boundary conditions, after which they are named. D-branes were discovered by Dai, Leigh and Polchinski, and independently by Hořava in 1989. In 1995, Polchinski identified D-branes with black p-brane solutions of supergravity, a discovery that triggered the Second Superstring Revolution and led to both holographic and M-theory dualities.

D-branes are typically classified by their spatial dimension, which is indicated by a number written after the *D*. A D0-brane is a single point, a D1-brane is a line (sometimes called a "D-string"), a D2-brane is a plane, and a D25-brane fills the highest-dimensional space considered in bosonic string theory. There are also instantonic D(-1)-branes, which are localized in both space and time.

8.1   Theoretical background

The equations of motion of string theory require that the endpoints of an open string (a string with endpoints) satisfy one of two types of boundary conditions: The Neumann boundary condition, corresponding to free endpoints moving through spacetime at the speed of light, or the Dirichlet boundary conditions, which pin the string endpoint. Each coordinate of the string must satisfy one or the other of these conditions. There can also exist strings with mixed boundary conditions, where the two endpoints satisfy NN, DD, ND and DN boundary conditions. If p spatial dimensions satisfy the Neumann boundary condition, then the string endpoint is confined to move within a p-dimensional hyperplane. This hyperplane provides one description of a Dp-brane.

Although rigid in the limit of zero coupling, the spectrum of open strings ending on a D-brane contains modes associated with its fluctuations, implying that D-branes are dynamical objects. When N D-branes are nearly coincident, the spectrum of strings stretching between them becomes very rich. One set of modes produce a non-abelian gauge theory on the world-volume. Another set of modes is an $N \times N$ dimensional matrix for each transverse dimension of the brane. If these matrices commute, they may be diagonalized, and the eigenvalues define the position of the N D-branes in space. More generally, the branes are described by non-commutative geometry, which allows exotic behavior such as the Myers effect, in which a collection of Dp-branes expand into a D(p+2)-brane.

Tachyon condensation is a central concept in this field. Ashoke Sen has argued that in Type IIB string theory, tachyon condensation allows (in the absence of Neveu-Schwarz 3-form flux) an arbitrary D-brane configuration to be obtained from a stack of D9 and anti D9-branes. Edward Witten has shown that such configurations will be classified by the K-theory of the spacetime. Tachyon condensation is still very poorly understood. This is due to the lack of an exact string field theory that would describe the off-shell evolution of the tachyon.

8.2   Braneworld cosmology

This has implications for physical cosmology. Because string theory implies that the Universe has more dimensions than we expect—26 for bosonic string theories and 10 for superstring theories—we have to find a reason why the extra dimensions are not apparent. One possibility would be that the visible Universe is in fact a very large D-brane extending over three spatial dimensions. Material objects, made of open strings, are bound to the D-brane, and cannot move "at right angles to reality" to explore the Universe outside the brane. This scenario is called a brane cosmology. The force of gravity is *not* due to open strings; the gravitons which carry gravitational forces are vibrational states of *closed* strings. Because closed strings do not have to be attached to D-branes, gravitational effects could depend upon the extra dimensions orthogonal to the brane.

8.3 D-brane scattering

When two D-branes approach each other the interaction is captured by the one loop annulus amplitude of strings between the two branes. The scenario of two parallel branes approaching each other at a constant velocity can be mapped to the problem of two stationary branes that are rotated relative to each other by some angle. The annulus amplitude yields singularities that correspond to the on-shell production of open strings stretched between the two branes. This is true irrespective of the charge of the D-branes. At non-relativistic scattering velocities the open strings may be described by a low-energy effective action that contains two complex scalar fields that are coupled via a term $\phi^2\chi^2$. Thus, as the field ϕ (separation of the branes) changes, the mass of the field χ changes. This induces open string production and as a result the two scattering branes will be trapped.

8.4 Gauge theories

The arrangement of D-branes constricts the types of string states which can exist in a system. For example, if we have two parallel D2-branes, we can easily imagine strings stretching from brane 1 to brane 2 or vice versa. (In most theories, strings are *oriented* objects: each one carries an "arrow" defining a direction along its length.) The open strings permissible in this situation then fall into two categories, or "sectors": those originating on brane 1 and terminating on brane 2, and those originating on brane 2 and terminating on brane 1. Symbolically, we say we have the [1 2] and the [2 1] sectors. In addition, a string may begin and end on the same brane, giving [1 1] and [2 2] sectors. (The numbers inside the brackets are called *Chan-Paton indices*, but they are really just labels identifying the branes.) A string in either the [1 2] or the [2 1] sector has a minimum length: it cannot be shorter than the separation between the branes. All strings have some tension, against which one must pull to lengthen the object; this pull does work on the string, adding to its energy. Because string theories are by nature relativistic, adding energy to a string is equivalent to adding mass, by Einstein's relation $E = mc^2$. Therefore, the separation between D-branes controls the minimum mass open strings may have.

Furthermore, affixing a string's endpoint to a brane influences the way the string can move and vibrate. Because particle states "emerge" from the string theory as the different vibrational states the string can experience, the arrangement of D-branes controls the types of particles present in the theory. The simplest case is the [1 1] sector for a Dp-brane, that is to say the strings which begin and end on any particular D-brane of p dimensions. Examining the conse-

quences of the Nambu-Goto action (and applying the rules of quantum mechanics to quantize the string), one finds that among the spectrum of particles is one resembling the photon, the fundamental quantum of the electromagnetic field. The resemblance is precise: a p-dimensional version of the electromagnetic field, obeying a p-dimensional analogue of Maxwell's equations, exists on every Dp-brane.

In this sense, then, one can say that string theory "predicts" electromagnetism: D-branes are a necessary part of the theory if we permit open strings to exist, and all D-branes carry an electromagnetic field on their volume.

Other particle states originate from strings beginning and ending on the same D-brane. Some correspond to massless particles like the photon; also in this group are a set of massless scalar particles. If a Dp-brane is embedded in a spacetime of d spatial dimensions, the brane carries (in addition to its Maxwell field) a set of $d - p$ massless scalars (particles which do not have polarizations like the photons making up light). Intriguingly, there are just as many massless scalars as there are directions perpendicular to the brane; the *geometry* of the brane arrangement is closely related to the *quantum field theory* of the particles existing on it. In fact, these massless scalars are Goldstone excitations of the brane, corresponding to the different ways the symmetry of empty space can be broken. Placing a D-brane in a universe breaks the symmetry among locations, because it defines a particular place, assigning a special meaning to a particular location along each of the $d - p$ directions perpendicular to the brane.

The quantum version of Maxwell's electromagnetism is only one kind of gauge theory, a $U(1)$ gauge theory where the gauge group is made of unitary matrices of order 1. D-branes can be used to generate gauge theories of higher order, in the following way:

Consider a group of N separate Dp-branes, arranged in parallel for simplicity. The branes are labeled 1,2,...,N for convenience. Open strings in this system exist in one of many sectors: the strings beginning and ending on some brane i give that brane a Maxwell field and some massless scalar fields on its volume. The strings stretching from brane i to another brane j have more intriguing properties. For starters, it is worthwhile to ask which sectors of strings can interact with one another. One straightforward mechanism for a string interaction is for two strings to join endpoints (or, conversely, for one string to "split down the middle" and make two "daughter" strings). Since endpoints are restricted to lie on D-branes, it is evident that a [1 2] string may interact with a [2 3] string, but not with a [3 4] or a [4 17] one. The masses of these strings will be influenced by the separation between the branes, as discussed above, so for simplicity's sake we can imagine the branes squeezed closer and closer together, until they lie atop one another. If we regard two overlapping branes as distinct objects, then we still have all the sectors we had before, but without the effects due to the brane separations.

The zero-mass states in the open-string particle spectrum for a system of N coincident D-branes yields a set of interacting quantum fields which is exactly a $U(N)$ gauge theory. (The string theory does contain other interactions, but they are only detectable at very high energies.) Gauge theories were not invented starting with bosonic or fermionic strings; they originated from a different area of physics, and have become quite useful in their own right. If nothing else, the relation between D-brane geometry and gauge theory offers a useful pedagogical tool for explaining gauge interactions, even if string theory fails to be the "theory of everything".

8.5   Black holes

Another important use of D-branes has been in the study of black holes. Since the 1970s, scientists have debated the problem of black holes having entropy. Consider, as a thought experiment, dropping an amount of hot gas into a black hole. Since the gas cannot escape from the hole's gravitational pull, its entropy would seem to have vanished from the universe. In order to maintain the second law of thermodynamics, one must postulate that the black hole gained whatever entropy the infalling gas originally had. Attempting to apply quantum mechanics to the study of black holes, Stephen Hawking discovered that a hole should emit energy with the characteristic spectrum of thermal radiation. The characteristic temperature of this Hawking radiation is given by

$$T_{\mathrm{H}} = \frac{\hbar c^3}{8\pi G M k_B} \quad \left(\approx \frac{1.227 \times 10^{23}\ kg}{M}\ K \right)$$

where G is Newton's gravitational constant, M is the black hole's mass and kB is Boltzmann's constant.

Using this expression for the Hawking temperature, and assuming that a zero-mass black hole has zero entropy, one can use thermodynamic arguments to derive the "Bekenstein entropy":

$$S_{\mathrm{B}} = \frac{k_B 4\pi G}{\hbar c} M^2.$$

The Bekenstein entropy is proportional to the black hole mass squared; because the Schwarzschild radius is proportional to the mass, the Bekenstein entropy is proportional to the black hole's *surface area*. In fact,

$$S_{\mathrm{B}} = \frac{A k_B}{4 l_{\mathrm{P}}^2},$$

where l_{P} is the Planck length.

The concept of black hole entropy poses some interesting conundra. In an ordinary situation, a system has entropy when a large number of different "microstates" can satisfy the same macroscopic condition. For example, given a box full of gas, many different arrangements of the gas atoms can have the same total energy. However, a black hole was believed to be a featureless object (in John Wheeler's catchphrase, "Black holes have no hair"). What, then, are the "degrees of freedom" which can give rise to black hole entropy?

String theorists have constructed models in which a black hole is a very long (and hence very massive) string. This model gives rough agreement with the expected entropy of a Schwarzschild black hole, but an exact proof has yet to be found one way or the other. The chief difficulty is that it is relatively easy to count the degrees of freedom quantum strings possess *if they do not interact with one another*. This is analogous to the ideal gas studied in introductory thermodynamics: the easiest situation to model is when the gas atoms do not have interactions among themselves. Developing the kinetic theory of gases in the case where the gas atoms or molecules experience inter-particle forces (like the van der Waals force) is more difficult. However, a world without interactions is an uninteresting place: most significantly for the black hole problem, gravity is an interaction, and so if the "string coupling" is turned off, no black hole could ever arise. Therefore, calculating black hole entropy requires working in a regime where string interactions exist.

Extending the simpler case of non-interacting strings to the regime where a black hole could exist requires supersymmetry. In certain cases, the entropy calculation done for zero string coupling remains valid when the strings interact. The challenge for a string theorist is to devise a situation in which a black hole can exist which does not "break" supersymmetry. In recent years, this has been done by building black holes out of D-branes. Calculating the entropies of these hypothetical holes gives results which agree with the expected Bekenstein entropy. Unfortunately, the cases studied so far all involve higher-dimensional spaces — D5-branes in nine-dimensional space, for example. They do not directly apply to the familiar case, the Schwarzschild black holes observed in our own universe.

8.6   History

Dirichlet boundary conditions and D-branes had a long "pre-history" before their full significance was recognized. Mixed Dirichlet/Neumann boundary conditions were first considered by Warren Siegel in 1976 as a means of lowering the critical dimension of open string theory from 26 or

10 to 4 (Siegel also cites unpublished work by Halpern, and a 1974 paper by Chodos and Thorn, but a reading of the latter paper shows that it is actually concerned with linear dilation backgrounds, not Dirichlet boundary conditions). This paper, though prescient, was little-noted in its time (a 1985 parody by Siegel, "The Super-g String," contains an almost dead-on description of braneworlds). Dirichlet conditions for all coordinates including Euclidean time (defining what are now known as D-instantons) were introduced by Michael Green in 1977 as a means of introducing point-like structure into string theory, in an attempt to construct a string theory of the strong interaction. String compactifications studied by Harvey and Minahan, Ishibashi and Onogi, and Pradisi and Sagnotti in 1987-89 also employed Dirichlet boundary conditions.

The fact that T-duality interchanges the usual Neumann boundary conditions with Dirichlet boundary conditions was discovered independently by Horava and by Dai, Leigh, and Polchinski in 1989; this result implies that such boundary conditions must necessarily appear in regions of the moduli space of any open string theory. The Dai et al. paper also notes that the locus of the Dirichlet boundary conditions is dynamical, and coins the term Dirichlet-brane (D-brane) for the resulting object (this paper also coins orientifold for another object that arises under string T-duality). A 1989 paper by Leigh showed that D-brane dynamics are governed by the Dirac-Born-Infeld action. D-instantons were extensively studied by Green in the early 1990s, and were shown by Polchinski in 1994 to produce the $e^{-1/g}$ nonperturbative string effects anticipated by Shenker. In 1995 Polchinski showed that D-branes are the sources of electric and magnetic Ramond–Ramond fields that are required by string duality, leading to rapid progress in the nonperturbative understanding of string theory.

8.7 See also

- Bogomol'nyi–Prasad–Sommerfield bound

- M-theory

8.8 References

- Bachas, C. P. "Lectures on D-branes" (1998). arXiv:hep-th/9806199.

- Giveon, A. and Kutasov, D. "Brane dynamics and gauge theory", *Rev. Mod. Phys.* **71**, 983 (1999). arXiv:hep-th/9802067.

- Hashimoto, Koji, *D-Brane: Superstrings and New Perspective of Our World.* Springer (2012). ISBN 978-3-642-23573-3

- Johnson, Clifford (2003). *D-branes.* Cambridge: Cambridge University Press. ISBN 0-521-80912-6.

- Polchinski, Joseph, *TASI Lectures on D-branes*, arXiv:hep-th/9611050. Lectures given at TASI '96.

- Polchinski, Joseph, *Phys. Rev. Lett.* **75**, 4724 (1995). An article which established D-branes' significance in string theory.

- Zwiebach, Barton. *A First Course in String Theory.* Cambridge University Press (2004). ISBN 0-521-83143-1.

Chapter 9

Brane

For other uses, see Brane (disambiguation).

In string theory and related theories such as supergravity theories, a **brane** is a physical object that generalizes the notion of a point particle to higher dimensions. For example, a point particle can be viewed as a brane of dimension zero, while a string can be viewed as a brane of dimension one. It is also possible to consider higher-dimensional branes. In dimension p, these are called p-branes. The word "brane" comes from the word "membrane" which refers to a two-dimensional brane.[1]

Branes are dynamical objects which can propagate through spacetime according to the rules of quantum mechanics. They have mass and can have other attributes such as charge. A p-brane sweeps out a $(p+1)$-dimensional volume in spacetime called its *worldvolume*. Physicists often study fields analogous to the electromagnetic field, which live on the worldvolume of a brane.[2]

In string theory, D-branes are an important class of branes that arise when one considers open strings. As an open string propagates through spacetime, its endpoints are required to lie on a D-brane. The letter "D" in D-brane refers to a certain mathematical condition on the system known as the Dirichlet boundary condition. The study of D-branes in string theory has led to important results such as the AdS/CFT correspondence, which has shed light on many problems in quantum field theory.

Branes are also frequently studied from a purely mathematical point of view since they are related to subjects such as homological mirror symmetry and noncommutative geometry. Mathematically, branes may be represented as objects of certain categories, such as the derived category of coherent sheaves on a Calabi–Yau manifold, or the Fukaya category.

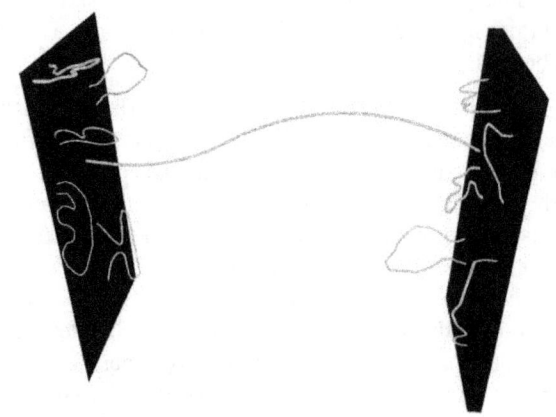

Open strings attached to a pair of D-branes

9.1 D-branes

Main article: D-brane

In string theory, a string may be open (forming a segment with two endpoints) or closed (forming a closed loop). D-branes are an important class of branes that arise when one considers open strings. As an open string propagates through spacetime, its endpoints are required to lie on a D-brane. The letter "D" in D-brane refers to a condition that it satisfies, the Dirichlet boundary condition.[3]

One crucial point about D-branes is that the dynamics on the D-brane worldvolume is described by a gauge theory, a kind of highly symmetric physical theory which is also used to describe the behavior of elementary particles in the standard model of particle physics. This connection has led to many important insights into gauge theory. For example, it led to the discovery of the AdS/CFT correspondence, a theoretical tool that physicists use to translate difficult problems in gauge theory into more mathematically tractable problems in string theory.[4]

9.2 Mathematical viewpoint

Mathematically, branes can be described using the notion of a category.[5] This is a mathematical structure consisting of *objects*, and for any pair of objects, a set of *morphisms* between them. In most examples, the objects are mathematical structures (such as sets, vector spaces, or topological spaces) and the morphisms are functions between these structures.[6] One can also consider categories where the objects are D-branes and the morphisms between two branes α and β are states of open strings stretched between α and β.[7]

A cross section of a Calabi–Yau manifold

In one version of string theory known as the topological B-model, the D-branes are complex submanifolds of certain six-dimensional shapes called Calabi–Yau manifolds, together with additional data that arise physically from having charges at the endpoints of strings.[8] Intuitively, one can think of a submanifold as a surface embedded inside of a Calabi–Yau manifold, although submanifolds can also exist in dimensions different from two.[9] In mathematical language, the category having these branes as its objects is known as the derived category of coherent sheaves on the Calabi–Yau.[10] In another version of string theory called the topological A-model, the D-branes can again be viewed as submanifolds of a Calabi–Yau manifold. Roughly speaking, they are what mathematicians call special Lagrangian submanifolds.[11] This means among other things that they have half the dimension of the space in which they sit, and they are length-, area-, or volume-minimizing.[12] The category having these branes as its objects is called the Fukaya category.[13]

The derived category of coherent sheaves is constructed using tools from complex geometry, a branch of mathematics that describes geometric curves in algebraic terms and solves geometric problems using algebraic equations.[14] On the other hand, the Fukaya category is constructed using symplectic geometry, a branch of mathematics that arose from studies of classical physics. Symplectic geometry studies spaces equipped with a symplectic form, a mathematical tool that can be used to compute area in two-dimensional examples.[15]

The homological mirror symmetry conjecture of Maxim Kontsevich states that the derived category of coherent sheaves on one Calabi–Yau manifold is equivalent in a certain sense to the Fukaya category of a completely different Calabi–Yau manifold.[16] This equivalence provides an unexpected bridge between two branches of geometry, namely complex and symplectic geometry.[17]

9.3 See also

- Black brane
- Brane cosmology
- Dirac membrane
- M2-brane
- M5-brane
- NS5-brane

9.4 Notes

[1] Moore 2005, p. 214

[2] Moore 2005, p. 214

[3] Moore 2005, p. 215

[4] Moore 2005, p. 215

[5] Aspinwall et al. 2009

[6] A basic reference on category theory is Mac Lane 1998.

[7] Zaslow 2008, p. 536

[8] Zaslow 2008, p. 536

[9] Yau and Nadis 2010, p. 165

[10] Aspinwal et al. 2009, p. 575

[11] Aspinwal et al. 2009, p. 575

[12] Yau and Nadis 2010, p. 175

[13] Aspinwal et al. 2009, p. 575

[14] Yau and Nadis 2010, pp. 180–1

[15] Zaslow 2008, p. 531

[16] Aspinwall et al. 2009, p. 616

[17] Yau and Nadis 2010, p. 181

9.5 References

- Aspinwall, Paul; Bridgeland, Tom; Craw, Alastair; Douglas, Michael; Gross, Mark; Kapustin, Anton; Moore, Gregory; Segal, Graeme; Szendröi, Balázs; Wilson, P.M.H., eds. (2009). *Dirichlet Branes and Mirror Symmetry*. American Mathematical Society. ISBN 978-0-8218-3848-8.

- Mac Lane, Saunders (1998). *Categories for the Working Mathematician*. ISBN 978-0-387-98403-2.

- Moore, Gregory (2005). "What is ... a Brane?" (PDF). *Notices of the AMS* **52**: 214. Retrieved June 2013.

- Yau, Shing-Tung; Nadis, Steve (2010). *The Shape of Inner Space: String Theory and the Geometry of the Universe's Hidden Dimensions*. Basic Books. ISBN 978-0-465-02023-2.

- Zaslow, Eric (2008). "Mirror Symmetry". In Gowers, Timothy. *The Princeton Companion to Mathematics*. ISBN 978-0-691-11880-2.

Chapter 10

Fundamental interaction

Fundamental interactions, also known as **fundamental forces**, are the interactions in physical systems that do not appear to be reducible to more basic interactions. There are four conventionally accepted fundamental interactions—gravitational, electromagnetic, strong nuclear, and weak nuclear. Each one is understood as the dynamics of a *field*. The gravitational force is modelled as a continuous classical field. The other three are each modelled as discrete quantum fields, and exhibit a measurable unit or *elementary particle*.

The two nuclear interactions produce strong forces at minuscule, subatomic distances. The strong nuclear interaction is responsible for the binding of atomic nuclei. The weak nuclear interaction also acts on the nucleus, mediating radioactive decay. Electromagnetism and gravity produce significant forces at macroscopic scales where the effects can be seen directly in every day life. Electrical and magnetic fields tend to cancel each other out when large collections of objects are considered, so over the largest distances (on the scale of planets and galaxies), gravity tends to be the dominant force.

Theoretical physicists working beyond the Standard Model seek to quantize the gravitational field toward predictions that particle physicists can experimentally confirm, thus yielding acceptance to a theory of quantum gravity (QG). (Phenomena suitable to model as a fifth force—perhaps an added gravitational effect—remain widely disputed.) Other theorists seek to unite the electroweak and strong fields within a Grand Unified Theory (GUT). While all four fundamental interactions are widely thought to align on a highly minuscule scale, particle accelerators cannot produce the massive energy levels required to experimentally probe at that Planck scale (which would experimentally confirm such theories.) Yet some theories, such as the string theory, seek both QG and GUT within one framework, unifying all four fundamental interactions along with mass generation within a theory of everything (ToE).

The four fundamental interactions of nature[1]

Property/Interaction	Gravitation	Weak (Electroweak)	Electromagnetic	Strong	
				Fundamental	Residual
Acts on:	Mass - Energy	Flavor	Electric charge	Color charge	Atomic nuclei
Particles experiencing:	All	Quarks, leptons	Electrically charged	Quarks, Gluons	Hadrons
Particles mediating:	Not yet observed (Graviton hypothesised)	W^+ W^- Z^0	γ (photon)	Gluons	Mesons
Strength at the scale of quarks:	10^{-41}	10^{-4}	1	60	Not applicable to quarks
Strength at the scale of protons/neutrons:	10^{-36}	10^{-7}	1	Not applicable to hadrons	20

10.1 General relativity

In his 1687 theory, Isaac Newton postulated space as an infinite and unalterable physical structure existing before, within, and around all objects while their states and relations unfold at a constant pace everywhere, thus absolute space and time. Inferring that all objects bearing mass approach at a constant rate, but collide by impact proportional to their masses, Newton inferred that matter exhibits an attractive force. His law of universal gravitation mathematically stated it to span the entire universe instantly (despite absolute time), or, if not actually a force, to be instant interaction among all objects (despite absolute space.) As conventionally interpreted, Newton's theory of motion modelled a *central force* without a communicating medium.[2] Thus Newton's theory violated the first principle of mechanical philosophy, as stated by Descartes, *No action at a distance*. Conversely, during the 1820s, when explaining magnetism, Michael Faraday inferred a *field* filling space and transmitting that force. Faraday conjectured that ultimately, all forces unified into one.

In the early 1870s, James Clerk Maxwell unified electricity and magnetism as effects of an electromagnetic field whose third consequence was light, travelling at constant speed in a vacuum. The electromagnetic field theory contradicted predictions of Newton's theory of motion, unless physical states of the luminiferous aether—presumed to fill all space whether within matter or in a vacuum and to manifest the electromagnetic field—aligned all phenomena and thereby held valid the Newtonian principle relativity or invariance. Disfavouring hypotheses at unobservables, Albert Einstein discarded the aether, and aligned electrodynamics with relativity by denying absolute space and time, and stating relative space and time. The two phenomena altered in the vicinity of an object measured to be in motion—length contraction and time dilation for the object experienced to be in relative motion—Einstein's principle special relativity, published in 1905.

Special relativity was accepted as a theory too. It rendered Newton's theory of motion apparently untenable, especially since Newtonian physics postulated an object's mass to be constant. A consequence of special relativity is mass being a variant form of energy, condensed into an object. By the equivalence principle, published by Einstein in 1907, gravitation is indistinguishable from acceleration, perhaps two phenomena sharing a mechanism. That year, Hermann Minkowski modelled special relativity to a unification of space and time, 4D spacetime. Stretching the three spatial dimensions onto the single dimension of time's arrow, Einstein arrived at the general theory of relativity in 1915.[3] Einstein interpreted space as a substance, *Einstein-aether*, whose physical properties receive motion from an object and transmit it to other objects while modulating events unfolding. Equivalent to energy, mass contracts space, which dilates time—events unfold more slowly—establishing local tension. The object relieves it in the likeness of a free fall at light speed along the pathway of least resistance, a straight line's equivalent on the curved surface of 4D spacetime, a pathway termed *worldline*.

Einstein abolished *action at a distance* by theorizing a gravitational field—4D spacetime—that waves while transmitting motion across the universe at light speed. All objects always travel at light speed in 4D spacetime. At zero relative speed, an object is observed to travel none through space, but age most rapidly. That is, an object at relative rest in 3D space exhibits its constant energy to an observer by exhibiting top speed along 1D time flow. Conversely, at highest relative speed, an object traverses 3D space at light speed, yet is ageless, none of its constant energy available to internal motion as flow along 1D time. Whereas Newtonian inertia is an idealized case of an object either keeping rest or holding constant velocity by its hypothetical existence in a universe otherwise devoid of matter, Einsteinian inertia is indistinguishable from an object experiencing no acceleration by existing in a gravitational field possibly full of matter distributed uniformly. Conversely, even massless energy manifests gravitation—which is acceleration—on local objects by "curving" the surface of 4D spacetime. Physicists renounced belief that motion must be mediated by a *force*.

10.2   Standard Model

Main article: Standard Model
See also: Lambda-CDM model

 The electromagnetic, strong, and weak interactions associate with elementary particles, whose behaviours are modelled in quantum mechanics (QM). For predictive success with QM's probabilistic outcomes, particle physics conventionally models QM events across a field set to special relativity, altogether relativistic quantum field theory (QFT).[4] Force particles, called gauge bosons—*force carriers* or *messenger particles* of underlying fields—interact with mat-

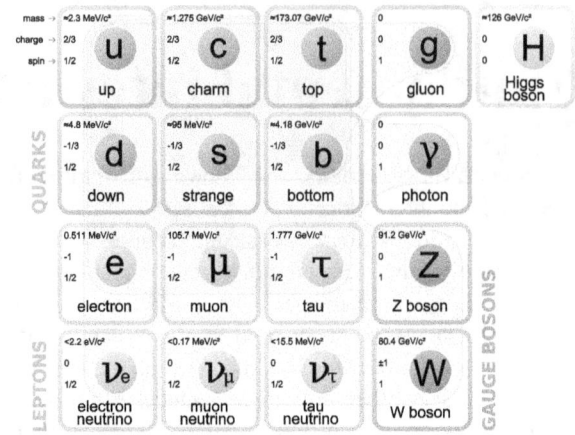

The Standard Model of elementary particles, with the fermions in the first three columns, the gauge bosons in the fourth column, and the Higgs boson in the fifth column

ter particles, called fermions. Everyday matter is atoms, composed of three fermion types: up-quarks and down-quarks constituting, as well as electrons orbiting, the atom's nucleus. Atoms interact, form molecules, and manifest further properties through electromagnetic interactions among their electrons absorbing and emitting photons, the electromagnetic field's force carrier, which if unimpeded traverse potentially infinite distance. Electromagnetism's QFT is quantum electrodynamics (QED).

The electromagnetic interaction was modelled with the weak interaction, whose force carriers are W and Z bosons, traversing the minuscule distance, in electroweak theory (EWT). Electroweak interaction would operate at such high temperatures as soon after the presumed Big Bang, but, as the early universe cooled, split into electromagnetic and weak interactions. The strong interaction, whose force carrier is the gluon, traversing minuscule distance among quarks, is modeled in quantum chromodynamics (QCD). EWT, QCD, and the Higgs mechanism, whereby the Higgs field manifests Higgs bosons that interact with some quantum particles and thereby endow those particles with mass comprise particle physics' Standard Model (SM). Predictions are usually made using calculational approximation methods, although such perturbation theory is inadequate to model some experimental observations (for instance bound states and solitons.) Still, physicists widely accept the Standard Model as science's most experimentally confirmed theory.

Beyond the Standard Model, some theorists work to unite the electroweak and strong interactions within a Grand Unified Theory (GUT). Some attempts at GUTs hypothesize "shadow" particles, such that every known matter particle associates with an undiscovered force particle, and vice versa, altogether supersymmetry (SUSY). Other theorists seek to quantize the gravitational field by the modelling behaviour of its hypothetical force carrier, the graviton and achieve quantum gravity (QG). One approach to QG is loop quantum gravity (LQG). Still other theorists seek both QG and GUT within one framework, reducing all four fundamental interactions to a Theory of Everything (ToE). The most prevalent aim at a ToE is string theory, although to model matter particles, it added SUSY to force particles—and so, strictly speaking, became superstring theory. Multiple, seemingly disparate superstring theories were unified on a backbone, M-theory. Theories beyond the Standard Model remain highly speculative, lacking great experimental support.

10.3 Overview of the fundamental interactions

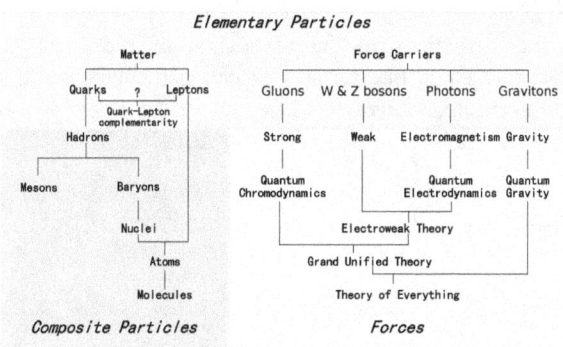

An overview of the various families of elementary and composite particles, and the theories describing their interactions. Fermions are on the left, and Bosons are on the right.

In the conceptual model of fundamental interactions, matter consists of fermions, which carry properties called charges and spin $\pm^1/_2$ (intrinsic angular momentum $\pm^\hbar/_2$, where $\hbar$ is the reduced Planck constant). They attract or repel each other by exchanging bosons.

The interaction of any pair of fermions in perturbation theory can then be modelled thus:

Two fermions go in → *interaction* by boson exchange → Two changed fermions go out.

The exchange of bosons always carries energy and momentum between the fermions, thereby changing their speed and direction. The exchange may also transport a charge between the fermions, changing the charges of the fermions in the process (e.g., turn them from one type of fermion to another). Since bosons carry one unit of angular momentum, the fermion's spin direction will flip from $+^1/_2$

to $-^1/_2$ (or vice versa) during such an exchange (in units of the reduced Planck's constant).

Because an interaction results in fermions attracting and repelling each other, an older term for "interaction" is force.

According to the present understanding, there are four fundamental interactions or forces: gravitation, electromagnetism, the weak interaction, and the strong interaction. Their magnitude and behaviour vary greatly, as described in the table below. Modern physics attempts to explain every observed physical phenomenon by these fundamental interactions. Moreover, reducing the number of different interaction types is seen as desirable. Two cases in point are the unification of:

- Electric and magnetic force into electromagnetism;

- The electromagnetic interaction and the weak interaction into the electroweak interaction; see below.

Both magnitude ("relative strength") and "range", as given in the table, are meaningful only within a rather complex theoretical framework. It should also be noted that the table below lists properties of a conceptual scheme that is still the subject of ongoing research.

Interaction	Current theory	Mediators	Relative strength[①]	Long-distance behavior	Range (m)[citation needed]
Strong	Quantum chromodynamics (QCD)	gluons	10^{38}	1 (see discussion below)	10^{-15}
Electromagnetic	Quantum electrodynamics (QED)	photons	10^{36}	$\dfrac{1}{r^2}$	∞
Weak	Electroweak Theory (EWT)	W and Z bosons	10^{25}	$\dfrac{1}{r} e^{-m_{W,Z}\,r}$	10^{-18}
Gravitation	General Relativity (GR)	gravitons (hypothetical)	1	$\dfrac{1}{r^2}$	∞

The modern (perturbative) quantum mechanical view of the fundamental forces other than gravity is that particles of matter (fermions) do not directly interact with each other, but rather carry a charge, and exchange virtual particles (gauge bosons), which are the interaction carriers or force mediators. For example, photons mediate the interaction of electric charges, and gluons mediate the interaction of color charges.

10.4 The interactions

10.4.1 Gravity

Gravitation is by far the weakest of the four interactions. The weakness of gravity can easily be demonstrated by suspending a pin using a simple magnet (such as a refrigerator magnet). The magnet is able to hold the pin against the gravitational pull of the entire Earth.

Yet gravitation is very important for macroscopic objects and over macroscopic distances for the following reasons. Gravitation:

- Is the only interaction that acts on all particles having mass, energy and/or momentum

- Has an infinite range, like electromagnetism but unlike strong and weak interaction

- Cannot be absorbed, transformed, or shielded against

- Always attracts and never repels

Even though electromagnetism is far stronger than gravitation, electrostatic attraction is not relevant for large celestial bodies, such as planets, stars, and galaxies, simply because such bodies contain equal numbers of protons and electrons and so have a net electric charge of zero. Nothing "cancels" gravity, since it is only attractive, unlike electric forces which can be attractive or repulsive. On the other hand, all objects having mass are subject to the gravitational force, which only attracts. Therefore, only gravitation matters on the large-scale structure of the universe.

The long range of gravitation makes it responsible for such large-scale phenomena as the structure of galaxies and black holes and it retards the expansion of the universe. Gravitation also explains astronomical phenomena on more modest scales, such as planetary orbits, as well as everyday experience: objects fall; heavy objects act as if they were glued to the ground, and animals can only jump so high.

Gravitation was the first interaction to be described mathematically. In ancient times, Aristotle hypothesized that objects of different masses fall at different rates. During the Scientific Revolution, Galileo Galilei experimentally determined that this was not the case — neglecting the friction due to air resistance, and buoyancy forces if an atmosphere is present (e.g. the case of a dropped air-filled balloon vs a water-filled balloon) all objects accelerate toward the Earth at the same rate. Isaac Newton's law of Universal Gravitation (1687) was a good approximation of the behaviour of gravitation. Our present-day understanding of gravitation stems from Albert Einstein's General Theory of Relativity of 1915, a more accurate (especially for cosmological masses and distances) description of gravitation in terms of the geometry of spacetime.

Merging general relativity and quantum mechanics (or quantum field theory) into a more general theory of quantum gravity is an area of active research. It is hypothesized that gravitation is mediated by a massless spin-2 particle called the graviton.

Although general relativity has been experimentally confirmed (at least for weak fields) on all but the smallest scales, there are rival theories of gravitation. Those taken seriously by [citation needed] the physics community all reduce to general relativity in some limit, and the focus of observational work is to establish limitations on what deviations from general relativity are possible.

Proposed extra dimensions could explain why the gravity force is so weak.[6]

10.4.2 Electroweak interaction

Main article: Electroweak interaction

Electromagnetism and weak interaction appear to be very different at everyday low energies. They can be modeled using two different theories. However, above unification energy, on the order of 100 GeV, they would merge into a single electroweak force.

Electroweak theory is very important for modern cosmology, particularly on how the universe evolved. This is because shortly after the Big Bang, the temperature was approximately above 10^{15} K. Electromagnetic force and weak force were merged into a combined electroweak force.

For contributions to the unification of the weak and electromagnetic interaction between elementary particles, Abdus Salam, Sheldon Glashow and Steven Weinberg were awarded the Nobel Prize in Physics in 1979.[7][8]

Electromagnetism

Main article: Electromagnetism

Electromagnetism is the force that acts between electrically charged particles. This phenomenon includes the electrostatic force acting between charged particles at rest, and the combined effect of electric and magnetic forces acting between charged particles moving relative to each other.

Electromagnetism is infinite-ranged like gravity, but vastly stronger, and therefore describes a number of macroscopic phenomena of everyday experience such as friction, rainbows, lightning, and all human-made devices using electric current, such as television, lasers, and computers. Electromagnetism fundamentally determines all macroscopic, and many atomic levels, properties of the chemical elements, including all chemical bonding.

In a four kilogram (~1 gallon) jug of water there are

$$4000 \text{ g } H_2O \cdot \frac{1 \text{ mol } H_2O}{18 \text{ g } H_2O} \cdot \frac{10 \text{ mol } e^-}{1 \text{ mol } H_2O} \cdot \frac{96,000 \text{ C}}{1 \text{ mol } e^-} =$$

$$2.1 \times 10^{8} C$$

of total electron charge. Thus, if we place two such jugs a meter apart, the electrons in one of the jugs repel those in the other jug with a force of

$$\frac{1}{4\pi\varepsilon_0} \frac{(2.1 \times 10^8 C)^2}{(1m)^2} = 4.1 \times 10^{26} N.$$

This is larger than the planet Earth would weigh if weighed on another Earth. The atomic nuclei in one jug also repel those in the other with the same force. However, these repulsive forces are canceled by the attraction of the electrons in jug A with the nuclei in jug B and the attraction of the nuclei in jug A with the electrons in jug B, resulting in no net force. Electromagnetic forces are tremendously stronger than gravity but cancel out so that for large bodies gravity dominates.

Electrical and magnetic phenomena have been observed since ancient times, but it was only in the 19th century that it was discovered that electricity and magnetism are two aspects of the same fundamental interaction. By 1864, Maxwell's equations had rigorously quantified this unified interaction. Maxwell's theory, restated using vector calculus, is the classical theory of electromagnetism, suitable for most technological purposes.

The constant speed of light in a vacuum (customarily described with the letter "c") can be derived from Maxwell's equations, which are consistent with the theory of special relativity. Einstein's 1905 theory of special relativity, however, which flows from the observation that the speed of light is constant no matter how fast the observer is moving, showed that the theoretical result implied by Maxwell's equations has profound implications far beyond electromagnetism on the very nature of time and space.

In another work that departed from classical electromagnetism, Einstein also explained the photoelectric effect by hypothesizing that light was transmitted in quanta, which we now call photons. Starting around 1927, Paul Dirac combined quantum mechanics with the relativistic theory of electromagnetism. Further work in the 1940s, by Richard Feynman, Freeman Dyson, Julian Schwinger, and Sin-Itiro Tomonaga, completed this theory, which is now called quantum electrodynamics, the revised theory of electromagnetism. Quantum electrodynamics and quantum mechanics provide a theoretical basis for electromagnetic behavior such as quantum tunneling, in which a certain percentage of electrically charged particles move in ways that would be impossible under the classical electromagnetic theory, that is necessary for everyday electronic devices such as transistors to function.

Weak interaction

Main article: Weak interaction

The *weak interaction* or *weak nuclear force* is responsible for some nuclear phenomena such as beta decay. Electromagnetism and the weak force are now understood to be two aspects of a unified electroweak interaction — this discovery was the first step toward the unified theory known as the Standard Model. In the theory of the electroweak interaction, the carriers of the weak force are the massive gauge bosons called the W and Z bosons. The weak interaction is the only known interaction which does not conserve parity; it is left-right asymmetric. The weak interaction even violates CP symmetry but does conserve CPT.

10.4.3 Strong interaction

Main article: Strong interaction

The *strong interaction*, or *strong nuclear force*, is the most complicated interaction, mainly because of the way it varies with distance. At distances greater than 10 femtometers, the strong force is practically unobservable. Moreover, it holds only inside the atomic nucleus.

After the nucleus was discovered in 1908, it was clear that a new force was needed to overcome the electrostatic repulsion, a manifestation of electromagnetism, of the positively charged protons. Otherwise, the nucleus could not exist. Moreover, the force had to be strong enough to squeeze the protons into a volume that is 10^{-15} of that of the entire atom. From the short range of this force, Hideki Yukawa predicted that it was associated with a massive particle, whose mass is approximately 100 MeV.

The 1947 discovery of the pion ushered in the modern era of particle physics. Hundreds of hadrons were discovered from the 1940s to 1960s, and an extremely complicated theory of hadrons as strongly interacting particles was developed. Most notably:

- The pions were understood to be oscillations of vacuum condensates;

- Jun John Sakurai proposed the rho and omega vector bosons to be force carrying particles for approximate symmetries of isospin and hypercharge;

- Geoffrey Chew, Edward K. Burdett and Steven Frautschi grouped the heavier hadrons into families that could be understood as vibrational and rotational excitations of strings.

While each of these approaches offered deep insights, no approach led directly to a fundamental theory.

Murray Gell-Mann along with George Zweig first proposed fractionally charged quarks in 1961. Throughout the 1960s, different authors considered theories similar to the modern fundamental theory of quantum chromodynamics (QCD) as simple models for the interactions of quarks. The first to hypothesize the gluons of QCD were Moo-Young Han and Yoichiro Nambu, who introduced the quark color charge and hypothesized that it might be associated with a force-carrying field. At that time, however, it was difficult to see how such a model could permanently confine quarks. Han and Nambu also assigned each quark color an integer electrical charge, so that the quarks were fractionally charged only on average, and they did not expect the quarks in their model to be permanently confined.

In 1971, Murray Gell-Mann and Harald Fritzsch proposed that the Han/Nambu color gauge field was the correct theory of the short-distance interactions of fractionally charged quarks. A little later, David Gross, Frank Wilczek, and David Politzer discovered that this theory had the property of asymptotic freedom, allowing them to make contact with experimental evidence. They concluded that QCD was the complete theory of the strong interactions, correct at all distance scales. The discovery of asymptotic freedom led most physicists to accept QCD since it became clear that even the long-distance properties of the strong interactions could be consistent with experiment if the quarks are permanently confined.

Assuming that quarks are confined, Mikhail Shifman, Arkady Vainshtein, and Valentine Zakharov were able to compute the properties of many low-lying hadrons directly from QCD, with only a few extra parameters to describe the vacuum. In 1980, Kenneth G. Wilson published computer calculations based on the first principles of QCD, establishing, to a level of confidence tantamount to certainty, that QCD will confine quarks. Since then, QCD has been the established theory of the strong interactions.

QCD is a theory of fractionally charged quarks interacting by means of 8 photon-like particles called gluons. The gluons interact with each other, not just with the quarks, and at long distances the lines of force collimate into strings. In this way, the mathematical theory of QCD not only explains how quarks interact over short distances but also the string-like behavior, discovered by Chew and Frautschi, which they manifest over longer distances.

10.4.4   Beyond the Standard Model

Main article: Physics beyond the Standard Model
See also: Elementary particle § Beyond the Standard Model

Numerous theoretical efforts have been made to systematize the existing four fundamental interactions on the model of electroweak unification.

Grand Unified Theories (GUTs) are proposals to show that all of the fundamental interactions, other than gravity, arise from a single interaction with symmetries that break down

at low energy levels. GUTs predict relationships among constants of nature that are unrelated in the SM. GUTs also predict gauge coupling unification for the relative strengths of the electromagnetic, weak, and strong forces, a prediction verified at the Large Electron–Positron Collider in 1991 for supersymmetric theories.

Theories of everything, which integrate GUTs with a quantum gravity theory face a greater barrier, because no quantum gravity theories, which include string theory, loop quantum gravity, and twistor theory, have secured wide acceptance. Some theories look for a graviton to complete the Standard Model list of force-carrying particles, while others, like loop quantum gravity, emphasize the possibility that time-space itself may have a quantum aspect to it.

Some theories beyond the Standard Model include a hypothetical fifth force, and the search for such a force is an ongoing line of experimental research in physics. In supersymmetric theories, there are particles that acquire their masses only through supersymmetry breaking effects and these particles, known as moduli can mediate new forces. Another reason to look for new forces is the recent discovery that the expansion of the universe is accelerating (also known as dark energy), giving rise to a need to explain a nonzero cosmological constant, and possibly to other modifications of general relativity. Fifth forces have also been suggested to explain phenomena such as CP violations, dark matter, and dark flow.

10.5   See also

- Standard Model

  - Strong interaction
  - Electroweak interaction
  - Weak interaction
  - Gravity

    - Quantum gravity
    - String Theory
    - Theory of Everything

- Grand Unified Theory

  - Gauge coupling unification
  - Unified Field Theory

- Quintessence, a hypothesized fifth force.

- *People*: Isaac Newton, James Clerk Maxwell, Albert Einstein, Richard Feynman, Sheldon Glashow, Abdus Salam, Steven Weinberg, Gerardus 't Hooft, David Gross, Edward Witten, Howard Georgi.

10.6 References

[1] http://www.pha.jhu.edu/~{ }dfehling/particle.gif

[2] Newton's absolute space was a medium, but not one transmitting gravitation.

[3] Special relativity holds for objects at vast speed but of negligible mass, for instance elementary particles. Yet by yielding gravitation, which is a manner of acceleration, notable mass breaks inertia—that is, constant speed and direction—and thereby violates special relativity. Special relativity could approximately predict a massive object's motion during barely an instant, however, and thus is a temporally limited case of general relativity.

[4] Meinard Kuhlmann, "Physicists debate whether the world is made of particles or fields—or something else entirely", *Scientific American*, 24 Jul 2013.

[5] Approximate. See Coupling constant for more exact strengths, depending on the particles and energies involved.

[6] CERN (20 January 2012). "Extra dimensions, gravitons, and tiny black holes".

[7] Bais, Sander (2005), *The Equations. Icons of knowledge*, ISBN 0-674-01967-9 p.84

[8] "The Nobel Prize in Physics 1979". The Nobel Foundation. Retrieved 2008-12-16.

Bibliography General:

- Davies, Paul (1986), *The Forces of Nature*, Cambridge Univ. Press 2nd ed.

- Feynman, Richard (1967), *The Character of Physical Law*, MIT Press, ISBN 0-262-56003-8

- Schumm, Bruce A. (2004), *Deep Down Things*, Johns Hopkins University Press While all interactions are discussed, discussion is especially thorough on the weak.

- Weinberg, Steven (1993), *The First Three Minutes: A Modern View of the Origin of the Universe*, Basic Books, ISBN 0-465-02437-8

- Weinberg, Steven (1994), *Dreams of a Final Theory*, Basic Books, ISBN 0-679-74408-8

Texts:

- Padmanabhan, T. (1998), *After The First Three Minutes: The Story of Our Universe*, Cambridge Univ. Press, ISBN 0-521-62972-1

- Perkins, Donald H. (2000), *Introduction to High Energy Physics*, Cambridge Univ. Press, ISBN 0-521-62196-8

- Riazuddin (December 29, 2009). "Non-standard interactions" (PDF). *NCP 5th Particle Physics Sypnoisis* (Islamabad: Riazuddin, Head of High-Energy Theory Group at National Center for Physics) **1** (1): 1–25. Retrieved March 19, 2011.

Chapter 11

Theory of everything

This article is about the physical concept. For other uses, see Theory of everything (disambiguation).

A **theory of everything** (**ToE**) or **final theory**, **ultimate theory**, or **master theory** is a hypothetical single, all-encompassing, coherent theoretical framework of physics that fully explains and links together all physical aspects of the universe.[1]:6 Finding a ToE is one of the major unsolved problems in physics. Over the past few centuries, two theoretical frameworks have been developed that, as a whole, most closely resemble a ToE. These two theories upon which all modern physics rests are general relativity (GR) and quantum field theory (QFT). GR is a theoretical framework that only focuses on gravity for understanding the universe in regions of both large-scale and high-mass: stars, galaxies, clusters of galaxies, etc. On the other hand, QFT is a theoretical framework that only focuses on three non-gravitational forces for understanding the universe in regions of both small scale and low mass: sub-atomic particles, atoms, molecules, etc. QFT successfully implemented the Standard Model and unified the interactions (so-called Grand Unified Theory) between the three non-gravitational forces: weak, strong, and electromagnetic force.[2]:122

Through years of research, physicists have experimentally confirmed with tremendous accuracy virtually every prediction made by these two theories when in their appropriate domains of applicability. In accordance with their findings, scientists also learned that GR and QFT, as they are currently formulated, are mutually incompatible – they cannot both be right. Since the usual domains of applicability of GR and QFT are so different, most situations require that only one of the two theories be used.[3][4]:842–844 As it turns out, this incompatibility between GR and QFT is only an apparent issue in regions of extremely small-scale and high-mass, such as those that exist within a black hole or during the beginning stages of the universe (i.e., the moment immediately following the Big Bang). To resolve this conflict, a theoretical framework revealing a deeper underlying reality, unifying gravity with the other three interactions, must be discovered to harmoniously integrate the realms of GR and QFT into a seamless whole: a single theory that, in principle, is capable of describing all phenomena. In pursuit of this goal, quantum gravity has recently become an area of active research.

Over the past few decades, a single explanatory framework, called "string theory", has emerged that intends to be the ultimate theory of the universe. Some physicists believe that, at the beginning of the universe (up to 10^{-43} seconds after the Big Bang), the four fundamental forces were once a single fundamental force. According to string theory, every particle in the universe, at its most microscopic level (Planck length), consists of varying combinations of vibrating strings (or strands) with preferred patterns of vibration. String theory claims that it is through these specific oscillatory patterns of strings that a particle of unique mass and force charge is created (that is to say, the electron is a type of string that vibrates one way, while the up-quark is a type of string vibrating another way, and so forth).

Initially, the term *theory of everything* was used with an ironic connotation to refer to various overgeneralized theories. For example, a grandfather of Ijon Tichy — a character from a cycle of Stanisław Lem's science fiction stories of the 1960s — was known to work on the "General Theory of Everything". Physicist John Ellis[5] claims to have introduced the term into the technical literature in an article in *Nature* in 1986.[6] Over time, the term stuck in popularizations of theoretical physics research.

11.1 Historical antecedents

11.1.1 From ancient Greece to Einstein

Archimedes was possibly the first scientist known to have described nature with axioms (or principles) and then deduce new results from them.[7] He thus tried to describe "everything" starting from a few axioms. Any "theory of everything" is similarly expected to be based on axioms and to deduce all observable phenomena from them.[8]:340

The concept of 'atom', introduced by Democritus, unified all phenomena observed in nature as the motion of atoms. In ancient Greek times philosophers speculated that the apparent diversity of observed phenomena was due to a single type of interaction, namely the collisions of atoms. Following atomism, the mechanical philosophy of the 17th century posited that all forces could be ultimately reduced to contact forces between the atoms, then imagined as tiny solid particles.[9]:184[10]

In the late 17th century, Isaac Newton's description of the long-distance force of gravity implied that not all forces in nature result from things coming into contact. Newton's work in his *Mathematical Principles of Natural Philosophy* dealt with this in a further example of unification, in this case unifying Galileo's work on terrestrial gravity, Kepler's laws of planetary motion and the phenomenon of tides by explaining these apparent actions at a distance under one single law: the law of universal gravitation.[11]

In 1814, building on these results, Laplace famously suggested that a sufficiently powerful intellect could, if it knew the position and velocity of every particle at a given time, along with the laws of nature, calculate the position of any particle at any other time:[12]:ch 7

> An intellect which at a certain moment would know all forces that set nature in motion, and all positions of all items of which nature is composed, if this intellect were also vast enough to submit these data to analysis, it would embrace in a single formula the movements of the greatest bodies of the universe and those of the tiniest atom; for such an intellect nothing would be uncertain and the future just like the past would be present before its eyes.
> — *Essai philosophique sur les probabilités*, Introduction. 1814

Laplace thus envisaged a combination of gravitation and mechanics as a theory of everything. Modern quantum mechanics implies that uncertainty is inescapable, and thus that Laplace's vision has to be amended: a theory of everything must include gravitation and quantum mechanics.

In 1820, Hans Christian Ørsted discovered a connection between electricity and magnetism, triggering decades of work that culminated in 1865, in James Clerk Maxwell's theory of electromagnetism. During the 19th and early 20th centuries, it gradually became apparent that many common examples of forces – contact forces, elasticity, viscosity, friction, and pressure – result from electrical interactions between the smallest particles of matter.

In his experiments of 1849–50, Michael Faraday was the first to search for a unification of gravity with electricity and magnetism.[13] However, he found no connection.

In 1900, David Hilbert published a famous list of mathematical problems. In Hilbert's sixth problem, he challenged researchers to find an axiomatic basis to all of physics. In this problem he thus asked for what today would be called a theory of everything.[14]

In the late 1920s, the new quantum mechanics showed that the chemical bonds between atoms were examples of (quantum) electrical forces, justifying Dirac's boast that "the underlying physical laws necessary for the mathematical theory of a large part of physics and the whole of chemistry are thus completely known".[15]

After 1915, when Albert Einstein published the theory of gravity (general relativity), the search for a unified field theory combining gravity with electromagnetism began with a renewed interest. In Einstein's day, the strong and the weak forces had not yet been discovered, yet, he found the potential existence of two other distinct forces -gravity and electromagnetism- far more alluring. This launched his thirty-year voyage in search of the so-called "unified field theory" that he hoped would show that these two forces are really manifestations of one grand underlying principle. During these last few decades of his life, this quixotic quest isolated Einstein from the mainstream of physics. Understandably, the mainstream was instead far more excited about the newly emerging framework of quantum mechanics. Einstein wrote to a friend in the early 1940s, "I have become a lonely old chap who is mainly known because he doesn't wear socks and who is exhibited as a curiosity on special occasions." Prominent contributors were Gunnar Nordström, Hermann Weyl, Arthur Eddington, Theodor Kaluza, Oskar Klein, and most notably, Albert Einstein and his collaborators. Einstein intensely searched for, but ultimately failed to find, a unifying theory.[16]:ch 17 (But see:Einstein–Maxwell–Dirac equations.) More than a half a century later, Einstein's dream of discovering a unified theory has become the Holy Grail of modern physics.

11.1.2 Twentieth century and the nuclear interactions

In the twentieth century, the search for a unifying theory was interrupted by the discovery of the strong and weak nuclear forces (or interactions), which differ both from gravity and from electromagnetism. A further hurdle was the acceptance that in a ToE, quantum mechanics had to be incorporated from the start, rather than emerging as a consequence of a deterministic unified theory, as Einstein had hoped.

Gravity and electromagnetism could always peacefully coexist as entries in a list of classical forces, but for many years it seemed that gravity could not even be incorporated into the quantum framework, let alone unified with the other fundamental forces. For this reason, work on unification, for much of the twentieth century, focused on understanding the three "quantum" forces: electromagnetism and the weak and strong forces. The first two were combined in 1967–68 by Sheldon Glashow, Steven Weinberg, and Abdus Salam into the "electroweak" force.[17] Electroweak unification is a broken symmetry: the electromagnetic and weak forces appear distinct at low energies because the particles carrying the weak force, the W and Z bosons, have non-zero masses of 80.4 GeV/c^2 and 91.2 GeV/c^2, whereas the photon, which carries the electromagnetic force, is massless. At higher energies Ws and Zs can be created easily and the unified nature of the force becomes apparent.

While the strong and electroweak forces peacefully coexist in the Standard Model of particle physics, they remain distinct. So far, the quest for a theory of everything is thus unsuccessful on two points: neither a unification of the strong and electroweak forces – which Laplace would have called 'contact forces' – has been achieved, nor has a unification of these forces with gravitation been achieved.

11.2   Modern physics

11.2.1   Conventional sequence of theories

A Theory of Everything would unify all the fundamental interactions of nature: gravitation, strong interaction, weak interaction, and electromagnetism. Because the weak interaction can transform elementary particles from one kind into another, the ToE should also yield a deep understanding of the various different kinds of possible particles. The usual assumed path of theories is given in the following graph, where each unification step leads one level up:

In this graph, electroweak unification occurs at around 100 GeV, grand unification is predicted to occur at 10^{16} GeV, and unification of the GUT force with gravity is expected at the Planck energy, roughly 10^{19} GeV.

Several Grand Unified Theories (GUTs) have been proposed to unify electromagnetism and the weak and strong forces. Grand unification would imply the existence of an electronuclear force; it is expected to set in at energies of the order of 10^{16} GeV, far greater than could be reached by any possible Earth-based particle accelerator. Although the simplest GUTs have been experimentally ruled out, the general idea, especially when linked with supersymmetry, remains a favorite candidate in the theoretical physics community. Supersymmetric GUTs seem plausible not only for

their theoretical "beauty", but because they naturally produce large quantities of dark matter, and because the inflationary force may be related to GUT physics (although it does not seem to form an inevitable part of the theory). Yet GUTs are clearly not the final answer; both the current standard model and all proposed GUTs are quantum field theories which require the problematic technique of renormalization to yield sensible answers. This is usually regarded as a sign that these are only effective field theories, omitting crucial phenomena relevant only at very high energies.[3]

The final step in the graph requires resolving the separation between quantum mechanics and gravitation, often equated with general relativity. Numerous researchers concentrate their efforts on this specific step; nevertheless, no accepted theory of quantum gravity – and thus no accepted theory of everything – has emerged yet. It is usually assumed that the ToE will also solve the remaining problems of GUTs.

In addition to explaining the forces listed in the graph, a ToE may also explain the status of at least two candidate forces suggested by modern cosmology: an inflationary force and dark energy. Furthermore, cosmological experiments also suggest the existence of dark matter, supposedly composed of fundamental particles outside the scheme of the standard model. However, the existence of these forces and particles has not been proven yet.

11.2.2   String theory and M-theory

Since the 1990s, some physicists believe that 11-dimensional M-theory, which is described in some limits by one of the five perturbative superstring theories, and in another by the maximally-supersymmetric 11-dimensional supergravity, is the theory of everything. However, there is no widespread consensus on this issue.

A surprising property of string/M-theory is that extra dimensions are required for the theory's consistency. In this regard, string theory can be seen as building on the insights of the Kaluza–Klein theory, in which it was realized that applying general relativity to a five-dimensional universe (with one of them small and curled up) looks from the four-dimensional perspective like the usual general relativity together with Maxwell's electrodynamics. This lent credence to the idea of unifying gauge and gravity interactions, and to extra dimensions, but did not address the detailed experimental requirements. Another important property of string theory is its supersymmetry, which together with extra dimensions are the two main proposals for resolving the hierarchy problem of the standard model, which is (roughly) the question of why gravity is so much weaker than any other force. The extra-dimensional solution involves allowing gravity to propagate into the other dimensions while keeping other forces confined to a four-dimensional spacetime, an idea that has been realized with explicit stringy mechanisms.[18]

Research into string theory has been encouraged by a variety of theoretical and experimental factors. On the experimental side, the particle content of the standard model supplemented with neutrino masses fits into a spinor representation of SO(10), a subgroup of E8 that routinely emerges in string theory, such as in heterotic string theory[19] or (sometimes equivalently) in F-theory.[20][21] String theory has mechanisms that may explain why fermions come in three hierarchical generations, and explain the mixing rates between quark generations.[22] On the theoretical side, it has begun to address some of the key questions in quantum gravity, such as resolving the black hole information paradox, counting the correct entropy of black holes[23][24] and allowing for topology-changing processes.[25][26][27] It has also led to many insights in pure mathematics and in ordinary, strongly-coupled gauge theory due to the Gauge/String duality.

In the late 1990s, it was noted that one major hurdle in this endeavor is that the number of possible four-dimensional universes is incredibly large. The small, "curled up" extra dimensions can be compactified in an enormous number of different ways (one estimate is 10^{500}) each of which leads to different properties for the low-energy particles and forces. This array of models is known as the string theory landscape.[8]:347

One proposed solution is that many or all of these possibilities are realised in one or another of a huge number of universes, but that only a small number of them are habitable, and hence the fundamental constants of the universe are ultimately the result of the anthropic principle rather than dictated by theory. This has led to criticism of string theory,[28] arguing that it cannot make useful (i.e., original, falsifiable, and verifiable) predictions and regarding it as a pseudoscience. Others disagree,[29] and string theory remains an extremely active topic of investigation in theoretical physics.

11.2.3 Loop quantum gravity

Current research on loop quantum gravity may eventually play a fundamental role in a ToE, but that is not its primary aim.[30] Also loop quantum gravity introduces a lower bound on the possible length scales.

There have been recent claims that loop quantum gravity may be able to reproduce features resembling the Standard Model. So far only the first generation of fermions (leptons and quarks) with correct parity properties have been modelled by Sundance Bilson-Thompson using preons consti-

tuted of braids of spacetime as the building blocks.[31] However, there is no derivation of the Lagrangian that would describe the interactions of such particles, nor is it possible to show that such particles are fermions, nor that the gauge groups or interactions of the Standard Model are realised. Utilization of quantum computing concepts made it possible to demonstrate that the particles are able to survive quantum fluctuations.[32]

This model leads to an interpretation of electric and colour charge as topological quantities (electric as number and chirality of twists carried on the individual ribbons and colour as variants of such twisting for fixed electric charge).

Bilson-Thompson's original paper suggested that the higher-generation fermions could be represented by more complicated braidings, although explicit constructions of these structures were not given. The electric charge, colour, and parity properties of such fermions would arise in the same way as for the first generation. The model was expressly generalized for an infinite number of generations and for the weak force bosons (but not for photons or gluons) in a 2008 paper by Bilson-Thompson, Hackett, Kauffman and Smolin.[33]

11.2.4 Other attempts

A recent development is the theory of causal fermion systems,[34] giving all three current physical theories (quantum mechanics, general relativity and quantum field theory) as limiting cases.

A recent and very prolific attempt is called Causal Sets. As some of the approaches mentioned above, its direct goal isn't necessarily to achieve a ToE but primarily a working theory of quantum gravity, which might eventually include the standard model and become a candidate for a ToE. Its founding principle is that spacetime is fundamentally discrete and that the spacetime events are related by a partial order. This partial order has the physical meaning of the causality relations between relative past and future distinguishing spacetime events.

Outside the previously mentioned attempts there is Garrett Lisi's E8 proposal. This theory provides an attempt of identifying general relativity and the standard model within the Lie group E8. The theory doesn't provide a novel quantization procedure and the author suggests its quantization might follow the Loop Quantum Gravity approach above mentioned.[35]

Christoph Schiller's Strand Model attempts to account for the gauge symmetry of the Standard Model of particle physics, U(1)×SU(2)×SU(3), with the three Reidemeister moves of knot theory by equating each elementary particle to a different tangle of one, two, or three strands (selectively a long prime knot or unknotted curve, a rational tangle, or a braided tangle respectively).

11.2.5 Present status

At present, there is no candidate theory of everything that includes the standard model of particle physics and general relativity. For example, no candidate theory is able to calculate the fine structure constant or the mass of the electron. Most particle physicists expect that the outcome of the ongoing experiments – the search for new particles at the large particle accelerators and for dark matter – are needed in order to provide further input for a ToE.

11.3 Theory of everything and philosophy

Main article: Theory of everything (philosophy)

The philosophical implications of a physical ToE are frequently debated. For example, if philosophical physicalism is true, a physical ToE will coincide with a philosophical theory of everything.

The "system building" style of metaphysics attempts to answer *all* the important questions in a coherent way, providing a complete picture of the world. Plato and Aristotle could be said to have created early examples of comprehensive systems. In the early modern period (17th and 18th centuries), the system-building *scope* of philosophy is often linked to the rationalist *method* of philosophy, which is the technique of deducing the nature of the world by pure *a priori* reason. Examples from the early modern period include the Leibniz's Monadology, Descarte's Dualism, and Spinoza's Monism. Hegel's Absolute idealism and Whitehead's Process philosophy were later systems.

11.4 Arguments against a theory of everything

In parallel to the intense search for a ToE, various scholars have seriously debated the possibility of its discovery.

11.4.1 Gödel's incompleteness theorem

A number of scholars claim that Gödel's incompleteness theorem suggests that any attempt to construct a ToE is bound to fail. Gödel's theorem, informally stated, asserts that any formal theory expressive enough for elementary arithmetical facts to be expressed and strong enough for them to be proved is either inconsistent (both a statement and its denial can be derived from its axioms) or incomplete, in the sense that there is a true statement that can't be derived in the formal theory.

Stanley Jaki, in his 1966 book *The Relevance of Physics*, pointed out that, because any "theory of everything" will certainly be a consistent non-trivial mathematical theory, it must be incomplete. He claims that this dooms searches for a deterministic theory of everything.[36] In a later reflection, Jaki states that it is wrong to say that a final theory is impossible, but rather that "when it is on hand one cannot know rigorously that it is a final theory."[37]

Freeman Dyson has stated that "Gödel's theorem implies that pure mathematics is inexhaustible. No matter how many problems we solve, there will always be other problems that cannot be solved within the existing rules. […] Because of Gödel's theorem, physics is inexhaustible too. The laws of physics are a finite set of rules, and include the rules for doing mathematics, so that Gödel's theorem applies to them."[38]

Stephen Hawking was originally a believer in the Theory of Everything but, after considering Gödel's Theorem, concluded that one was not obtainable: "Some people will be very disappointed if there is not an ultimate theory, that can be formulated as a finite number of principles. I used to belong to that camp, but I have changed my mind."[39]

Jürgen Schmidhuber (1997) has argued against this view; he points out that Gödel's theorems are irrelevant for computable physics.[40] In 2000, Schmidhuber explicitly constructed limit-computable, deterministic universes whose pseudo-randomness based on undecidable, Gödel-like halting problems is extremely hard to detect but does not at all prevent formal ToEs describable by very few bits of information.[41]

Related critique was offered by Solomon Feferman,[42] among others. Douglas S. Robertson offers Conway's game of life as an example:[43] The underlying rules are simple and complete, but there are formally undecidable questions about the game's behaviors. Analogously, it may (or may not) be possible to completely state the underlying rules of physics with a finite number of well-defined laws, but there is little doubt that there are questions about the behavior of physical systems which are formally undecidable on the basis of those underlying laws.

Since most physicists would consider the statement of the underlying rules to suffice as the definition of a "theory of everything", most physicists argue that Gödel's Theorem does *not* mean that a ToE cannot exist. On the other hand, the scholars invoking Gödel's Theorem appear, at least in some cases, to be referring not to the underlying rules, but to the understandability of the behavior of all physical systems, as when Hawking mentions arranging blocks into rectangles, turning the computation of prime numbers into a physical question.[44] This definitional discrepancy may explain some of the disagreement among researchers.

11.4.2 Fundamental limits in accuracy

No physical theory to date is believed to be precisely accurate. Instead, physics has proceeded by a series of "successive approximations" allowing more and more accurate predictions over a wider and wider range of phenomena. Some physicists believe that it is therefore a mistake to confuse theoretical models with the true nature of reality, and hold that the series of approximations will never terminate in the "truth". Einstein himself expressed this view on occasions.[45] Following this view, we may reasonably hope for *a* theory of everything which self-consistently incorporates all currently known forces, but we should not expect it to be the final answer.

On the other hand, it is often claimed that, despite the apparently ever-increasing complexity of the mathematics of each new theory, in a deep sense associated with their underlying gauge symmetry and the number of fundamental physical constants, the theories are becoming simpler. If this is the case, the process of simplification cannot continue indefinitely.

11.4.3 Lack of fundamental laws

There is a philosophical debate within the physics community as to whether a theory of everything deserves to be called *the* fundamental law of the universe.[46] One view is the hard reductionist position that the ToE is the fundamental law and that all other theories that apply within the universe are a consequence of the ToE. Another view is that emergent laws, which govern the behavior of complex systems, should be seen as equally fundamental. Examples of emergent laws are the second law of thermodynamics and the theory of natural selection. The advocates of emergence argue that emergent laws, especially those describing complex or living systems are independent of the low-level, microscopic laws. In this view, emergent laws are as fundamental as a ToE.

The debates do not make the point at issue clear. Possibly the only issue at stake is the right to apply the high-status term "fundamental" to the respective subjects of research. A well-known one took place between Steven Weinberg and Philip Anderson

11.4.4 Impossibility of being "of everything"

Although the name "theory of everything" suggests the determinism of Laplace's quotation, this gives a very misleading impression. Determinism is frustrated by the probabilistic nature of quantum mechanical predictions, by the extreme sensitivity to initial conditions that leads to mathematical chaos, by the limitations due to event horizons, and by the extreme mathematical difficulty of applying the theory. Thus, although the current standard model of particle physics "in principle" predicts almost all known non-gravitational phenomena, in practice only a few quantitative results have been derived from the full theory (e.g., the masses of some of the simplest hadrons), and these results (especially the particle masses which are most relevant for low-energy physics) are less accurate than existing experimental measurements. The ToE would almost certainly be even harder to apply for the prediction of experimental results, and thus might be of limited use.

A motive for seeking a ToE, apart from the pure intellectual satisfaction of completing a centuries-long quest, is that prior examples of unification have predicted new phenomena, some of which (e.g., electrical generators) have proved of great practical importance. And like in these prior examples of unification, the ToE would probably allow us to confidently define the domain of validity and residual error of low-energy approximations to the full theory.

11.4.5 Infinite number of onion layers

Lee Smolin regularly argues that the layers of nature may be like the layers of an onion, and that the number of layers might be infinite. This would imply an infinite sequence of physical theories.

The argument is not universally accepted, because it is not obvious that infinity is a concept that applies to the foundations of nature.

11.4.6 Impossibility of calculation

Weinberg[47] points out that calculating the precise motion of an actual projectile in the Earth's atmosphere is impossible. So how can we know we have an adequate theory for describing the motion of projectiles? Weinberg suggests that we know *principles* (Newton's laws of motion and gravitation) that work "well enough" for simple examples, like the motion of planets in empty space. These principles have worked so well on simple examples that we can be reasonably confident they will work for more complex examples. For example, although general relativity includes equations that do not have exact solutions, it is widely accepted as a valid theory because all of its equations with exact solutions have been experimentally verified. Likewise, a ToE must work for a wide range of simple examples in such a way that we can be reasonably confident it will work for every situation in physics.

11.5 See also

- Absolute (philosophy)

- An Exceptionally Simple Theory of Everything

- Argument from beauty

- Attractor

- Beyond black holes

- Beyond the standard model

- Big Bang

- Brownian motion

- Chaos theory

- Chronology of the universe

- Electroweak interaction

- Holographic principle

- Mathematical beauty

- Mathematical universe hypothesis

- Multiverse

- Standard Model (mathematical formulation)

- Superfluid vacuum theory (SVT)

- *The Theory of Everything (2014 film)* – a feature film about Prof. Stephen Hawking and his first wife Jane Hawking

- Timeline of the Big Bang

- Zero-energy universe

11.6 References

11.6.1 Footnotes

[1] Steven Weinberg. *Dreams of a Final Theory: The Scientist's Search for the Ultimate Laws of Nature.* Knopf Doubleday Publishing Group. ISBN 978-0-307-78786-6.

[2] Stephen W. Hawking (28 February 2006). *The Theory of Everything: The Origin and Fate of the Universe.* Phoenix Books; Special Anniv. ISBN 978-1-59777-508-3.

[3] Carlip, Steven (2001). "Quantum Gravity: a Progress Report". *Reports on Progress in Physics* **64** (8): 885. arXiv:gr-qc/0108040. Bibcode:2001RPPh...64..885C. doi:10.1088/0034-4885/64/8/301.

[4] Susanna Hornig Priest (14 July 2010). *Encyclopedia of Science and Technology Communication.* SAGE Publications. ISBN 978-1-4522-6578-0.

[5] Ellis, John (2002). "Physics gets physical (correspondence)". *Nature* **415** (6875): 957. Bibcode:2002Natur.415..957E. doi:10.1038/415957b.

[6] Ellis, John (1986). "The Superstring: Theory of Everything, or of Nothing?". *Nature* **323** (6089): 595–598. Bibcode:1986Natur.323..595E. doi:10.1038/323595a0.

[7] Rorres, Chris (2009). "ARCHIMEDES AND THE QUEST FOR THE THEORY OF EVERYTHING".

[8] Chris Impey (26 March 2012). *How It Began: A Time-Traveler's Guide to the Universe.* W. W. Norton. ISBN 978-0-393-08002-5.

[9] William E. Burns (1 January 2001). *The Scientific Revolution: An Encyclopedia.* ABC-CLIO. ISBN 978-0-87436-875-8.

[10] Shapin, Steven (1996). *The Scientific Revolution.* University of Chicago Press. ISBN 0-226-75021-3.

[11] Newton, Sir Isaac (1729). *The Mathematical Principles of Natural Philosophy* **II**. p. 255.

[12] Sean Carroll (7 January 2010). *From Eternity to Here: The Quest for the Ultimate Theory of Time.* Penguin Group US. ISBN 978-1-101-15215-7.

[13] Faraday, M. (1850). "Experimental Researches in Electricity. Twenty-Fourth Series. On the Possible Relation of Gravity to Electricity". *Abstracts of the Papers Communicated to the Royal Society of London* **5**: 994–995. doi:10.1098/rspl.1843.0267.

[14] Gorban, Alexander N.; Karlin, Ilya (2013). "Hilbert's 6th Problem: Exact and approximate hydrodynamic manifolds for kinetic equations". *Bulletin of the American Mathematical Society* **51** (2): 187. doi:10.1090/S0273-0979-2013-01439-3.

[15] Dirac, P.A.M. (1929). "Quantum mechanics of many-electron systems". *Proceedings of the Royal Society of London A* **123** (792): 714. Bibcode:1929RSPSA.123..714D. doi:10.1098/rspa.1929.0094.

[16] Abraham Pais (23 September 1982). *Subtle is the Lord : The Science and the Life of Albert Einstein: The Science and the Life of Albert Einstein*. Oxford University Press. ISBN 978-0-19-152402-8.

[17] Weinberg (1993), Ch. 5

[18] Holloway, M (2005). "The Beauty of Branes" (PDF). *Scientific American* (Scientific American) **293** (4): 38. Bibcode:2005SciAm.293d..38H. doi:10.1038/scientificamerican1005-38. PMID 16196251. Retrieved August 13, 2012.

[19] Nilles, Hans Peter; Ramos-Sánchez, Saúl; Ratz, Michael; Vaudrevange, Patrick K. S. (2008). "From strings to the MSSM". *The European Physical Journal C* **59** (2): 249. arXiv:0806.3905. Bibcode:2009EPJC...59..249N. doi:10.1140/epjc/s10052-008-0740-1.

[20] Beasley, Chris; Heckman, Jonathan J; Vafa, Cumrun (2009). "GUTs and exceptional branes in F-theory — I". *Journal of High Energy Physics* **2009**: 058. arXiv:0802.3391. Bibcode:2009JHEP...01..058B. doi:10.1088/1126-6708/2009/01/058.

[21] Donagi, Ron; Wijnholt, Martijn (2008). "Model Building with F-Theory". arXiv:0802.2969v3 [hep-th].

[22] Heckman, Jonathan J.; Vafa, Cumrun (2008). "Flavor Hierarchy from F-theory". *Nuclear Physics B* **837**: 137–151. arXiv:0811.2417v3. doi:10.1016/j.nuclphysb.2010.05.009.

[23] Strominger, Andrew; Vafa, Cumrun (1996). "Microscopic origin of the Bekenstein-Hawking entropy". *Physics Letters B* **379**: 99. arXiv:hep-th/9601029. Bibcode:1996PhLB..379...99S. doi:10.1016/0370-2693(96)00345-0.

[24] Horowitz, Gary (1996). "Gravitational Wave Astronomy". *The Origin of Black Hole Entropy in String Theory*. Astrophysics and Space Science Library **211**. p. 95. arXiv:gr-qc/9604051. doi:10.1007/978-94-011-5812-1_7. ISBN 978-94-010-6455-2.

[25] Greene, Brian R.; Morrison, David R.; Strominger, Andrew (1995). "Black hole condensation and the unification of string vacua". *Nuclear Physics B* **451**: 109. arXiv:hep-th/9504145. Bibcode:1995NuPhB.451..109G. doi:10.1016/0550-3213(95)00371-X.

[26] Aspinwall, Paul S.; Greene, Brian R.; Morrison, David R. (1994). "Calabi-Yau moduli space, mirror manifolds and spacetime topology change in string theory". *Nuclear Physics B* **416** (2): 414. arXiv:hep-th/9309097. Bibcode:1994NuPhB.416..414A. doi:10.1016/0550-3213(94)90321-2.

[27] Adams, Allan; Liu, Xiao; McGreevy, John; Saltman, Alex; Silverstein, Eva (2005). "Things fall apart: Topology change from winding tachyons". *Journal of High Energy Physics* **2005** (10): 033. arXiv:hep-th/0502021. Bibcode:2005JHEP...10..033A. doi:10.1088/1126-6708/2005/10/033.

[28] Smolin, Lee (2006). *The Trouble With Physics: The Rise of String Theory, the Fall of a Science, and What Comes Next*. Houghton Mifflin. ISBN 978-0-618-55105-7.

[29] Duff, M. J. (2011). "String and M-Theory: Answering the Critics". *Foundations of Physics* **43**: 182. arXiv:1112.0788. Bibcode:2013FoPh...43..182D. doi:10.1007/s10701-011-9618-4.

[30] Potter, Franklin (15 February 2005). "Leptons And Quarks In A Discrete Spacetime" (PDF). *Frank Potter's Science Gems*. Retrieved 2009-12-01.

[31] Bilson-Thompson, Sundance O.; Markopoulou, Fotini; Smolin, Lee (2007). "Quantum gravity and the standard model". *Classical and Quantum Gravity* **24** (16): 3975–3994. arXiv:hep-th/0603022. Bibcode:2007CQGra..24.3975B. doi:10.1088/0264-9381/24/16/002.

[32] Castelvecchi, Davide; Valerie Jamieson (August 12, 2006). "You are made of space-time". *New Scientist* (2564).

[33] Sundance Bilson-Thompson; Jonathan Hackett; Lou Kauffman; Lee Smolin (2008). "Particle Identifications from Symmetries of Braided Ribbon Network Invariants". arXiv:0804.0037 [hep-th].

[34] F. Finster; J. Kleiner (2015). "Causal fermion systems as a candidate for a unified physical theory". *Journal of Physics: Conference Series* **626** (2015): 012020. arXiv:1502.03587. doi:10.1088/1742-6596/626/1/012020.

[35] A. G. Lisi (2007). "An Exceptionally Simple Theory of Everything". arXiv:0711.0770 [hep-th].

[36] Jaki, S.L. (1966). *The Relevance of Physics*. Chicago Press. pp. 127–130.

[37] Stanley L. Jaki (2004) "A Late Awakening to Gödel in Physics", pp. 8–9.

[38] Freeman Dyson, NYRB, May 13, 2004

[39] Stephen Hawking, Gödel and the end of physics, July 20, 2002

[40] Schmidhuber, Jürgen (1997). *A Computer Scientist's View of Life, the Universe, and Everything*. *Lecture Notes in Computer Science*. Springer. pp. 201–208. doi:10.1007/BFb0052071. ISBN 978-3-540-63746-2.

[41] Schmidhuber, Jürgen (2002). "Hierarchies of generalized Kolmogorov complexities and nonenumerable universal measures computable in the limit". *Sections in: Hierarchies of generalized Kolmogorov complexities and nonenumerable universal measures computable in the limit. International Journal of Foundations of Computer Science ():587-612 (2002). Section 6 in: the Speed Prior: A New Simplicity Measure Yielding Near-Optimal Computable Predictions. in J. Kivinen and R. H. Sloan, editors, Proceedings of the 15th Annual Conference on Computational Learning Theory (COLT 2002), Sydney, Australia, Lecture Notes in Artificial Intelligence, pages 216-–228. Springer, 2002* **13** (4): 1–5. arXiv:quant-ph/0011122.

[42] Feferman, Solomon (17 November 2006). "The nature and significance of Gödel's incompleteness theorems" (PDF). Institute for Advanced Study. Retrieved 2009-01-12.

[43] Robertson, Douglas S. (2007). "Goedel's Theorem, the Theory of Everything, and the Future of Science and Mathematics". *Complexity* **5** (5): 22–27. doi:10.1002/1099-0526(200005/06)5:5<22::AID-CPLX4>3.0.CO;2-0.

[44] Hawking, Stephen (20 July 2002). "Gödel and the end of physics". Retrieved 2009-12-01.

[45] Einstein, letter to Felix Klein, 1917. (On determinism and approximations.) Quoted in Pais (1982), Ch. 17.

[46] Weinberg (1993), Ch 2.

[47] Weinberg (1993) p. 5

11.6.2 Bibliography

- Pais, Abraham (1982) *Subtle is the Lord...: The Science and the Life of Albert Einstein* (Oxford University Press, Oxford, . Ch. 17, ISBN 0-19-853907-X

- Weinberg, Steven (1993) *Dreams of a Final Theory: The Search for the Fundamental Laws of Nature*, Hutchinson Radius, London, ISBN 0-09-177395-4

11.7 External links

- The Elegant Universe, *Nova* episode about the search for the theory of everything and string theory.

- Theory of Everything, freeview video by the Vega Science Trust, BBC and Open University.

- The Theory of Everything: Are we getting closer, or is a final theory of matter and the universe impossible? Debate between John Ellis (physicist), Frank Close and Nicholas Maxwell.

- Why The World Exists, a discussion between physicist Laura Mersini-Houghton, cosmologist George Francis Rayner Ellis and philosopher David Wallace about dark matter, parallel universes and explaining why these and the present Universe exist.

Chapter 12

An Exceptionally Simple Theory of Everything

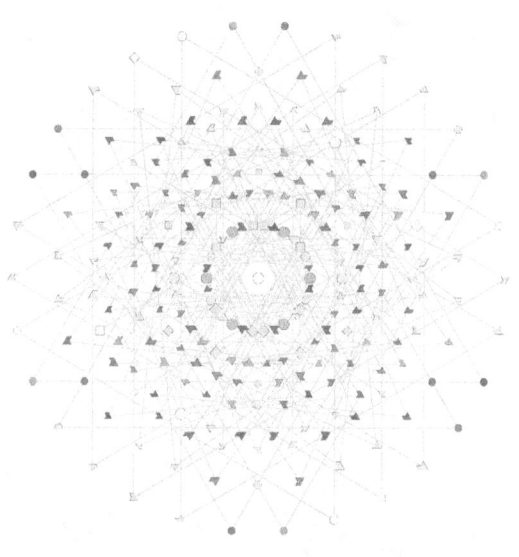

Elementary particle states assigned to E_8 roots corresponding to their spin, electroweak, and strong charges according to E_8 Theory, with particles related by triality. This eight-dimensional root diagram is shown projected onto a Coxeter plane.

"An Exceptionally Simple Theory of Everything"[1] is a physics preprint proposing a basis for a unified field theory, often referred to as "**E_8 Theory**",[2] which attempts to describe all known fundamental interactions in physics and to stand as a possible theory of everything. The paper was posted to the physics arXiv by Antony Garrett Lisi on November 6, 2007, and was not submitted to a peer-reviewed scientific journal.[3] The title is a pun on the algebra used, the Lie algebra of the largest "simple", "exceptional" Lie group, E_8. The paper's goal is to describe how the combined structure and dynamics of all gravitational and Standard Model particle fields, including fermions, are part of the E_8 Lie algebra.[2]

The theory is presented as an extension of the grand unified theory program, incorporating gravity and fermions. In the paper, Lisi states that all three generations of fermions do not directly embed in E_8 with correct quantum numbers and spins, but that they must be described via a triality transformation, noting that the theory is incomplete and that a correct description of the relationship between triality and generations, if it exists, awaits a better understanding.

The theory received accolades from a few physicists amid a flurry of media coverage, but also met with widespread skepticism.[4] *Scientific American* reported in March 2008 that the theory was being "largely but not entirely ignored" by the mainstream physics community, with a few physicists picking up the work to develop it further.[5] In a follow-up paper, Lee Smolin proposed a spontaneous symmetry breaking mechanism for obtaining the classical action in Lisi's model, and speculated on the path to its quantization.[6] In July 2009, Jacques Distler and Skip Garibaldi published a critical paper in *Communications in Mathematical Physics* called "There is no 'Theory of Everything' inside E_8",[7] arguing that Lisi's theory, and a large class of related models, cannot work. They offer a direct proof that it is impossible to embed all three generations of fermions in E_8, or to obtain even the one-generation Standard Model without the presence of an antigeneration. In response to Distler and Garibaldi's paper, Lisi argued in a new paper, "An Explicit Embedding of Gravity and the Standard Model in E_8",[8] peer reviewed and published in a conference proceedings, that Distler and Garibaldi's assumptions about fermion embeddings are incorrect and that the antigeneration is not by itself a problem sufficient to rule out the one-generation Standard Model.[8][9] In July 2010, a group of mathematicians and physicists, including David Vogan, Garibaldi, and Lisi, met for a week-long conference in Banff to discuss the mathematics and physics related to the exceptional groups.[10] In December 2010, Scientific American published a feature article on "A Geometric Theory of Everything", authored by Lisi and James

Owen Weatherall.[2] In May 2011, Lisi wrote an entry in the blog section of *Scientific American* addressing some of the criticism of his theory and how it had progressed, noting that the theory was still incomplete and made only tenuous predictions, with a precise description of the three generations of fermions and their masses remaining as the largest outstanding problem.[9] In June 2015, Lisi posted a paper, "Lie Group Cosmology", describing the geometry of E_8 Theory as an extension of Cartan geometry, and providing a description of the three generations of fermions via triality, while not predicting their masses.[11][12]

12.1   Overview

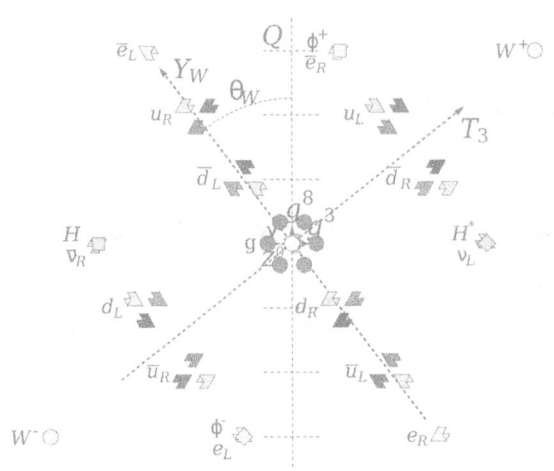

The pattern of weak isospin, T_3, and weak hypercharge, Y_W, and color charge of all known elementary particles, rotated by the weak mixing angle to show electric charge, Q, roughly along the vertical. The neutral Higgs field (gray square) breaks the electroweak symmetry and interacts with other particles to give them mass.

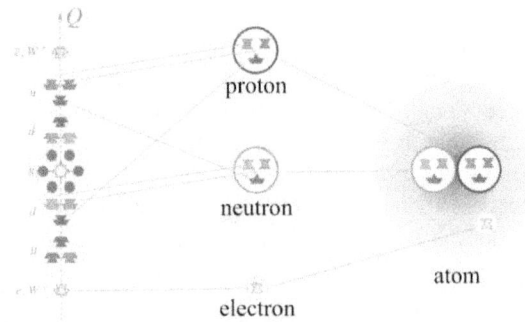

Electrons and quarks, with electric (Q) and color (g) charges, make up color-neutral protons (with total electric charge $Q=+1$) and neutrons (with electric charge $Q=0$), which make up atoms.

The goal of E_8 Theory is to describe all elementary particles and their interactions, including gravitation, as quantum excitations of a single Lie group geometry—specifically, excitations of the noncompact quaternionic real form of the largest simple exceptional Lie group, E_8. A Lie group, such as a one-dimensional circle, may be understood as a smooth manifold with a fixed, highly symmetric geometry. Larger Lie groups, as higher-dimensional manifolds, may be imagined as smooth surfaces composed of many circles (and hyperbolas) twisting around one another. At each point in a N-dimensional Lie group there can be N different orthogonal circles, tangent to N different orthogonal directions in the Lie group, spanning the N-dimensional Lie algebra of the Lie group. For a Lie group of rank R, one can choose at most R orthogonal circles that do not twist around each other, and so form a *maximal torus* within the Lie group, corresponding to a collection of R mutually-commuting Lie algebra generators, spanning a *Cartan subalgebra*. Each elementary particle state can be thought of as a different orthogonal direction, having an integral number of twists around each of the R directions of a chosen maximal torus. These R twist numbers (each multiplied by a scaling factor) are the R different kinds of elementary charge that each

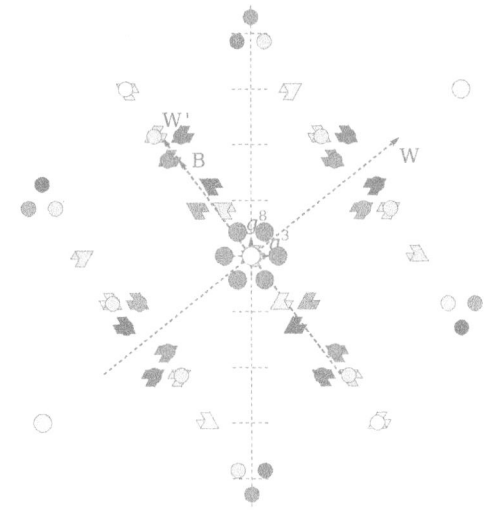

The pattern of weak isospin, W, weaker isospin, W', strong g_3 and g_8, and baryon minus lepton, B, charges for particles in the SO(10) model, rotated to show the embedding of the Georgi-Glashow model and Standard Model, with electric charge roughly along the vertical. In addition to Standard Model particles, the theory includes thirty colored X bosons, responsible for proton decay, and three W' and Z' bosons.

particle has. Mathematically, these charges are eigenvalues of the Cartan subalgebra generators, and are called roots or

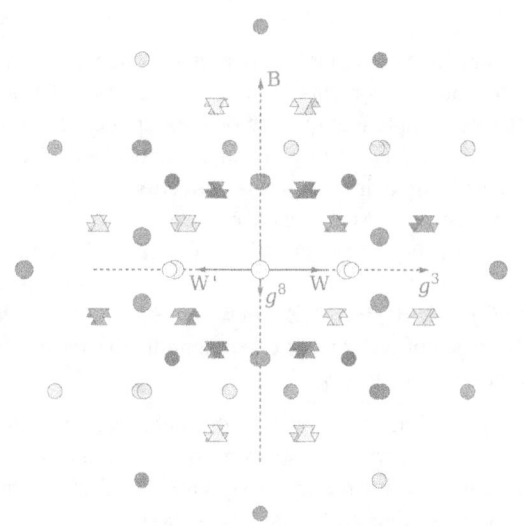

The pattern of weak isospin, W, weaker isospin, W', strong g₃ and g₈, and baryon minus lepton, B, charges for particles in the SO(10) Grand Unified Theory, rotated to show the embedding in E6.

weights of a representation.

In the Standard Model of particle physics, each different kind of elementary particle has four different charges, corresponding to twists along directions of a four-dimensional maximal torus in the twelve-dimensional Standard Model Lie group, $SU(3) \times SU(2) \times U(1)$. The two strong "color" charges, g^3 and g^8, correspond to twists along directions in the two-dimensional maximal torus of the eight-dimensional $SU(3)$ Lie group of the strong interaction. The weak isospin, T_3 (or W), and weak hypercharge, YW (or Y), correspond to twists along directions in the two-dimensional maximal torus of the four-dimensional $SU(2) \times U(1)$ Lie group of the electroweak interaction, with W and Y combining as electric charge, Q. Whenever an interaction occurs between elementary particles, with two coming together and becoming a third, or one particle becoming two, each type of charge must be conserved. For example, a red up quark, having charges ($g^3 = \frac{1}{2}$, $g^8 = \frac{1}{2\sqrt{3}}$, $W = \frac{1}{2}$, $Y = \frac{1}{3}$) can interact with a weak boson, W⁻, having charges ($g^3 = 0$, $g^3 = 0$, $W = -1$, $Y = 0$), to produce a red down quark, having charges ($g^3 = \frac{1}{2}$, $g^8 = \frac{1}{2\sqrt{3}}$, $W = \frac{-1}{2}$, $Y = \frac{1}{3}$). The complete pattern of all Standard Model particle charges in four dimensions may be projected down to two dimensions and plotted in a charge diagram.

In grand unified theories (GUTs), the 12-dimensional Standard Model Lie group, $SU(3) \times SU(2) \times U(1)$ (modded by $\mathbf{Z}_6$), is considered as a subgroup of a higher-dimensional Lie group, such as of 24-dimensional SU(5) in the Georgi-Glashow model or of 45-dimensional Spin(10) in the

SO(10) model (Spin(10) being the double cover of SO(10), and having the same Lie algebra). Since there is a different elementary particle for each dimension of the Lie group, in addition to the 12 Standard Model gauge bosons there are 12 X and Y bosons in the SU(5) Model and 18 more X bosons and 3 W' and Z' bosons in Spin(10). In Spin(10) there is a five-dimensional maximal torus, and the Standard Model hypercharge, Y, is a combination of two new Spin(10) charges: "weaker charge", W', and baryon minus lepton number, B. In the Spin(10) model, one generation of 16 fermions (including left-handed electrons, neutrinos, three colors of up quarks, three colors of down quarks, and their anti-particles) lives neatly in the 16-complex-dimensional spinor representation space of Spin(10). The combination of these 32 real fermions and 45 bosons, along with another U(1) Lie group (corresponding to Peccei–Quinn symmetry), constitute the 78-dimensional real compact exceptional Lie group, E6. (This unusual algebraic structure, reminiscent of supersymmetry, of gauge fields and spinors combined in a simple Lie group, is characteristic of the exceptional groups.)

As well as being in some representation space of the Standard Model or Grand Unified Theory Lie group, each physical fermion is a spinor under the gravitational noncompact Spin(1,3) Lie group of rotations and boosts. This six-dimensional Lie group has a two-dimensional maximal torus (technically a hyperboloid) and thus two kinds of charge, spin, S_z, and boost, S_t. A Dirac fermion (consisting of fermion and anti-fermion) has eight real degrees of freedom corresponding to its real vs. imaginary parts, left or right **chirality**, and being spin up or down. Using the Lie group equivalence of Spin(1,3) and SL(2,**C**), and the chirality of Standard Model weak force fermion interactions, each fermion (and each anti-fermion) can be described as a two-complex-dimensional left-chiral Weyl spinor under gravitational SL(2,**C**). Accounting for the up or down spin for each of the 16 left-chiral fermions of one generation (or 15 fermions if neutrinos are Majorana), each fermion generation corresponds to 64 (or 60) real degrees of freedom.

In GraviGUT unification, the gravitational Spin(1,3) and Spin(10) GUT Lie groups are combined (modded by $\mathbf{Z}_2$) as parts of a Spin(11,3) Lie group, acting on each generation of fermions in a real 64-dimensional spinor representation. The remaining parts of Spin(11,3) include the 4-dimensional spacetime frame and a Higgs field transforming as a 10 under Spin(10). The resulting gauge theory of gravity, Higgs, and gauge bosons is an extension of the MacDowell-Mansouri formalism to higher dimensions. Several physicists objected to the apparent violation of the Coleman-Mandula theorem, which states the impossibility of mixing gravity and gauge fields in a unified Lie group over spacetime, given reasonable assumptions. Proponents of GraviGUT unification and E_8 Theory claim that the Coleman-Mandula theorem is not violated because the assumptions are not met.[13]

In E_8 Theory, it is observed that the GraviGUT algebra of spin(11,3) acting on one generation of fermions in a real positive-chiral 64-spinor, 64_+, can be part of the 248-dimensional real quaternionic e8 Lie algebra,

$$e8 = \text{spin}(12,4) + 128_+$$

The strongest criticism of E_8 Theory, stated by Distler, Garibaldi, and others, including Lisi in the original paper, is that given an embedding of gravitational spin(1,3) in the spin(12,4) subalgebra of e8, the 128_+ includes not only the 64_+ of a generation of fermions, but a 64- "anti-generation" of mirror fermions with non-physical chirality. Since we do not see mirror fermions in nature, Distler and Garibaldi consider this to be a disproof of E_8 Theory. Lisi has voiced two responses to this criticism. The first response is that these mirror fermions might exist and have very large masses. The second response, stated in the original paper and in his latest work, is that there is not a single embedding of gravitational spin(1,3) in e8, but three embeddings related by triality, with respect to which the 64- contains a second generation of physical fermions, and the third generation of fermions is contained within spin(12,4).[14]

The algebraic breakdown of the 248-dimensional e8 Lie algebra relevant to E8 Theory is

$$e8 = \text{spin}(4,4) + \text{spin}(8) + 8V \otimes 8V + 8_+ \otimes 8_+ + 8_- \otimes 8_-$$

This decomposition, attributed to Bertram Kostant, relies on the triality isomorphism between eight-dimensional vectors, 8_v, positive-chiral spinors, 8_+, and negative-chiral spinors, 8_-, relating to the division algebra of the octonions.[15] Within this decomposition, the strong force su(3) embeds in spin(8), three triality-related gravitational spin(1,3)'s embed in spin(4,4), the three generations of 60 fermions embed in $8V \otimes 8V + 8_+ \otimes 8_+ + 8_- \otimes 8_-$, and the gravitational frame, Higgs, and electroweak bosons embed throughout, with 18 colored X bosons remaining as new predicted particles.[16]

In E_8 Theory's current state, it is not possible to calculate masses for the existing or predicted particles. Lisi states the theory is young and incomplete, requiring a better understanding of the three fermion generations and their masses, and places a low confidence in its predictions. However, the discovery of new particles that do not fit in Lisi's classification, such as superpartners or new fermions, would fall outside the model and falsify the theory.

12.2 Technical Overview

The fundamental geometric idea of E_8 Theory is that our universe and its contents exists as quantum excitations of the largest simple real quaternionic exceptional Lie group, $E_{8(-24)}$. This is described via an extension of Cartan geometry employing a **superconnection**. The relevant Cartan geometry is modeled on Klein geometry, beginning with a homogeneous space, G/H, in which the initial Lie group is $G = E_{8(-24)}$ and the subgroup is $H = SL(2,\mathbf{C}) \times S(U(3) \times U(2)) \times \mathbf{Z}_3$, in which $\mathbf{Z}_3 = \{1, T, T^2\}$ is the cyclic group of order three corresponding to a triality automorphism, T, of $E_{8(-24)}$.[11]

Usually, in Cartan geometry, the deformation of a Lie group, G, preserving the structure of a subgroup, H, is described by allowing the Lie group's Maurer-Cartan form, θ, to vary, becoming the Cartan connection,

$$C = W + \mathcal{E}$$

The resulting geometry, $\tilde{G}$, is that of a principal bundle, with W the principal H-connection, a 1-form valued in Lie(H) over a base manifold, B, modeled on G/H, with the **frame**, $\mathcal{E}$, a 1-form valued in Lie(G/H). If H is a reductive subgroup of G, the curvature of the Cartan connection is

$$FC = dC + CC = (dW + WW + \mathcal{E}\mathcal{E}) + (d\mathcal{E} + W\mathcal{E} + \mathcal{E}W)$$

In Lisi's extension of Cartan geometry, the Cartan connection over B is interpreted as a superconnection,

$$G = W + E + \Psi$$

over spacetime, M (a subspace of B), in which, Ψ, a Lie(G/H) valued 1-form over B assumed to be orthogonal to M, is interpreted as a set of three fermionic (Grassmann number) fields over M, valued in Lie(G/H), related by triality, T.[11] According to Lisi, the description of Grassmann number fermions as 1-forms orthogonal to spacetime, valued in a spinor representation, provides a clear geometric understanding of what fermions are.[12] In E_8 Theory, the H-connection is physically the gravitational spin connection, $\frac{1}{2}\omega$, plus the Standard Model and X boson gauge fields, $H = g + W + B + X$, while E is the 1-form frame over M, assumed to equal the gravitational frame 1-form, e, times (in the Clifford algebra sense) the Higgs, $E = e\Phi$. The fermionic part of the superconnection, Ψ, is interpreted as the three generation multiplets of Standard Model fermions. The curvature of the resulting superconnection,

$$G = \frac{1}{2}\omega + H + e\Phi + \Psi$$

is

$$F = dG + GG = (\tfrac{1}{2}R - e\,e\,\Phi^2) + FH + (T\Phi + eD\Phi) + D\Psi + \Psi\Psi$$

in which $R = d\omega + \tfrac{1}{2}\omega\omega$ is the gravitational Riemann curvature 2-form, $FH = dH + HH$ is the gauge field curvature, $T = de + \tfrac{1}{2}\omega e + \tfrac{1}{2}e\omega$ is the gravitational torsion, $D\Phi = d\Phi + [H,\Phi]$ is the covariant derivative of the Higgs, and $D\Psi = d\Psi + [\tfrac{1}{2}\omega + H + e\Phi, \Psi]$ is the covariant Dirac derivative of the fermions in curved spacetime.

In this geometric description, physical four-dimensional spacetime, M, is considered as a sheaf of gauge-related subspaces of $\tilde{G}$. For the case in which the curvature vanishes, $F = 0$, there is no excitation of the Lie group, G, and the Higgs field has a vacuum expectation value, $\Phi=\Phi_0$, corresponding to a positive cosmological constant, $\Lambda = -12\,\Phi_0^{\,2}$, with the vacuum spacetime, as a subspace of G, identified as de Sitter spacetime, satisfying $R = -6\Lambda ee$.

Within a Lie group, the Maurer-Cartan form, θ, is the natural frame and determines the Haar measure for integration over the group manifold. With the Killing form of the Lie algebra, this also determines a natural metric and Hodge duality operator on the group manifold. For a deforming Lie group, the Maurer-Cartan form is replaced by the superconnection, G, defined over the entire deforming Lie group manifold via gauge transformation. This superconnection, G, determines the Hodge duality operator, $\star$, and the curvature, F, of the deforming Lie group. The action for E_8 Theory is the Yang-Mills action, integrated over the entire deforming Lie group.

$$S = \tfrac{1}{2}\int (F, \star F)$$

Since the structure of the H subgroup and fermionic directions of B are preserved, this action reduces to an integral over spacetime,

$$S = \tfrac{1}{2}\,V\int \{(ee, \star ee)\,\Phi^4 - (R, \star ee)\,\Phi^2 + \tfrac{1}{4}(R, \star R) + (D\Phi, \star D\Phi) - (T, \star T)\,\Phi^2 + (FH, \star FH) + (D\Psi, \star D\Psi)\}$$

in which V is a constant volume factor, and the Hodge star, $\star$, is now the Hodge star for M determined by the gravitational frame, e. As well as the usual Einstein-Hilbert action for gravity, Yang-Mills action, and Higgs action, this action includes a quadratic torsion term, a quadratic curvature term, and a quadratic spinor Lagrangian.[11]

12.3 Chronology and reaction

Three previous arXiv preprints by Lisi deal with mathematical physics related to the theory. "Clifford Geometrodynamics",[17] in 2002, endeavors to describe fermions geometrically as BRST ghosts. "Clifford bundle formulation of BF gravity generalized to the standard model",[18] in 2005, describes the algebra of gravitational and Standard Model fields acting on a generation of fermions, but does not mention E_8. "Quantum mechanics from a universal action reservoir",[19] in 2006, attempts to derive quantum mechanics using information theory.

Before writing his 2007 paper, Lisi discussed his work on an Foundational Questions Institute (FQXi) forum,[20] at an FQXi conference,[21] and for an FQXi article.[22] Lisi gave his first talk on E_8 Theory at the Loops '07 conference in Morelia, Mexico,[23] soon followed by a talk at the Perimeter Institute.[24] John Baez commented on Lisi's work in "This Week's Finds in Mathematical Physics (Week 253)",[25] and Lisi was interviewed on Sabine Hossenfelder's "Backreaction" blog.[26] Lisi's arXiv preprint, "An Exceptionally Simple Theory of Everything", appeared on November 6, 2007, and immediately attracted a great deal of attention. Lisi made a further presentation for the International Loop Quantum Gravity Seminar on November 13, 2007,[27] and responded to press inquiries on an FQXi forum.[28] He presented his work at the TED Conference on February 28, 2008.[29]

Numerous news sites from all over the world reported on the new theory in 2007 and 2008, noting Lisi's personal history and the controversy in the physics community. The first mainstream and scientific press coverage began with articles in *The Daily Telegraph* and *New Scientist*,[30] with articles soon following in many other newspapers and magazines.

Lisi's paper spawned a variety of reactions and debates across various physics blogs and online discussion groups. The first to comment was Sabine Hossenfelder, summarizing the paper and noting the lack of a dynamical symmetry breaking mechanism.[31] Luboš Motl offered a colorful critique, objecting to the addition of bosons and fermions in Lisi's superconnection, and to the violation of the Coleman-Mandula theorem.[32] In the presentation "What's new at the arXiv?" on May 20, 2008, Simeon Warner stated that Lisi's paper is the most downloaded article on the arXiv.[33][34] Among the physicists early to comment on E_8 Theory, Sabine Hossenfelder, Peter Woit and Lee Smolin were generally supportive, while Luboš Motl and Jacques Distler were critical.

On his blog, *Musings*, Jacques Distler offered one of the strongest criticisms of Lisi's approach, claiming to demonstrate that, unlike in the Standard Model, Lisi's model is nonchiral — consisting of a generation and an antigeneration — and to prove that any alternative embedding in E_8 must be similarly nonchiral.[35][36][37] These arguments were distilled in a paper written jointly with Skip Garibaldi, "There is no 'Theory of Everything' inside E_8",[7] published in *Communications in Mathematical Physics*. In this paper, Distler and Garibaldi offer a proof that it is impossible to embed all three generations of fermions in E_8, or to obtain even the one-generation Standard Model. In a press release from his university, "Rock climber takes on surfer's theory",[38][39] Garibaldi states that his article with Distler is a rebuttal of Lisi's theory. In response, Lisi argues that Distler and Garibaldi made unnecessary assumptions about how the embedding needs to happen.[9] Addressing the one generation case, in June 2010 Lisi posted a new paper on E_8 Theory, "An Explicit Embedding of Gravity and the Standard Model in E_8",[8] peer reviewed and published in a conference proceedings, describing how the algebra of gravity and the Standard Model with one generation of fermions embeds in the E_8 Lie algebra explicitly using matrix representations. When this embedding is done, Lisi agrees that there is an antigeneration of fermions (also known as "mirror fermions") remaining in E_8; but while Distler and Garibaldi state that these mirror fermions make the theory nonchiral, Lisi states that these mirror fermions might have high masses, making the theory chiral, or that they might be related to the other generations.[9]

The group blog, *The n-Category Cafe*, provides some of the more technical discussions, with posts by Lisi, Urs Schreiber,[40] Kea,[41] and Jacques Distler.[41]

Thirty-eight arXiv preprints have cited Lisi's work. Lee Smolin's "The Plebanski action extended to a unification of gravity and Yang–Mills theory", December 6, 2007, proposes a symmetry breaking mechanism to go from an E_8 symmetric action to Lisi's action for the Standard Model and gravity.[6] Roberto Percacci's "Mixing internal and spacetime transformations: some examples and counterexamples"[13] addresses a general loophole in the Coleman-Mandula theorem also thought to work in E_8 Theory.[9] Percacci and Fabrizio Nesti's "Chirality in unified theories of gravity"[42] confirms the embedding of the algebra of gravitational and Standard Model forces acting on a generation of fermions in $\text{spin}(3,11) + 64_+$, mentioning that Lisi's "ambitious attempt to unify all known fields into a single representation of E_8 stumbled into chirality issues".[42] Mathematician Bertram Kostant discussed Lisi's work in a colloquium presentation at UC Riverside.[43] In a joint paper with Lee Smolin and Simone Speziale,[44] published in *Journal of Physics A*, Lisi proposes a new action and symmetry breaking mechanism. In "An Explicit Embedding of Gravity and the Standard Model in E_8",[8] Lisi describes E_8 Theory using explicit matrix representations. In "Lie Group Cosmology",[11] Lisi describes the extension of Cartan geometry underlying E_8 Theory.

On August 4, 2008, FQXi awarded Lisi a grant for further development of E_8 Theory.[45][46]

In September 2010, *Scientific American* reported on a conference inspired by Lisi's work.[47]

In December 2010 *Scientific American* published a feature article on E_8 Theory, "A Geometric Theory of Everything",[2] written by Lisi and James Owen Weatherall.

In December 2011, in his paper, "String and M-theory: answering the critics",[48] for a Special Issue of Foundations of Physics: "Forty Years Of String Theory: Reflecting On the Foundations", Michael Duff argues against Lisi's theory and the attention it has received in the popular press.[49] Duff states that Lisi's paper was incorrect, citing Distler and Garibaldi's proof, and criticizes the press for giving too much positive attention to an "outsider" scientist and theory.

12.4   References

[1] A. G. Lisi (2007). "An Exceptionally Simple Theory of Everything". arXiv:0711.0770 [hep-th].

[2] A. G. Lisi; J. O. Weatherall (2010). "A Geometric Theory of Everything". *Scientific American* **303** (6): 54–61. doi:10.1038/scientificamerican1210-54. PMID 21141358.

[3] Greg Boustead (2008-11-17). "Garrett Lisi's Exceptional Approach to Everything". *SEED Magazine*.

[4] Amber Dance (2008-04-01). "Outsider Science". *Symmetry Magazine*. Archived from the original on 5 July 2008. Retrieved 2008-06-15.

[5] Collins, Graham P. (March 2008). "Wipeout?". *Scientific American*: 30–32. Retrieved 2008-06-18.

[6] Lee Smolin (2007). "The Plebanski action extended to a unification of gravity and Yang-Mills theory". arXiv:0712.0977 [hep-th].

[7] Jacques Distler; Skip Garibaldi (2009). "There is no 'Theory of Everything' inside E_8". arXiv:0905.2658 [math.RT].

[8] A. G. Lisi (2010). "An Explicit Embedding of Gravity and the Standard Model in E_8". arXiv:1006.4908 [gr-qc].

[9] A G Lisi (2011-05-11). "Garrett Lisi Responds to Criticism of his Proposed Unified Theory of Physics". *Scientific American*. Archived from the original on 2011-07-02. Retrieved 2011-07-30.

[10] David Vogan (2011-12-07). "Structure and representations of exceptional groups" (PDF). *BIRS reports*. Retrieved 2011-12-21.

[11] A. G. Lisi (2015). "Lie Group Cosmology". arXiv:1506.08073 [gr-qc].

[12] John Horgan (2014-10-20). "Surfer-Physicist Offers Alternative to String Theory, Academia". *Scientific American*. Retrieved 2015-12-13.

[13] Roberto Percacci (2008). "Mixing internal and space-time transformations: some examples and counterexamples". arXiv:0803.0303 [hep-th].

[14] "A Geometric Theory of Everything" (PDF).

[15] Baez, John C. (2002). "The Octonions". *Bulletin of the American Mathematical Society* **39** (2): 145–205. arXiv:math/0105155v4. doi:10.1090/S0273-0979-01-00934-X. ISSN 0273-0979. MR 1886087.

[16] "The Big Bang: what will we find?". *The Daily Telegraph*. 2008-03-25. Retrieved 2008-06-15.

[17] A. G. Lisi (2002). "Clifford Geometrodynamics". arXiv:gr-qc/0212041.

[18] A. G. Lisi (2005). "Clifford bundle formulation of BF gravity generalized to the standard model". arXiv:gr-qc/0511120.

[19] A. G. Lisi (2006). "Quantum mechanics from a universal action reservoir". arXiv:physics/0605068.

[20] A. G. Lisi (2007-06-09). "Pieces of E_8". *FQXi forum*. Archived from the original on 2 June 2008. Retrieved 2008-06-15.

[21] A. G. Lisi (2007-07-21). "Standard model and gravity". *inaugural FQXi conference*. Retrieved 2008-06-15.

[22] Scott Dodd (2007-10-26). "Surfing the Folds of Spacetime" (PDF). *FQXi article*. Retrieved 2008-06-15.

[23] A. G. Lisi (2007-06-25). "Deferential Geometry". *Loops '07 conference*. Retrieved 2008-06-15.

[24] A. G. Lisi (2007-10-04). "An Exceptionally Simple Theory of Everything". *Perimeter Institute talk*. Retrieved 2008-06-15.

[25] John Baez (2007-06-27). "This Week's Finds in Mathematical Physics (Week 253)". Archived from the original on 30 June 2008. Retrieved 2008-06-15.

[26] Sabine Hossenfelder (2007-08-06). "Garrett Lisi's Inspiration". *Backreaction*. Retrieved 2008-06-15.

[27] A. G. Lisi (2007-11-13). "A Connection With Everything". *International Loop Quantum Gravity Seminar*. Archived from the original on 22 May 2008. Retrieved 2008-06-15.

[28] A. G. Lisi (2007-11-20). "An Exceptionally Simple FAQ". *FQXi forum*. Archived from the original on 2 June 2008. Retrieved 2008-06-15.

[29] A. G. Lisi (2008-02-28). "Garrett Lisi: A beautiful new theory of everything". *TED talks*. Archived from the original on 18 October 2008. Retrieved 2008-10-17.

[30] Zeeya Merali (2007-11-15). "Is mathematical pattern the theory of everything?". *New Scientist*. Archived from the original on 12 May 2008. Retrieved 2008-06-15.

[31] Sabine Hossenfelder (2007-11-06). "A Theoretically Simple Exception of Everything". *Backreaction*. Archived from the original on 26 May 2008. Retrieved 2008-06-15.

[32] Luboš Motl (2007-11-07). "Garrett Lisi: An exceptionally simple theory of everything". *The Reference Frame*. Retrieved 2008-06-15.

[33] Peter Woit (2008-05-28). "INSPIRE". *Not Even Wrong*. Retrieved 2008-08-05.

[34] Simeon Warner (2008-05-20). "What's new at the arXiv?". *HEP Information Resource Summit*. Retrieved 2008-07-22. (The slide containing this statement was subsequently removed from the presentation file.)

[35] Jacques Distler (2007-11-21). "A Little Group Theory". *Musings*. Archived from the original on 12 May 2008. Retrieved 2008-06-15.

[36] Jacques Distler (2007-12-09). "A Little More Group Theory". *Musings*. Retrieved 2008-11-15.

[37] Jacques Distler (2008-09-14). "My Dinner with Garrett". *Musings*. Archived from the original on 2008-11-19. Retrieved 2008-11-15.

[38] Carol Clark (2010-03-18). "Rock climber takes on surfer's theory". *esciencecommons*. Retrieved 2011-07-30.

[39] "No 'Simple Theory of Everything' Inside the Enigmatic E_8, Researcher Says". *ScienceDaily*. 2010-03-26. Retrieved 2011-07-30.

[40] Urs Schreiber (2008-05-10). "E_8 Quillen Superconnection". *The n-Category Cafe*. Archived from the original on 2008-06-19. Retrieved 2008-06-15.

[41] http://golem.ph.utexas.edu/category/2008/05/e8_quillen_superconnection.html#c016877

[42] R. Percacci; F. Nesti (2009). "Chirality in unified theories of gravity". arXiv:0909.4537 [hep-th].

[43] Bertram Kostant (2008-02-12). "On Some Mathematics in Garrett Lisi's 'E_8 Theory of Everything'". *UC Riverside mathematics colloquium*. Archived from the original on 28 June 2008. Retrieved 2008-06-15.

[44] A. G. Lisi; Lee Smolin; Simone Speziale (2010). "Unification of gravity, gauge fields, and Higgs bosons". arXiv:1004.4866 [gr-qc].

[45] "E_8 Theory". *FQXi*. 2008-08-04. Archived from the original on 2008-08-09. Retrieved 2008-08-05.

[46] "FQXi Grants". *FQXi*. Archived from the original on 3 July 2008. Retrieved 2008-08-08.

[47] Merali, Zeeya (September 2010). "Rummaging for a Final Theory". *Scientific American*. Retrieved 2010-08-25.

[48] M. J. Duff (2011). "String and M-theory: answering the critics". arXiv:1112.0788v1.

[49] Peter Woit (2011-12-07). "String and M-theory: answering the critics". *Not Even Wrong*. Retrieved 2011-12-21.

12.5   External links

- Deferential Geometry - Lisi's wiki, containing some mathematical background

- The Elementary Particle Explorer - an online tool for rotating and examining the particle charges and interactions in the Standard Model, GUTs, and E_8 Theory

- An 8-dimensional model of the universe - Lisi presents his theory at TED

Chapter 13

Mathematical universe hypothesis

In physics and cosmology, the **mathematical universe hypothesis** (**MUH**), also known as the **Ultimate Ensemble**, is a speculative "theory of everything" (TOE) proposed by the cosmologist Max Tegmark.[1][2]

13.1 Description

Tegmark's MUH is: *Our external physical reality is a mathematical structure*. That is, the physical universe *is mathematics* in a well-defined sense, and "in those [worlds] complex enough to contain self-aware substructures [they] will subjectively perceive themselves as existing in a physically 'real' world".[3][4] The hypothesis suggests that worlds corresponding to different sets of initial conditions, physical constants, or altogether different equations may be considered equally real. Tegmark elaborates the MUH into the **Computable Universe Hypothesis** (**CUH**), which posits that all computable mathematical structures (in Gödel's sense) exist.[5]

The theory can be considered a form of Pythagoreanism or Platonism in that it posits the existence of mathematical entities; a form of mathematical monism in that it denies that anything exists except mathematical objects; and a formal expression of ontic structural realism.

Tegmark claims that the hypothesis has no free parameters and is not observationally ruled out. Thus, he reasons, it is preferred over other theories-of-everything by Occam's Razor. He suggests conscious experience would take the form of mathematical "self-aware substructures" that exist in a physically "real" world.

The hypothesis is related to the anthropic principle and to Tegmark's categorization of four levels of the multiverse.[6]

Andreas Albrecht of Imperial College in London called it a "provocative" solution to one of the central problems facing physics. Although he "wouldn't dare" go so far as to say he believes it, he noted that "it's actually quite difficult to construct a theory where everything we see is all there is".[7]

13.2 Criticisms and responses

13.2.1 Definition of the Ensemble

Jürgen Schmidhuber[8] argues that "Although Tegmark suggests that '... all mathematical structures are a priori given equal statistical weight,' there is no way of assigning equal non-vanishing probability to all (infinitely many) mathematical structures." Schmidhuber puts forward a more restricted ensemble which admits only universe representations describable by constructive mathematics, that is, computer programs. He explicitly includes universe representations describable by non-halting programs whose output bits converge after finite time, although the convergence time itself may not be predictable by a halting program, due to Kurt Gödel's limitations.[9]

In response, Tegmark notes[3] (sec. V.E) that the measure over all universes has not yet been constructed for the String theory landscape either, so this should not be regarded as a "show-stopper".

13.2.2 Consistency with Gödel's theorem

See also: Consistency and Gödel's completeness theorem

It has also been suggested that the MUH is inconsistent with Gödel's incompleteness theorem. In a three-way debate between Tegmark and fellow physicists Piet Hut and Mark Alford,[10] the "secularist" (Alford) states that "the methods allowed by formalists cannot prove all the theorems in a sufficiently powerful system... The idea that math is 'out there' is incompatible with the idea that it consists of formal systems."

Tegmark's response in[10] (sec VI.A.1) is to offer a new hypothesis "that only Gödel-complete (fully decidable) mathematical structures have physical existence. This drastically shrinks the Level IV multiverse, essentially placing an upper limit on complexity, and may have the attractive side effect of explaining the relative simplicity of our universe." Tegmark goes on to note that although conventional theories in physics are Gödel-undecidable, the actual mathematical structure describing our world could still be Gödel-complete, and "could in principle contain observers capable of thinking about Gödel-incomplete mathematics, just as finite-state digital computers can prove certain theorems about Gödel-incomplete formal systems like Peano arithmetic." In[3] (sec. VII) he gives a more detailed response, proposing as an alternative to MUH the more restricted "Computable Universe Hypothesis" (CUH) which only includes mathematical structures that are simple enough that Gödel's theorem does not require them to contain any undecidable or uncomputable theorems. Tegmark admits that this approach faces "serious challenges", including (a) it excludes much of the mathematical landscape; (b) the measure on the space of allowed theories may itself be uncomputable; and (c) "virtually all historically successful theories of physics violate the CUH".

13.2.3   Observability

Stoeger, Ellis, and Kircher[11] (sec. 7) note that in a true multiverse theory, "the universes are then completely disjoint and nothing that happens in any one of them is causally linked to what happens in any other one. This lack of any causal connection in such multiverses really places them beyond any scientific support". Ellis[12] (p29) specifically criticizes the MUH, stating that an infinite ensemble of completely disconnected universes is "completely untestable, despite hopeful remarks sometimes made, see, e.g., Tegmark (1998)." Tegmark maintains that MUH is testable, stating that it predicts (a) that "physics research will uncover mathematical regularities in nature", and (b) by assuming that we occupy a typical member of the multiverse of mathematical structures, one could "start testing multiverse predictions by assessing how typical our universe is" ([3] sec. VIII.C).

13.2.4   Plausibility of Radical Platonism

See also: Philosophy of mathematics § Platonism

The MUH is based on the Radical Platonist view that math is an external reality ([3] sec V.C). However, Jannes[13] argues that "mathematics is at least in part a human construction", on the basis that if it is an external reality, then it should be found in some other animals as well: "Tegmark argues that, if we want to give a complete description of reality, then we will need a language independent of us humans, understandable for non-human sentient entities, such as aliens and future supercomputers. Brian Greene ([14] p.

299) argues similarly: "The deepest description of the universe should not require concepts whose meaning relies on human experience or interpretation. Reality transcends our existence and so shouldn't, in any fundamental way, depend on ideas of our making."

However there are many non-human entities, plenty of which are intelligent, and many of which can apprehend, memorise, compare and even approximately add numerical quantities. Several animals have also passed the mirror test of self-consciousness. But a few surprising examples of mathematical abstraction notwithstanding (for example, chimpanzees can be trained to carry out symbolic addition with digits, or the report of a parrot understanding a "zero-like concept"), all examples of animal intelligence with respect to mathematics are limited to basic counting abilities. He adds, "non-human intelligent beings should exist that understand the language of advanced mathematics. However, none of the non-human intelligent beings that we know of confirm the status of (advanced) mathematics as an objective language." In the paper "On Math, Matter and Mind"[10] the secularist viewpoint examined argues (sec. VI.A) that math is evolving over time, there is "no reason to think it is converging to a definite structure, with fixed questions and established ways to address them", and also that "The Radical Platonist position is just another metaphysical theory like solipsism... In the end the metaphysics just demands that we use a different language for saying what we already knew." Tegmark responds (sec VI.A.1) that "The notion of a mathematical structure is rigorously defined in any book on Model Theory", and that non-human mathematics would only differ from our own "because we are uncovering a different part of what is in fact a consistent and unified picture, so math is converging in this sense." In his 2014 book on the MUH,[15] Tegmark argues that the resolution is that we invent the language of mathematics but discover the structure of mathematics.

13.2.5   Coexistence of all mathematical structures

Don Page has argued[16] (sec 4) that "At the ultimate level, there can be only one world and, if mathematical structures are broad enough to include all possible worlds or at least our own, there must be one unique mathematical structure that describes ultimate reality. So I think it is logical nonsense to talk of Level 4 in the sense of the co-existence of all mathematical structures." Tegmark responds ([3] sec. V.E) that "this is less inconsistent with Level IV than it may sound, since many mathematical structures decompose into unrelated substructures, and separate ones can be unified."

13.2.6 Consistency with our "simple universe"

Alexander Vilenkin comments[17] (Ch. 19, p. 203) that "the number of mathematical structures increases with increasing complexity, suggesting that 'typical' structures should be horrendously large and cumbersome. This seems to be in conflict with the beauty and simplicity of the theories describing our world". He goes on to note (footnote 8, p. 222) that Tegmark's solution to this problem, the assigning of lower "weights" to the more complex structures ([6] sec. V.B) seems arbitrary ("Who determines the weights?") and may not be logically consistent ("It seems to introduce an additional mathematical structure, but all of them are supposed to be already included in the set").

13.2.7 Occam's razor

Tegmark has been criticized as misunderstanding the nature and application of Occam's razor; Massimo Pigliucci reminds that "Occam's razor is just a useful heuristic, it should never be used as the final arbiter to decide which theory is to be favored".[18]

13.3 Major books

- *Our Mathematical Universe*: written by Max Tegmark and published on January 7, 2014, this book describes Tegmark's theory.

13.4 See also

- Church–Turing thesis

- Cosmology

- Digital physics

- Impossible world

- Modal realism

- Multiverse

- Ontology

- String theory

- The Unreasonable Effectiveness of Mathematics in the Natural Sciences

- Theory of everything

13.5 References

[1] Tegmark, Max (November 1998). "Is "the Theory of Everything" Merely the Ultimate Ensemble Theory?". *Annals of Physics* **270** (1): 1–51. arXiv:gr-qc/9704009. Bibcode:1998AnPhy.270....1T. doi:10.1006/aphy.1998.5855.

[2] M. Tegmark 2014, "Our Mathematical Universe", Knopf

[3] Tegmark, Max (February 2008). "The Mathematical Universe". *Foundations of Physics* **38** (2): 101–150. arXiv:0704.0646. Bibcode:2008FoPh...38..101T. doi:10.1007/s10701-007-9186-9.

[4] Tegmark (1998), p. 1.

[5] "[0704.0646] The Mathematical Universe".

[6] Tegmark, Max (2003). "Parallel Universes". In Barrow, J.D.; Davies, P.C.W.' & Harper, C.L. *"Science and Ultimate Reality: From Quantum to Cosmos" honoring John Wheeler's 90th birthday*. Cambridge University Press. arXiv:astro-ph/0302131.

[7] Chown, Markus (June 1998). "Anything goes". *New Scientist* **158** (2157).

[8] J. Schmidhuber (2000) "Algorithmic Theories of Everything."

[9] Schmidhuber, J. (2002). "Hierarchies of generalized Kolmogorov complexities and nonenumerable universal measures computable in the limit". *International Journal of Foundations of Computer Science* **13** (4): 587–612. doi:10.1142/S0129054102001291.

[10] Hut, P.; Alford, M.; Tegmark, M. (2006). "On Math, Matter and Mind". *Foundations of Physics* **36**: 765–94. arXiv:physics/0510188. Bibcode:2006FoPh...36..765H. doi:10.1007/s10701-006-9048-x.

[11] W. R. Stoeger, G. F. R. Ellis, U. Kirchner (2006) "Multiverses and Cosmology: Philosophical Issues."

[12] G.F.R. Ellis, "83 years of general relativity and cosmology: Progress and problems", Class. Quant. Grav. 16, A37-A75, 1999

[13] Gil Jannes, "Some comments on 'The Mathematical Universe'", Found. Phys. 39, 397-406, 2009 arXiv:0904.0867

[14] B. Greene 2011, "*The Hidden Reality* "

[15] M. Tegmark 2014, "Our Mathematical Universe"

[16] D. Page, "Predictions and Tests of Multiverse Theories."

[17] A. Vilenkin (2006) *Many Worlds in One: The Search for Other Universes*. Hill and Wang, New York.

[18] "Mathematical Universe? I Ain't Convinced". *Science 2.0.*

13.6   Further reading

- Schmidhuber, J. (1997) "A Computer Scientist's View of Life, the Universe, and Everything" in C. Freksa, ed., *Foundations of Computer Science: Potential - Theory - Cognition*. Lecture Notes in Computer Science, Springer: p.201-08.

- Tegmark, Max (1998). "Is the 'theory of everything' merely the ultimate ensemble theory?". *Annals of Physics* **270**: 1–51. arXiv:gr-qc/9704009. Bibcode:1998AnPhy.270....1T. doi:10.1006/aphy.1998.5855.

- Tegmark, Max (2008), "The Mathematical Universe", *Foundations of Physics* **38**:101–50.

- Tegmark, Max (2014), *Our Mathematical Universe: My Quest for the Ultimate Nature of Reality*, ISBN 978-0-307-59980-3

- Woit, P. (17 January 2014), "Book Review: 'Our Mathematical Universe' by Max Tegmark", *The Wall Street Journal*.

13.7   External links

- Jürgen Schmidhuber "The ensemble of universes describable by constructive mathematics."

- Page maintained by Max Tegmark with links to his technical and popular writings.

- "The 'Everything' mailing list" (and archives). Discusses the idea that all possible universes exist.

- "Is the universe actually made of math?" Interview with Max Tegmark in *Discover Magazine*.

Chapter 14

Compactification (physics)

For the concept of compactification in mathematics, see compactification (mathematics).

In physics, **compactification** means changing a theory with respect to one of its space-time dimensions. Instead of having a theory with this dimension being infinite, one changes the theory so that this dimension has a finite length, and may also be periodic.

Compactification plays an important part in thermal field theory where one compactifies time, in string theory where one compactifies the extra dimensions of the theory, and in two- or one-dimensional solid state physics, where one considers a system which is limited in one of the three usual spatial dimensions.

At the limit where the size of the compact dimension goes to zero, no fields depend on this extra dimension, and the theory is dimensionally reduced.

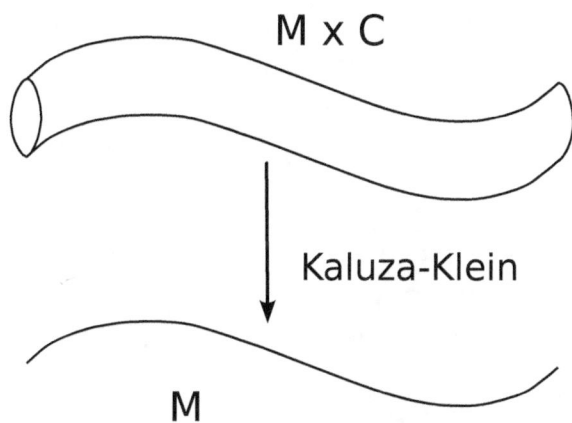

The space $M \times C$ is compactified over the compact C and after Kaluza–Klein decomposition, we have an effective field theory over M.

14.1 Compactification in string theory

In string theory, compactification is a generalization of Kaluza–Klein theory. It tries to conciliate the gap between the conception of our universe based on its four observable dimensions with the ten, eleven, or twenty-six dimensions which theoretical equations lead us to suppose the universe is made with.

For this purpose it is assumed the extra dimensions are "wrapped" up on themselves, or "curled" up on Calabi–Yau spaces, or on orbifolds. Models in which the compact directions support fluxes are known as *flux compactifications*. The coupling constant of string theory, which determines the probability of strings to split and reconnect, can be described by a field called dilaton. This in turn can be described as the size of an extra (eleventh) dimension which is compact. In this way, the ten-dimensional type IIA string theory can be described as the compactification of M-theory in eleven dimensions. Furthermore, different versions of string theory are related by different compactifications in a procedure known as T-duality.

The formulation of more precise versions of the meaning of compactification in this context has been promoted by discoveries such as the mysterious duality.

14.2 Flux compactification

A **flux compactification** is a particular way to deal with additional dimensions required by string theory.

It assumes that the shape of the internal manifold is a Calabi–Yau manifold or generalized Calabi–Yau manifold which is equipped with non-zero values of fluxes, i.e. differential forms that generalize the concept of an electromagnetic field (see p-form electrodynamics).

The hypothetical concept of the anthropic landscape in string theory follows from a large number of possibilities in which the integers that characterize the fluxes can be chosen without violating rules of string theory. The flux compactifications can be described as F-theory vacua or type IIB string theory vacua with or without D-branes.

14.3   See also

- Dimensional reduction

- Kaluza–Klein theory

14.4   References

- Chapter 16 of Michael Green, John H. Schwarz and Edward Witten (1987) *Superstring theory.* Cambridge University Press. *Vol. 2: Loop amplitudes, anomalies and phenomenology.* ISBN 0-521-35753-5.

- Brian R. Greene, "String Theory on Calabi–Yau Manifolds". arXiv:hep-th/9702155.

- Mariana Graña, "Flux compactifications in string theory: A comprehensive review", *Physics Reports* **423**, 91–158 (2006). arXiv:hep-th/0509003.

- Michael R. Douglas and Shamit Kachru "Flux compactification", *Rev. Mod. Phys.* **79**, 733 (2007). arXiv:hep-th/0610102.

- Ralph Blumenhagen, Boris Körs, Dieter Lüst, Stephan Stieberger, "Four-dimensional string compactifications with D-branes, orientifolds and fluxes", *Physics Reports* **445**, 1–193 (2007). arXiv:hep-th/0610327.

14.5   External links

- Flux compactification on arxiv.org

Chapter 15

Extra dimensions

In physics, **extra dimensions** are proposed additional space or time dimensions beyond the (3 + 1) typical of our observed space-time, such as the first attempts based on the Kaluza–Klein theory. Among theories proposing extra dimension are:[1]

1. Large extra dimension, mostly motivated by the ADD model, by Nima Arkani-Hamed, Savas Dimopoulos, and Gia Dvali in 1998, in an attempt to solve the hierarchy problem. This theory requires that the fields of the Standard Model are confined to a four-dimensional membrane, while gravity propagates in several additional spatial dimensions that are large compared to the Planck scale.[2]

2. Warped extra dimensions, such as those proposed by the Randall–Sundrum model (RS), based on warped geometry where our universe is a five-dimensional anti-de Sitter space and the elementary particles except for the graviton are localized on a (3 + 1)-dimensional brane or branes.[3]

3. Universal extra dimension, proposed and first studied in 2000, assume, at variance with the ADD and RS approaches, that all fields propagate universally in the extra dimensions.

4. Multiple time dimensions, i.e. the possibility that there might be more than one dimension of time, has occasionally been discussed in physics and philosophy, although those models have to deal with the problem of causality.

15.1 References

[1] Rizzo, Thomas G. (2004). "Pedagogical Introduction to Extra Dimensions". *SLAC Summer Institute*. Retrieved 2016.

[2] For a pedagogical introduction, see M. Shifman (2009). *Large Extra Dimensions: Becoming acquainted with an alternative paradigm*. Crossing the boundaries: Gauge dynamics at strong coupling. Singapore: World Scientific. arXiv:0907.3074.

[3] Randall, Lisa; Sundrum, Raman (1999). "Large Mass Hierarchy from a Small Extra Dimension". *Physical Review Letters* **83** (17): 3370–3373. arXiv:hep-ph/9905221. Bibcode:1999PhRvL..83.3370R. doi:10.1103/PhysRevLett.83.3370.

Chapter 16

Quantum gravity

Quantum gravity (**QG**) is a field of theoretical physics that seeks to describe the force of gravity according to the principles of quantum mechanics, and where quantum effects cannot be ignored.[1]

The current understanding of gravity is based on Albert Einstein's general theory of relativity, which is formulated within the framework of classical physics. On the other hand, the nongravitational forces are described within the framework of quantum mechanics, a radically different formalism for describing physical phenomena based on the wave-like nature of matter.[2] The necessity of a quantum mechanical description of gravity follows from the fact that one cannot consistently couple a classical system to a quantum one.[3]

Although a quantum theory of gravity is needed in order to reconcile general relativity with the principles of quantum mechanics, difficulties arise when one attempts to apply the usual prescriptions of quantum field theory to the force of gravity.[4] From a technical point of view, the problem is that the theory one gets in this way is not renormalizable and therefore cannot be used to make meaningful physical predictions. As a result, theorists have taken up more radical approaches to the problem of quantum gravity, the most popular approaches being string theory and loop quantum gravity.[5] A recent development is the theory of causal fermion systems which gives quantum mechanics, general relativity, and quantum field theory as limiting cases.[6][7][8][9][10][11]

Strictly speaking, the aim of quantum gravity is only to describe the quantum behavior of the gravitational field and should not be confused with the objective of unifying all fundamental interactions into a single mathematical framework. While any substantial improvement into the present understanding of gravity would aid further work towards unification, study of quantum gravity is a field in its own right with various branches having different approaches to unification. Although some quantum gravity theories, such as string theory, try to unify gravity with the other fundamental forces, others, such as loop quantum gravity,

make no such attempt; instead, they make an effort to quantize the gravitational field while it is kept separate from the other forces. A theory of quantum gravity that is also a grand unification of all known interactions is sometimes referred to as a theory of everything (TOE).

One of the difficulties of quantum gravity is that quantum gravitational effects are only expected to become apparent near the Planck scale, a scale far smaller in distance (equivalently, far larger in energy) than what is currently accessible at high energy particle accelerators. As a result, quantum gravity is a mainly theoretical enterprise, although there are speculations about how quantum gravity effects might be observed in existing experiments.[12]

16.1 Overview

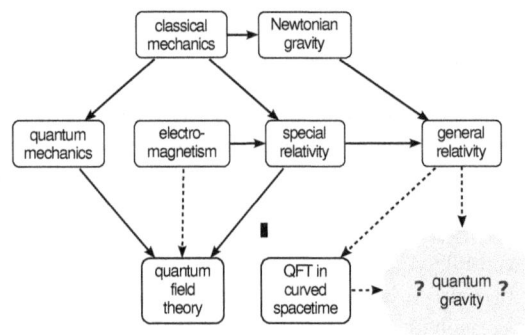

Diagram showing where quantum gravity sits in the hierarchy of physics theories

Much of the difficulty in meshing these theories at all energy scales comes from the different assumptions that these theories make on how the universe works. Quantum field theory depends on particle fields embedded in the flat spacetime of special relativity. General relativity models gravity as a curvature within space-time that changes as a gravitational mass moves. Historically, the most obvious way of combining the two (such as treating gravity as simply another particle field) ran quickly into what is known as the renormalization problem. In the old-fashioned understanding of renormalization, gravity particles would attract each other and adding together all of the interactions results in many infinite values which cannot easily be cancelled out mathematically to yield sensible, finite results. This is in contrast with quantum electrodynamics where, given that the series still do not converge, the interactions sometimes evaluate to infinite results, but those are few enough in number to be removable via renormalization.

16.1.1 Effective field theories

Quantum gravity can be treated as an effective field theory. Effective quantum field theories come with some high-energy cutoff, beyond which we do not expect that the theory provides a good description of nature. The "infinities" then become large but finite quantities depending on this finite cutoff scale, and correspond to processes that involve very high energies near the fundamental cutoff. These quantities can then be absorbed into an infinite collection of coupling constants, and at energies well below the fundamental cutoff of the theory, to any desired precision; only a finite number of these coupling constants need to be measured in order to make legitimate quantum-mechanical predictions. This same logic works just as well for the highly successful theory of low-energy pions as for quantum gravity. Indeed, the first quantum-mechanical corrections to graviton-scattering and Newton's law of gravitation have been explicitly computed[13] (although they are so infinitesimally small that we may never be able to measure them). In fact, gravity is in many ways a much better quantum field theory than the Standard Model, since it appears to be valid all the way up to its cutoff at the Planck scale.

While confirming that quantum mechanics and gravity are indeed consistent at reasonable energies, it is clear that near or above the fundamental cutoff of our effective quantum theory of gravity (the cutoff is generally assumed to be of the order of the Planck scale), a new model of nature will be needed. Specifically, the problem of combining quantum mechanics and gravity becomes an issue only at very high energies, and may well require a totally new kind of model.

16.1.2 Quantum gravity theory for the highest energy scales

The general approach to deriving a quantum gravity theory that is valid at even the highest energy scales is to assume that such a theory will be simple and elegant and, accordingly, to study symmetries and other clues offered by current theories that might suggest ways to combine them into a comprehensive, unified theory. One problem with this approach is that it is unknown whether quantum gravity will actually conform to a simple and elegant theory, as it should resolve the dual conundrums of special relativity with regard to the uniformity of acceleration and gravity, and general relativity with regard to spacetime curvature.

Such a theory is required in order to understand problems involving the combination of very high energy and very small dimensions of space, such as the behavior of black holes, and the origin of the universe.

16.2 Quantum mechanics and general relativity

16.2.1 The graviton

Main article: Graviton

At present, one of the deepest problems in theoretical physics is harmonizing the theory of general relativity, which describes gravitation, and applications to large-scale structures (stars, planets, galaxies), with quantum mechanics, which describes the other three fundamental forces acting on the atomic scale. This problem must be put in the proper context, however. In particular, contrary to the popular claim that quantum mechanics and general relativity are fundamentally incompatible, one can demonstrate that the structure of general relativity essentially follows inevitably from the quantum mechanics of interacting theoretical spin-2 massless particles (called gravitons).[14][15][16][17][18]

While there is no concrete proof of the existence of gravitons, quantized theories of matter may necessitate their existence. Supporting this theory is the observation that all fundamental forces except gravity have one or more known messenger particles, leading researchers to believe that at least one most likely does exist; they have dubbed this hypothetical particle the *graviton*. The predicted find would result in the classification of the graviton as a "force particle" similar to the photon of the electromagnetic field. Many of the accepted notions of a unified theory of physics since the 1970s assume, and to some degree depend upon, the existence of the graviton. These include string theory, superstring theory, M-theory, and loop quantum gravity. Detection of gravitons is thus vital to the validation of various lines of research to unify quantum mechanics and relativity theory.

Gravity Probe B (GP-B) has measured spacetime curvature near Earth to test related models in application of Einstein's general theory of relativity.

16.2.2   The dilaton

Main article: Dilaton

The dilaton made its first appearance in Kaluza–Klein theory, a five-dimensional theory that combined gravitation and electromagnetism. Generally, it appears in string theory. More recently, however, it's become central to the lower-dimensional many-bodied gravity problem[19] based on the field theoretic approach of Roman Jackiw. The impetus arose from the fact that complete analytical solutions for the metric of a covariant N-body system have proven elusive in general relativity. To simplify the problem, the

number of dimensions was lowered to *(1+1)*, i.e., one spatial dimension and one temporal dimension. This model problem, known as $R=T$ theory[20] (as opposed to the general $G=T$ theory) was amenable to exact solutions in terms of a generalization of the Lambert W function. It was also found that the field equation governing the dilaton (derived from differential geometry) was the Schrödinger equation and consequently amenable to quantization.[21]

Thus, one had a theory which combined gravity, quantization, and even the electromagnetic interaction, promising ingredients of a fundamental physical theory. It is worth noting that this outcome revealed a previously unknown and already existing *natural link* between general relativity and quantum mechanics. For some time, a generalization of this theory to *(3+1)* dimensions was unclear. However, a recent derivation in *(3+1)* dimensions under the right coordinate conditions yields a formulation similar to the earlier *(1+1)* namely a dilaton field governed by the logarithmic Schrödinger equation[22] which is seen in condensed matter physics and superfluids. The field equations are indeed amenable to such a generalization (as shown with the inclusion of a one-graviton process[23]) and yield the correct Newtonian limit in *d* dimensions but only if a dilaton is included. Furthermore, the results become even more tantalizing in view of the apparent resemblance between the dilaton and the Higgs boson.[24] However, more experimentation is needed to resolve the relationhip between these two particles.

Since this theory can combine gravitational, electromagnetic and quantum effects, their coupling could potentially lead to a means of vindicating the theory, through cosmology and even, perhaps, experimentally. When the equation E=mC^2 was solved through the Lorentz derivative, and applied to the velocity of the electron in relationship to micro time dilation, a gravitational force was discovered as space was bent due to near C velocity of the electron and the discovery that going light speed and gravity were linked unquestionably by Einstein. It has come into question how light speed is linked to gravity, and experiment was done with atomic clocks in a centrifuge where the acceleration in the disk was causing macroscopic time dilation according to Einsteins well known works. when cesium clocks were used in the 50s on the ground and a commercial airliner, the difference in the two clocks was measured to be about 45.9 microseconds time dilation for 24 hours of flight, the clock at sea level being slower.

16.2.3   Nonrenormalizability of gravity

Further information: Renormalization

General relativity, like electromagnetism, is a classical field theory. One might expect that, as with electromagnetism, the gravitational force should also have a corresponding quantum field theory.

However, gravity is perturbatively nonrenormalizable.[25][26] For a quantum field theory to be well-defined according to this understanding of the subject, it must be asymptotically free or asymptotically safe. The theory must be characterized by a choice of *finitely many* parameters, which could, in principle, be set by experiment. For example, in quantum electrodynamics these parameters are the charge and mass of the electron, as measured at a particular energy scale.

On the other hand, in quantizing gravity there are, in perturbation theory, *infinitely many independent parameters* (counterterm coefficients) needed to define the theory. For a given choice of those parameters, one could make sense of the theory, but since it's impossible to conduct infinite experiments to fix the values of every parameter, it has been argued that one does not, in perturbation theory, have a meaningful physical theory:

- At low energies, the logic of the renormalization group tells us that, despite the unknown choices of these infinitely many parameters, quantum gravity will reduce to the usual Einstein theory of general relativity.

- On the other hand, if we could probe very high energies where quantum effects take over, then *every one* of the infinitely many unknown parameters would begin to matter, and we could make no predictions at all.

If we treat QG as an effective field theory, there is a way around this problem.

That is, the meaningful theory of quantum gravity (that makes sense and is predictive at all energy levels) inherently implies some deep principle that reduces the infinitely many unknown parameters to a finite number that can then be measured:

- One possibility is that normal perturbation theory is not a reliable guide to the renormalizability of the theory, and that there really *is* a UV fixed point for gravity. Since this is a question of non-perturbative quantum field theory, it is difficult to find a reliable answer, but some people still pursue this option.

- Another possibility is that there are new unfound symmetry principles that constrain the parameters and reduce them to a finite set. This is the route taken by string theory, where all of the excitations of the string essentially manifest themselves as new symmetries.

16.2.4   QG as an effective field theory

Main article: Effective field theory

In an effective field theory, all but the first few of the infinite set of parameters in a non-renormalizable theory are suppressed by huge energy scales and hence can be neglected when computing low-energy effects. Thus, at least in the low-energy regime, the model is indeed a predictive quantum field theory.[13] (A very similar situation occurs for the very similar effective field theory of low-energy pions.) Furthermore, many theorists agree that even the Standard Model should really be regarded as an effective field theory as well, with "nonrenormalizable" interactions suppressed by large energy scales and whose effects have consequently not been observed experimentally.

Recent work[13] has shown that by treating general relativity as an effective field theory, one can actually make legitimate predictions for quantum gravity, at least for low-energy phenomena. An example is the well-known calculation of the tiny first-order quantum-mechanical correction to the classical Newtonian gravitational potential between two masses.

16.2.5   Spacetime background dependence

Main article: Background independence

A fundamental lesson of general relativity is that there is no fixed spacetime background, as found in Newtonian mechanics and special relativity; the spacetime geometry is dynamic. While easy to grasp in principle, this is the hardest idea to understand about general relativity, and its consequences are profound and not fully explored, even at the classical level. To a certain extent, general relativity can be seen to be a relational theory,[27] in which the only physically relevant information is the relationship between different events in space-time.

On the other hand, quantum mechanics has depended since its inception on a fixed background (non-dynamic) structure. In the case of quantum mechanics, it is time that is given and not dynamic, just as in Newtonian classical mechanics. In relativistic quantum field theory, just as in classical field theory, Minkowski spacetime is the fixed background of the theory.

String theory

String theory can be seen as a generalization of quantum field theory where instead of point particles, string-like objects propagate in a fixed spacetime background, although

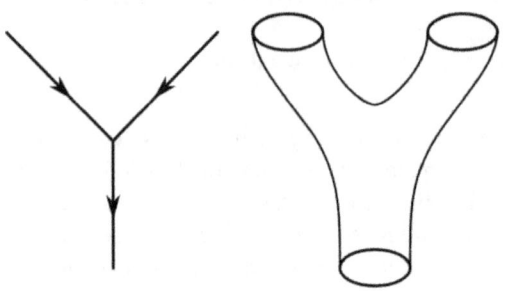

Interaction in the subatomic world: world lines of point-like particles in the Standard Model or a world sheet swept up by closed strings in string theory

the interactions among closed strings give rise to space-time in a dynamical way. Although string theory had its origins in the study of quark confinement and not of quantum gravity, it was soon discovered that the string spectrum contains the graviton, and that "condensation" of certain vibration modes of strings is equivalent to a modification of the original background. In this sense, string perturbation theory exhibits exactly the features one would expect of a perturbation theory that may exhibit a strong dependence on asymptotics (as seen, for example, in the AdS/CFT correspondence) which is a weak form of background dependence.

Background independent theories

Loop quantum gravity is the fruit of an effort to formulate a background-independent quantum theory.

Topological quantum field theory provided an example of background-independent quantum theory, but with no local degrees of freedom, and only finitely many degrees of freedom globally. This is inadequate to describe gravity in 3+1 dimensions, which has local degrees of freedom according to general relativity. In 2+1 dimensions, however, gravity is a topological field theory, and it has been successfully quantized in several different ways, including spin networks.

16.2.6   Semi-classical quantum gravity

Quantum field theory on curved (non-Minkowskian) backgrounds, while not a full quantum theory of gravity, has shown many promising early results. In an analogous way to the development of quantum electrodynamics in the early part of the 20th century (when physicists considered quantum mechanics in classical electromagnetic fields), the consideration of quantum field theory on a curved background

has led to predictions such as black hole radiation.

Phenomena such as the Unruh effect, in which particles exist in certain accelerating frames but not in stationary ones, do not pose any difficulty when considered on a curved background (the Unruh effect occurs even in flat Minkowskian backgrounds). The vacuum state is the state with the least energy (and may or may not contain particles). See Quantum field theory in curved spacetime for a more complete discussion.

16.2.7   Points of tension

There are other points of tension between quantum mechanics and general relativity.

- First, classical general relativity breaks down at singularities, and quantum mechanics becomes inconsistent with general relativity in the neighborhood of singularities (however, no one is certain that classical general relativity applies near singularities in the first place).

- Second, it is not clear how to determine the gravitational field of a particle, since under the Heisenberg uncertainty principle of quantum mechanics its location and velocity cannot be known with certainty. The resolution of these points may come from a better understanding of general relativity.[28]

- Third, there is the problem of time in quantum gravity. Time has a different meaning in quantum mechanics and general relativity and hence there are subtle issues to resolve when trying to formulate a theory which combines the two.[29]

16.3   Candidate theories

There are a number of proposed quantum gravity theories.[30] Currently, there is still no complete and consistent quantum theory of gravity, and the candidate models still need to overcome major formal and conceptual problems. They also face the common problem that, as yet, there is no way to put quantum gravity predictions to experimental tests, although there is hope for this to change as future data from cosmological observations and particle physics experiments becomes available.[31][32]

16.3.1   String theory

Main article: String theory

One suggested starting point is ordinary quantum field theories which, after all, are successful in describing the other

Projection of a Calabi–Yau manifold, one of the ways of compactifying the extra dimensions posited by string theory

three basic fundamental forces in the context of the standard model of elementary particle physics. However, while this leads to an acceptable effective (quantum) field theory of gravity at low energies,[33] gravity turns out to be much more problematic at higher energies. For ordinary field theories such as quantum electrodynamics, a technique known as renormalization is an integral part of deriving predictions which take into account higher-energy contributions,[34] but gravity turns out to be nonrenormalizable: at high energies, applying the recipes of ordinary quantum field theory yields models that are devoid of all predictive power.[35]

One attempt to overcome these limitations is to replace ordinary quantum field theory, which is based on the classical concept of a point particle, with a quantum theory of one-dimensional extended objects: string theory.[36] At the energies reached in current experiments, these strings are indistinguishable from point-like particles, but, crucially, different modes of oscillation of one and the same type of fundamental string appear as particles with different (electric and other) charges. In this way, string theory promises to be a unified description of all particles and interactions.[37] The theory is successful in that one mode will always correspond to a graviton, the messenger particle of gravity; however, the price of this success are unusual features such as six extra dimensions of space in addition to the usual three for space and one for time.[38]

In what is called the second superstring revolution, it was conjectured that both string theory and a unification of general relativity and supersymmetry known as supergravity[39] form part of a hypothesized eleven-dimensional model

known as M-theory, which would constitute a uniquely defined and consistent theory of quantum gravity.[40][41] As presently understood, however, string theory admits a very large number (10^{500} by some estimates) of consistent vacua, comprising the so-called "string landscape". Sorting through this large family of solutions remains a major challenge.

16.3.2 Loop quantum gravity

Main article: Loop quantum gravity

Loop quantum gravity seriously considers general relativ-

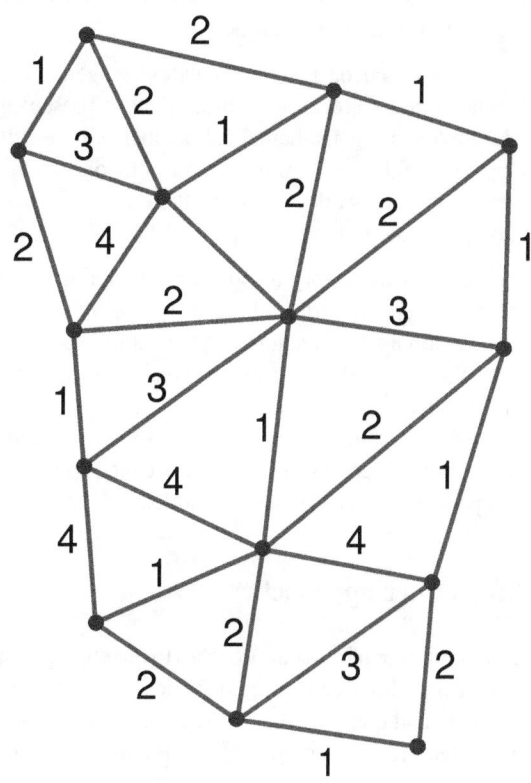

Simple spin network of the type used in loop quantum gravity

ity's insight that spacetime is a dynamical field and is therefore a quantum object. Its second idea is that the quantum discreteness that determines the particle-like behavior of other field theories (for instance, the photons of the electromagnetic field) also affects the structure of space.

The main result of loop quantum gravity is the derivation of a granular structure of space at the Planck length. This is derived from following considerations: In the case of electromagnetism, the quantum operator representing the energy of each frequency of the field has a discrete spectrum. Thus the energy of each frequency is quantized, and the quanta are the photons. In the case of gravity, the operators representing the area and the volume of each surface or space region likewise have discrete spectrum. Thus area and volume of any portion of space are also quantized, where the quanta are elementary quanta of space. It follows, then, that spacetime has an elementary quantum granular structure at the Planck scale, which cuts off the ultraviolet infinities of quantum field theory.

The quantum state of spacetime is described in the theory by means of a mathematical structure called spin networks. Spin networks were initially introduced by Roger Penrose in abstract form, and later shown by Carlo Rovelli and Lee Smolin to derive naturally from a non-perturbative quantization of general relativity. Spin networks do not represent quantum states of a field in spacetime: they represent directly quantum states of spacetime.

The theory is based on the reformulation of general relativity known as Ashtekar variables, which represent geometric gravity using mathematical analogues of electric and magnetic fields.[42][43] In the quantum theory, space is represented by a network structure called a spin network, evolving over time in discrete steps.[44][45][46][47]

The dynamics of the theory is today constructed in several versions. One version starts with the canonical quantization of general relativity. The analogue of the Schrödinger equation is a Wheeler–DeWitt equation, which can be defined within the theory.[48] In the covariant, or spinfoam formulation of the theory, the quantum dynamics is obtained via a sum over discrete versions of spacetime, called spinfoams. These represent histories of spin networks.

16.3.3   Other approaches

There are a number of other approaches to quantum gravity. The approaches differ depending on which features of general relativity and quantum theory are accepted unchanged, and which features are modified.[49][50] Examples include:

- Asymptotic safety in quantum gravity

- Euclidean quantum gravity

- Causal dynamical triangulation[51]

- Causal fermion systems,[6][7][8][9][10][11] giving quantum mechanics, general relativity and quantum field theory as limiting cases.

- Causal sets[52]

- Covariant Feynman path integral approach

- Group field theory[53]

- Wheeler-DeWitt equation

- Geometrodynamics

- Hořava–Lifshitz gravity

- MacDowell–Mansouri action

- Path-integral based models of quantum cosmology[54]

- Regge calculus

- Shape Dynamics

- String-nets giving rise to gapless helicity ±2 excitations with no other gapless excitations[55]

- Superfluid vacuum theory a.k.a. theory of BEC vacuum

- Supergravity

- Twistor theory[56]

- Canonical quantum gravity

- E8 Theory

- Quantum holonomy theory[57]

16.4   Weinberg–Witten theorem

In quantum field theory, the Weinberg–Witten theorem places some constraints on theories of composite gravity/emergent gravity. However, recent developments attempt to show that if locality is only approximate and the holographic principle is correct, the Weinberg–Witten theorem would not be valid.

16.5   Experimental tests

As was emphasized above, quantum gravitational effects are extremely weak and therefore difficult to test. For this reason, the possibility of experimentally testing quantum gravity had not received much attention prior to the late 1990s. However, in the past decade, physicists have realized that evidence for quantum gravitational effects can guide the development of the theory. Since theoretical development has been slow, the field of phenomenological quantum gravity, which studies the possibility of experimental tests, has obtained increased attention.[58][59]

The most widely pursued possibilities for quantum gravity phenomenology include violations of Lorentz invariance, imprints of quantum gravitational effects in the cosmic microwave background (in particular its polarization), and decoherence induced by fluctuations in the space-time foam.

The BICEP2 experiment detected what was initially thought to be primordial B-mode polarization caused by gravitational waves in the early universe. If truly primordial, these waves were born as quantum fluctuations in gravity itself. Cosmologist Ken Olum (Tufts University) stated: "I think this is the only observational evidence that we have that actually shows that gravity is quantized....It's probably the only evidence of this that we will ever have."[60]

16.6 See also

16.7 References

[1] Rovelli, Carlo. "Quantum gravity - Scholarpedia". *www.scholarpedia.org*. Retrieved 2016-01-09.

[2] Griffiths, David J. (2004). *Introduction to Quantum Mechanics*. Pearson Prentice Hall. OCLC 803860989.

[3] Wald, Robert M. (1984). *General Relativity*. University of Chicago Press. p. 382. OCLC 471881415.

[4] Zee, Anthony (2010). *Quantum Field Theory in a Nutshell* (2nd ed.). Princeton University Press. p. 172. OCLC 659549695.

[5] Penrose, Roger (2007). *The road to reality : a complete guide to the laws of the universe*. Vintage. p. 1017. OCLC 716437154.

[6] F. Finster, J. Kleiner, Causal fermion systems as a candidate for a unified physical theory, arXiv:1502.03587 [math-ph] (2015)

[7] F. Finster, The Principle of the Fermionic Projector, hep-th/0001048, hep-th/0202059, hep- th/0210121, AMS/IP Studies in Advanced Mathematics, vol. **35**, American Mathematical Society, Providence, RI, 2006.

[8] F. Finster, A formulation of quantum field theory realizing a sea of interacting Dirac particles, arXiv:0911.2102 [hep-th], Lett. Math. Phys. **97** (2011), no. 2, 165–183.

[9] F. Finster, An action principle for an interacting fermion system and its analysis in the continuum limit, arXiv:0908.1542 [math-ph] (2009).

[10] F. Finster, The continuum limit of a fermion system involving neutrinos: Weak and gravitational interactions, arXiv: 1211.3351 [math-ph] (2012).

[11] F. Finster, Perturbative quantum field theory in the framework of the fermionic projector, arXiv:1310.4121 [math-ph], J. Math. Phys. **55** (2014), no. 4, 042301.

[12] Quantum effects in the early universe might have an observable effect on the structure of the present universe, for example, or gravity might play a role in the unification of the other forces. Cf. the text by Wald cited above.

[13] Donoghue (1995). "Introduction to the Effective Field Theory Description of Gravity". arXiv:gr-qc/9512024. (verify against ISBN 9789810229085)

[14] Kraichnan, R. H. (1955). "Special-Relativistic Derivation of Generally Covariant Gravitation Theory". *Physical Review* **98** (4): 1118–1122. Bibcode:1955PhRv...98.1118K. doi:10.1103/PhysRev.98.1118.

[15] Gupta, S. N. (1954). "Gravitation and Electromagnetism". *Physical Review* **96** (6): 1683–1685. Bibcode:1954PhRv...96.1683G. doi:10.1103/PhysRev.96.1683.

[16] Gupta, S. N. (1957). "Einstein's and Other Theories of Gravitation". *Reviews of Modern Physics* **29** (3): 334–336. Bibcode:1957RvMP...29..334G. doi:10.1103/RevModPhys.29.334.

[17] Gupta, S. N. (1962). "Quantum Theory of Gravitation". *Recent Developments in General Relativity*. Pergamon Press. pp. 251–258.

[18] Deser, S. (1970). "Self-Interaction and Gauge Invariance". *General Relativity and Gravitation* **1**: 9–18. arXiv:gr-qc/0411023. Bibcode:1970GReGr...1....9D. doi:10.1007/BF00759198.

[19] Ohta, Tadayuki; Mann, Robert (1996). "Canonical reduction of two-dimensional gravity for particle dynamics". *Classical and Quantum Gravity* **13** (9): 2585–2602. arXiv:gr-qc/9605004. Bibcode:1996CQGra..13.2585O. doi:10.1088/0264-9381/13/9/022.

[20] Sikkema, A E; Mann, R B (1991). "Gravitation and cosmology in (1+1) dimensions". *Classical and Quantum Gravity* **8**: 219–235. Bibcode:1991CQGra...8..219S. doi:10.1088/0264-9381/8/1/022.

[21] Farrugia; Mann; Scott (2007). "N-body Gravity and the Schroedinger Equation". *Classical and Quantum Gravity* **24** (18): 4647–4659. arXiv:gr-qc/0611144. Bibcode:2007CQGra..24.4647F. doi:10.1088/0264-9381/24/18/006.

[22] Scott, T.C.; Zhang, Xiangdong; Mann, Robert; Fee, G.J. (2016). "Canonical reduction for dilatonic gravity in 3 + 1 dimensions". *Physical Review D* **93** (8): 084017. doi:10.1103/PhysRevD.93.084017.

[23] Mann, R B; Ohta, T (1997). "Exact solution for the metric and the motion of two bodies in (1+1)-dimensional gravity". *Phys. Rev. D* **55** (8): 4723–4747. arXiv:gr-qc/9611008. Bibcode:1997PhRvD..55.4723M. doi:10.1103/PhysRevD.55.4723.

[24] Bellazzini, B.; Csaki, C.; Hubisz, J.; Serra, J.; Terning, J. (2013). "A higgs-like dilaton". *Eur. Phys. J. C* **73** (2): 2333.

[25] Feynman, R. P.; Morinigo, F. B.; Wagner, W. G.; Hatfield, B. (1995). *Feynman lectures on gravitation*. Addison-Wesley. ISBN 0-201-62734-5.

[26] Hamber, H. W. (2009). *Quantum Gravitation - The Feynman Path Integral Approach*. Springer Publishing. ISBN 978-3-540-85292-6.

[27] Smolin, Lee (2001). *Three Roads to Quantum Gravity*. Basic Books. pp. 20–25. ISBN 0-465-07835-4. Pages 220–226 are annotated references and guide for further reading.

[28] Hunter Monroe (2005). "Singularity-Free Collapse through Local Inflation". arXiv:astro-ph/0506506.

[29] Edward Anderson (2010). "The Problem of Time in Quantum Gravity". arXiv:1009.2157 [gr-qc]. (also published as chapter 4 of ISBN 9781611229578)

[30] A timeline and overview can be found in Rovelli, Carlo (2000). "Notes for a brief history of quantum gravity". arXiv:gr-qc/0006061. (verify against ISBN 9789812777386)

[31] Ashtekar, Abhay (2007). "Loop Quantum Gravity: Four Recent Advances and a Dozen Frequently Asked Questions". *11th Marcel Grossmann Meeting on Recent Developments in Theoretical and Experimental General Relativity*. p. 126. arXiv:0705.2222. Bibcode:2008mgm..conf..126A. doi:10.1142/9789812834300_0008.

[32] Schwarz, John H. (2007). "String Theory: Progress and Problems". *Progress of Theoretical Physics Supplement* **170**: 214–226. arXiv:hep-th/0702219. Bibcode:2007PThPS.170..214S. doi:10.1143/PTPS.170.214.

[33] Donoghue, John F. (editor) (1995). "Introduction to the Effective Field Theory Description of Gravity". In Cornet, Fernando. *Effective Theories: Proceedings of the Advanced School, Almunecar, Spain, 26 June–1 July 1995*. Singapore: World Scientific. arXiv:gr-qc/9512024. ISBN 981-02-2908-9.

[34] Weinberg, Steven (1996). "Chapters 17–18". *The Quantum Theory of Fields II: Modern Applications*. Cambridge University Press. ISBN 0-521-55002-5.

[35] Goroff, Marc H.; Sagnotti, Augusto; Sagnotti, Augusto (1985). "Quantum gravity at two loops". *Physics Letters B* **160**: 81–86. Bibcode:1985PhLB..160...81G. doi:10.1016/0370-2693(85)91470-4.

[36] An accessible introduction at the undergraduate level can be found in Zwiebach, Barton (2004). *A First Course in String Theory*. Cambridge University Press. ISBN 0-521-83143-1., and more complete overviews in Polchinski, Joseph (1998). *String Theory Vol. I: An Introduction to the Bosonic String*. Cambridge University Press. ISBN 0-521-63303-6. and Polchinski, Joseph (1998b). *String Theory Vol. II: Superstring Theory and Beyond*. Cambridge University Press. ISBN 0-521-63304-4.

[37] Ibanez, L. E. (2000). "The second string (phenomenology) revolution". *Classical & Quantum Gravity* **17** (5): 1117–1128. arXiv:hep-ph/9911499.

Bibcode:2000CQGra..17.1117I.   doi:10.1088/0264-9381/17/5/321.

[38] For the graviton as part of the string spectrum, e.g. Green, Schwarz & Witten 1987, sec. 2.3 and 5.3; for the extra dimensions, ibid sec. 4.2.

[39] Weinberg, Steven (2000). "Chapter 31". *The Quantum Theory of Fields II: Modern Applications*. Cambridge University Press. ISBN 0-521-55002-5.

[40] Townsend, Paul K. (1996). *Four Lectures on M-Theory*. ICTP Series in Theoretical Physics. p. 385. arXiv:hep-th/9612121. Bibcode:1997hepcbconf..385T.

[41] Duff, Michael (1996). "M-Theory (the Theory Formerly Known as Strings)". *International Journal of Modern Physics A* **11** (32): 5623–5642. arXiv:hep-th/9608117.   Bibcode:1996IJMPA..11.5623D. doi:10.1142/S0217751X96002583.

[42] Ashtekar, Abhay (1986). "New variables for classical and quantum gravity". *Physical Review Letters* **57** (18): 2244–2247. Bibcode:1986PhRvL..57.2244A. doi:10.1103/PhysRevLett.57.2244. PMID 10033673.

[43] Ashtekar, Abhay (1987). "New Hamiltonian formulation of general relativity". *Physical Review D* **36** (6): 1587–1602. Bibcode:1987PhRvD..36.1587A. doi:10.1103/PhysRevD.36.1587.

[44] Thiemann, Thomas (2006). "Loop Quantum Gravity: An Inside View". *Approaches to Fundamental Physics*. Lecture Notes in Physics **721**: 185. arXiv:hep-th/0608210. Bibcode:2007LNP...721..185T. doi:10.1007/978-3-540-71117-9_10. ISBN 978-3-540-71115-5.

[45] Rovelli, Carlo (1998). "Loop Quantum Gravity". *Living Reviews in Relativity* **1**. Retrieved 2008-03-13.

[46] Ashtekar, Abhay; Lewandowski, Jerzy (2004). "Background Independent Quantum Gravity: A Status Report". *Classical & Quantum Gravity* **21** (15): R53–R152. arXiv:gr-qc/0404018. Bibcode:2004CQGra..21R..53A. doi:10.1088/0264-9381/21/15/R01.

[47] Thiemann, Thomas (2003). "Lectures on Loop Quantum Gravity". *Lecture Notes in Physics*. Lecture Notes in Physics **631**: 41–135. arXiv:gr-qc/0210094. Bibcode:2003LNP...631...41T. doi:10.1007/978-3-540-45230-0_3. ISBN 978-3-540-40810-9.

[48] Rovelli, Carlo (2004). *Quantum Gravity*. Cambridge University Press. ISBN 0521715962.

[49] Isham, Christopher J. (1994). "Prima facie questions in quantum gravity". In Ehlers, Jürgen; Friedrich, Helmut. *Canonical Gravity: From Classical to Quantum*. Springer. arXiv:gr-qc/9310031. ISBN 3-540-58339-4.

[50] Sorkin, Rafael D. (1997). "Forks in the Road, on the Way to Quantum Gravity". *International Journal of Theoretical Physics* **36** (12): 2759–2781. arXiv:gr-qc/9706002. Bibcode:1997IJTP...36.2759S. doi:10.1007/BF02435709.

[51] Loll, Renate (1998). "Discrete Approaches to Quantum Gravity in Four Dimensions". *Living Reviews in Relativity* **1**: 13. arXiv:gr-qc/9805049. Bibcode:1998LRR.....1...13L. doi:10.12942/lrr-1998-13. Retrieved 2008-03-09.

[52] Sorkin, Rafael D. (2005). "Causal Sets: Discrete Gravity". In Gomberoff, Andres; Marolf, Donald. *Lectures on Quantum Gravity*. Springer. arXiv:gr-qc/0309009. ISBN 0-387-23995-2.

[53] See Daniele Oriti and references therein.

[54] Hawking, Stephen W. (1987). "Quantum cosmology". In Hawking, Stephen W.; Israel, Werner. *300 Years of Gravitation*. Cambridge University Press. pp. 631–651. ISBN 0-521-37976-8.

[55] Wen 2006

[56] See ch. 33 in Penrose 2004 and references therein.

[57] "Quantum Holonomy Theory by J. Aastrup and J. M. Grimstrup" (PDF).

[58] Hossenfelder, Sabine (2011). "Experimental Search for Quantum Gravity". In V. R. Frignanni. *Classical and Quantum Gravity: Theory, Analysis and Applications*. Chapter 5: Nova Publishers. ISBN 978-1-61122-957-8.

[59] Hossenfelder, Sabine (2010-10-17). V. R. Frignanni, ed. "Experimental Search for Quantum Gravity Chapter 5". *Classical and Quantum Gravity: Theory, Analysis and Applications* (Nova Publishers) **5** (2011). arXiv:1010.3420. Bibcode:2010arXiv1010.3420H.

[60] Camille Carlisle. "First Direct Evidence of Big Bang Inflation". SkyandTelescope.com. Retrieved March 18, 2014.

16.8 Further reading

- Ahluwalia, D. V. (2002). "Interface of Gravitational and Quantum Realms". *Modern Physics Letters A* **17** (15–17): 1135. arXiv:gr-qc/0205121. Bibcode:2002MPLA...17.1135A. doi:10.1142/S021773230200765X.

- Ashtekar, Abhay (2005). "The winding road to quantum gravity" (PDF). *Current Science* **89**: 2064–2074.

- Carlip, Steven (2001). "Quantum Gravity: a Progress Report". *Reports on Progress in Physics* **64** (8): 885–942. arXiv:gr-qc/0108040. Bibcode:2001RPPh...64..885C. doi:10.1088/0034-4885/64/8/301.

- Herbert W. Hamber (2009). *Quantum Gravitation*. Springer Publishing. doi:10.1007/978-3-540-85293-3. ISBN 978-3-540-85292-6.

- Kiefer, Claus (2007). *Quantum Gravity*. Oxford University Press. ISBN 0-19-921252-X.

- Kiefer, Claus (2005). "Quantum Gravity: General Introduction and Recent Developments". *Annalen der Physik* **15**: 129–148. arXiv:gr-qc/0508120. Bibcode:2006AnP...518..129K. doi:10.1002/andp.200510175.

- Lämmerzahl, Claus, ed. (2003). *Quantum Gravity: From Theory to Experimental Search*. Lecture Notes in Physics. Springer. ISBN 3-540-40810-X.

- Rovelli, Carlo (2004). *Quantum Gravity*. Cambridge University Press. ISBN 0-521-83733-2.

- Quantum gravity Carlo Rovelli, Scholarpedia, 3(5):7117. doi:10.4249/scholarpedia.7117

- Trifonov, Vladimir (2008). "GR-friendly description of quantum systems". *International Journal of Theoretical Physics* **47** (2): 492–510. arXiv:math-ph/0702095. Bibcode:2008IJTP...47..492T. doi:10.1007/s10773-007-9474-3.

Chapter 17

String (physics)

This article is about quantum strings, phenomena in string theory. For a classical string, such as a guitar string, see vibrating string. For other uses, see string (disambiguation).

In physics, a **string** is a physical phenomenon that appears in string theory and related subjects. Unlike elementary particles, which are zero-dimensional or point-like by definition, strings are one-dimensional extended objects. Theories in which the fundamental objects are strings rather than point particles automatically have many properties that some physicists expect to hold in a fundamental theory of physics. Most notably, a theory of strings that evolve and interact according to the rules of quantum mechanics will automatically describe quantum gravity.

In string theory, the strings may be open (forming a segment with two endpoints) or closed (forming a loop like a circle) and may have other special properties. Prior to 1995, there were five known versions of string theory incorporating the idea of supersymmetry, which differed in the type of strings and in other aspects. Today these different string theories are thought to arise as different limiting cases of a single theory called M-theory.

In theories of particle physics based on string theory, the characteristic length scale of strings is typically on the order of the Planck length, the scale at which the effects of quantum gravity are believed to become significant. On much larger length scales, such as the scales visible in physics laboratories, such objects would be indistinguishable from zero-dimensional point particles, and the vibrational state of the string would determine the type of particle. Strings are also sometimes studied in nuclear physics where they are used to model flux tubes.

As it propagates through spacetime, a string sweeps out a two-dimensional surface called its worldsheet. This is analogous to the one-dimensional worldline traced out by a point particle. The physics of a string is described by means of a two-dimensional conformal field theory associated with the worldsheet. The formalism of two dimensional conformal field theory also has many applications outside of string theory, for example in condensed matter physics and parts of pure mathematics.

17.1 Types of strings

17.1.1 Closed and open strings

Strings can be either open or closed. A **closed string** is a string that has no end-points, and therefore is topologically equivalent to a circle. An **open string**, on the other hand, has two end-points and is topologically equivalent to a line interval. Not all string theories contain open strings, but every theory must contain closed strings, as interactions between open strings can always result in closed strings.

The oldest superstring theory containing open strings was type I string theory. However, the developments in string theory in the 1990s have shown that the open strings should always be thought of as ending on a new type of objects called D-branes, and the spectrum of possibilities for open strings has increased greatly.

Open and closed strings are generally associated with characteristic vibrational modes. One of the vibration modes of a closed string can be identified as the graviton. In certain string theories the lowest-energy vibration of an open string is a tachyon and can undergo tachyon condensation. Other vibrational modes of open strings exhibit the properties of photons and gluons.

17.1.2 Orientation

Strings can also possess an **orientation**, which can be thought of as an internal "arrow" which distinguishes the string from one with the opposite orientation. By contrast, an **unoriented string** is one with no such arrow on it.

17.2   See also

- Elementary particle

- Brane

- D-brane

17.3   References

- Schwarz, John (2000). "Introduction to Superstring Theory". Retrieved Dec. 12, 2005.

- "NOVA's strings homepage"

Chapter 18

Type I string theory

In theoretical physics, **type I string theory** is one of five consistent supersymmetric string theories in ten dimensions. It is the only one whose strings are unoriented (both orientations of a string are equivalent) and which contains not only closed strings, but also open strings.

The classic 1976 work of Ferdinando Gliozzi, Joel Scherk and David Olive paved the way to a systematic understanding of the rules behind string spectra in cases where only closed strings are present via modular invariance but did not lead to similar progress for models with closed strings, despite the fact that the original discussion was based on the type I string theory.

As first proposed by Augusto Sagnotti in 1987, the type I string theory can be obtained as an orientifold of type IIB string theory, with 32 half-D9-branes added in the vacuum to cancel various anomalies.

At low energies, type I string theory is described by the N=1 supergravity (type I supergravity) in ten dimensions coupled to the SO(32) supersymmetric Yang–Mills theory. The discovery in 1984 by Michael Green and John H. Schwarz that anomalies in type I string theory cancel sparked the first superstring revolution. However, a key property of these models, shown by A. Sagnotti in 1992, is that in general the Green-Schwarz mechanism takes a more general form, and involves several two forms in the cancellation mechanism.

The relation between the type-IIB string theory and the type-I string theory has a large number of surprising consequences, both in ten and in lower dimensions, that were first displayed by the String Theory group at the University of Rome "Tor Vergata" in the early 1990s. It opened the way to the construction of entire new classes of string spectra with or without supersymmetry. Joseph Polchinski's work on D-branes provided a geometrical interpretation for these results in terms of extended objects (D-brane, orientifold).

In the 1990s it was first argued by Edward Witten that type I string theory with the string coupling constant g is equivalent to the SO(32) heterotic string with the coupling $1/g$. This equivalence is known as S-duality.

18.1 References

- F. Gliozzi, J. Scherk and D.I. Olive, "Supersymmetry, Supergravity Theories And The Dual Spinor Model", *Nucl. Phys. B* **122** (1977) 253.

- E. Witten, "String theory dynamics in various dimensions", *Nucl. Phys. B* **443** (1995) 85. arXiv:hep-th/9503124.

- J. Polchinski, S. Chaudhuri and C.V. Johnson, "Notes on D-Branes", arXiv:hep-th/9602052.

- C. Angelantonj and A. Sagnotti, "Open strings", *Phys. Rept.* **1** [(Erratum-ibid.) 339] arXiv:hep-th/0204089.

Chapter 19

Type II string theory

In theoretical physics, **type II string theory** is a unified term that includes both **type IIA strings** and **type IIB strings** theories. Type II string theory accounts for two of the five consistent superstring theories in ten dimensions. Both theories have the maximal amount of supersymmetry — namely 32 supercharges — in ten dimensions. Both theories are based on oriented closed strings. On the worldsheet, they differ only in the choice of GSO projection.

19.1   Type IIA string

At low energies, type IIA string theory is described by type IIA supergravity in ten dimensions which is a non-chiral theory (i.e. left-right symmetric) with $(1,1)$ $d=10$ supersymmetry; the fact that the anomalies in this theory cancel is therefore trivial.

In the 1990s it was realized by Edward Witten (building on previous insights by Michael Duff, Paul Townsend, and others) that the limit of type IIA string theory in which the string coupling goes to infinity becomes a new 11-dimensional theory called M-theory.

The mathematical treatment of type IIA string theory belongs to symplectic topology and algebraic geometry, particularly Gromov–Witten invariants.

19.2   Type IIB string

At low energies, type IIB string theory is described by type IIB supergravity in ten dimensions which is a chiral theory (left-right asymmetric) with $(2,0)$ $d=10$ supersymmetry; the fact that the anomalies in this theory cancel is therefore non-trivial.

In the 1990s it was realized that type II string theory with the string coupling constant g is equivalent to the same theory with the coupling $1/g$. This equivalence is known as S-duality.

Orientifold of type IIB string theory leads to type I string theory.

The mathematical treatment of type IIB string theory belongs to algebraic geometry, specifically the deformation theory of complex structures originally studied by Kunihiko Kodaira and Donald C. Spencer.

In 1997 Juan Maldacena gave some arguments indicating that type IIB string theory is equivalent to a Supersymmetric Yang Mills theory with 4 supersymmetries and gauge group $SU(N)$, in the 't Hooft limit; it was the first suggestion concerning the AdS/CFT correspondence.[1]

19.3   Relationship between the type II theories

In the late 1980s, it was realized that type IIA string theory is related to type IIB string theory by T-duality.

19.4   See also

- Superstring theory
- Type I string
- Heterotic string

19.5   References

[1] J. Maldacena, "The Large N Limit of Superconformal Field Theories and Supergravity" arXiv:hep-th/9711200

Chapter 20

Heterotic string theory

This article is about string theory. For heterosis in biology, see Heterosis.

In string theory, a **heterotic string** is a closed string (or loop) which is a hybrid ('heterotic') of a superstring and a bosonic string. There are two kinds of heterotic string, the heterotic SO(32) and the heterotic $E_8 \times E_8$, abbreviated to **HO** and **HE**. Heterotic string theory was first developed in 1985 by David Gross, Jeffrey Harvey, Emil Martinec, and Ryan Rohm (the so-called "Princeton String Quartet"[1]), in one of the key papers that fueled the first superstring revolution.

In string theory, the left-moving and the right-moving excitations are completely decoupled,[2] and it is possible to construct a string theory whose left-moving (counterclockwise) excitations are treated as a bosonic string propagating in $D = 26$ dimensions, while the right-moving (clockwise) excitations are treated as a superstring in $D = 10$ dimensions.

The mismatched 16 dimensions must be compactified on an even, self-dual lattice (a discrete subgroup of a linear space). There are two possible even self-dual lattices in 16 dimensions, and it leads to two types of the heterotic string. They differ by the gauge group in 10 dimensions. One gauge group is SO(32) (the HO string) while the other is $E_8 \times E_8$ (the HE string).[3]

These two gauge groups also turned out to be the only two anomaly-free gauge groups that can be coupled to the $N = 1$ supergravity in 10 dimensions other than $U(1)^{496}$ and $E_8 \times U(1)^{248}$, which is suspected to lie in the swampland.

Every heterotic string must be a closed string, not an open string; it is not possible to define any boundary conditions that would relate the left-moving and the right-moving excitations because they have a different character.

A heterotic string is embedded in the membrane that creates harmonics on the string which translate into mass and energy through mechanisms discussed above.

20.1 String duality

String duality is a class of symmetries in physics that link different string theories. In the 1990s, it was realized that the strong coupling limit of the HO theory is type I string theory — a theory that also contains open strings; this relation is called S-duality. The HO and HE theories are also related by T-duality.

Because the various superstring theories were shown to be related by dualities, it was proposed that that each type of string was a different aspect of a single underlying theory called M-theory.

20.2 References

[1] Dennis Overbye, "String theory, at 20, explains it all (or not)". *NY Times*, 2004-12-07

[2] *String Theory and M-Theory* by Becker, Becker and Schwarz (2006), p. 253

[3] Joseph Polchinski (1998). *String Theory: Volume 2*, p. 45.

Chapter 21

Orthogonal group

"Rotation group" redirects here. For other uses, see Rotation group (disambiguation).

In mathematics, the **orthogonal group** in dimension n, denoted O(n), is the group of distance-preserving transformations of a Euclidean space of dimension n that preserve a fixed point, where the group operation is given by composing transformations. Equivalently, it is the group of $n{\times}n$ orthogonal matrices, where the group operation is given by matrix multiplication, and an orthogonal matrix is a real matrix whose inverse equals its transpose.

The determinant of an orthogonal matrix being either 1 or -1, an important subgroup of O(n) is the **special orthogonal group**, denoted SO(n), of the orthogonal matrices of determinant 1. This group is also called the **rotation group**, because, in dimensions 2 and 3, its elements are the usual rotations around a point (in dimension 2) or a line (in dimension 3). In low dimension, these groups have been widely studied, see SO(2), SO(3) and SO(4).

The term "orthogonal group" may also refer to a generalization of the above case: the group of invertible linear operators that preserve a non-degenerate symmetric bilinear form or quadratic form[1] on a vector space over a field. In particular, when the bilinear form is the scalar product on the vector space F^n of dimension n over a field F, with quadratic form the sum of squares, then the corresponding orthogonal group, denoted O(n, F), is the set of $n \times n$ orthogonal matrices with entries from F, with the group operation of matrix multiplication. This is a subgroup of the general linear group GL(n, F) given by

$$O(n, F) = \{Q \in \mathrm{GL}(n, F) \mid Q^\mathsf{T} Q = Q Q^\mathsf{T} = I\}$$

where Q^T is the transpose of Q and I is the identity matrix.

This article mainly discusses the orthogonal groups of quadratic forms that may be expressed over some bases as the dot product; over the reals, they are the positive definite quadratic forms. Over the reals, for any non-degenerate quadratic form, there is a basis, on which the matrix of the

form is a diagonal matrix such that the diagonal entries are either 1 or -1. Thus the orthogonal group depends only on the numbers of 1 and of -1, and is denoted O(p, q), where p is the number of ones and q the number of negative ones. For details, see indefinite orthogonal group.

The derived subgroup $\Omega(n, F)$ of O(n, F) is an often studied object because, when F is a finite field, $\Omega(n, F)$ is often a central extension of a finite simple group.

Both O(n, F) and SO(n, F) are algebraic groups, because the condition that a matrix be orthogonal, i.e. have its own transpose as inverse, can be expressed as a set of polynomial equations in the entries of the matrix. The Cartan–Dieudonné theorem describes the structure of the orthogonal group for a non-singular form.

21.1 Name

The determinant of any orthogonal matrix is either 1 or -1. The orthogonal n-by-n matrices with determinant 1 form a normal subgroup of O(n, F) known as the **special orthogonal group** SO(n, F), consisting of all proper rotations. (More precisely, SO(n, F) is the kernel of the Dickson invariant, discussed below.). By analogy with GL–SL (general linear group, special linear group), the orthogonal group is sometimes called the *general* **orthogonal group** and denoted GO, though this term is also sometimes used for *indefinite* orthogonal groups O(p, q). The term **rotation group** can be used to describe either the special or general orthogonal group.

21.2 In even and odd dimension

The structure of the orthogonal group differs in certain respects between even and odd dimensions; for example, over ordered fields (such as **R**) the $-I$ element is orientation-preserving in even dimensions, but orientation-reversing in odd dimensions. When this distinction is to be emphasized, the groups may be denoted O(2k) and O(2k + 1), reserving n for the dimension of the space (n = 2k or n = 2k + 1). The letters p or r are also used, indicating the rank of the corresponding Lie algebra; in odd dimension the corresponding Lie algebra is $\mathfrak{so}(2r + 1)$, while in even dimension the Lie algebra is $\mathfrak{so}(2r)$.

21.2.1 Difference between O(n) and SO(n) in even dimensions

In two dimensions, O(2) is the group of all rotations about the origin and all reflections along a line through the origin. SO(2) is the group of all rotations about the origin.

These groups are closely related: SO(2) is a subgroup of O(2), since any two reflections gives a rotation.

More generally, in any number of dimensions an even number of reflections gives a rotation, and a rotation followed by reflection (or vice versa) produces a reflection. Therefore, the rotations define a subgroup of O(2), but the reflections do not define a subgroup.

A "reflection through the origin" may be generated as a combination of one reflection along each of the axes. The 'reflection through the origin' is not a reflection in the usual sense in even dimensions, but rather a rotation. In two dimensions it is the only nontrivial rotation that when applied twice gives the identity. It is its own inverse in any number of dimensions. In 4D it is isoclinic, and if that classification were generalised it would be isoclinic in every even number of dimensions.

21.3   Over the real number field

Over the field **R** of real numbers, the orthogonal group O(n, **R**) and the special orthogonal group SO(n, **R**) are often simply denoted by O(n) and SO(n) if no confusion is possible. They form real compact Lie groups of dimension n(n − 1)/2. O(n, **R**) has two connected components, with SO(n, **R**) being the identity component, i.e., the connected component containing the identity matrix.

21.3.1   Geometric interpretation

The real orthogonal and real special orthogonal groups have the following geometric interpretations:

O(n, **R**) is a subgroup of the Euclidean group E(n), the group of isometries of **R**^{n}; it contains those that leave the origin fixed – O(n, **R**) = E(n) ∩ GL(n, **R**). It is the symmetry group of the sphere (n = 3) or (n − 1)-sphere and all objects with spherical symmetry, if the origin is chosen at the center.

SO(n, **R**) is a subgroup of E⁺(n), which consists of *direct* isometries, i.e., isometries preserving orientation; it contains those that leave the origin fixed – SO(n, **R**) = E⁺(n) ∩ GL(n, **R**) = E(n) ∩ GL⁺(n, **R**). It is the rotation group of the sphere and all objects with spherical symmetry, if the origin is chosen at the center.

{±I} is a normal subgroup and even a characteristic subgroup of O(n, **R**), and, if n is even, also of SO(n, **R**). If n is odd, O(n, **R**) is the internal direct product of SO(n, **R**) and {±I}. For every positive integer k the cyclic group Ck of k-fold rotations is a normal subgroup of O(2, **R**) and SO(2, **R**).

Relative to suitable orthogonal bases, the isometries are of the form:

$$\begin{bmatrix} R_1 & & & & & \\ & \ddots & & & 0 & \\ & & R_k & & & \\ & & & \pm 1 & & \\ & 0 & & & \ddots & \\ & & & & & \pm 1 \end{bmatrix}$$

where the matrices R_1, ..., Rk are 2-by-2 rotation matrices in orthogonal planes of rotation. As a special case, known as Euler's rotation theorem, any (non-identity) element of SO(3, **R**) is rotation about a uniquely defined axis.

The orthogonal group is generated by reflections (two reflections give a rotation), as in a Coxeter group,[note 1] and elements have length at most n (require at most n reflections to generate; this follows from the above classification, noting that a rotation is generated by 2 reflections, and is true more generally for indefinite orthogonal groups, by the Cartan–Dieudonné theorem). A longest element (element needing the most reflections) is reflection through the origin (the map v ↦ −v), though so are other maximal combinations of rotations (and a reflection, in odd dimension).

The symmetry group of a circle is O(2, **R**). The orientation preserving subgroup SO(2, **R**) is isomorphic (as a *real* Lie group) to the circle group, also known as U(1). This isomorphism sends the complex number exp(φ i) = cos φ + i sin φ of absolute value 1 to the special orthogonal matrix

$$\begin{bmatrix} \cos(\phi) & -\sin(\phi) \\ \sin(\phi) & \cos(\phi) \end{bmatrix}.$$

The group SO(3, **R**), understood as the set of rotations of 3-dimensional space, is of major importance in the sciences and engineering, and there are numerous charts on SO(3).

21.3.2 Maximal tori and Weyl groups

A maximal torus T for SO($2n$), of rank n, is given by the block-diagonal matrices

$$\begin{bmatrix} R_1 & & 0 \\ & \ddots & \\ 0 & & R_n \end{bmatrix},$$

where the Rj are 2-by-2 rotation matrices. The image $T \times \{1\}$ of the same torus under the block-diagonal inclusion

$$\text{SO}(2n) \cong \text{SO}(2n) \times \{1\} < \text{SO}(2n+1)$$

is a maximal torus for SO($2n+1$). The Weyl group of SO($2n+1$) is the semidirect product $\{\pm 1\}^n \rtimes S_n$ of a normal elementary abelian 2-subgroup and a symmetric group, where the nontrivial element of each $\{\pm 1\}$ factor of $\{\pm 1\}^n$ acts on the corresponding circle factor of $T \times \{1\}$ by inversion, and the symmetric group Sn acts on both $\{\pm 1\}^n$ and $T \times \{1\}$ by permuting factors. The elements of the Weyl group are represented by matrices in O($2n$) $\times \{\pm 1\}$. The Sn factor is represented by block permutation matrices with 2-by-2 blocks, and a final 1 on the diagonal. The $\{\pm 1\}^n$ component is represented by block-diagonal matrices with 2-by-2 blocks either

$$\begin{bmatrix} 1 & 0 \\ 0 & 1 \end{bmatrix} \quad \text{or} \quad \begin{bmatrix} 0 & 1 \\ 1 & 0 \end{bmatrix},$$

with the last component ± 1 chosen to make the determinant 1.

The Weyl group of SO($2n$) is the subgroup $H_{n-1} \rtimes S_n < \{\pm 1\}^n \rtimes S_n$ of that of SO($2n + 1$), where $Hn-1 < \{\pm 1\}^n$ is the kernel of the product homomorphism $\{\pm 1\}^n \to \{\pm 1\}$ given by $(\epsilon_1, \ldots, \epsilon_n) \mapsto \epsilon_1 \cdots \epsilon_n$; that is $Hn-1 < \{\pm 1\}^n$ is the subgroup with an even number of minus signs. The Weyl group of SO($2n$) is represented in SO($2n$) by the preimages under the standard injection SO($2n$) $\to$ SO($2n+1$) of the representatives for the Weyl group of SO($2n + 1$). Those matrices with an odd number of $\begin{bmatrix} 0 & 1 \\ 1 & 0 \end{bmatrix}$ blocks have no remaining final -1 coordinate to make their determinants positive, and hence cannot be represented in SO($2n$).

21.3.3 Low-dimensional topology

The low-dimensional (real) orthogonal groups are familiar spaces:

- O(1) = S^0, a two-point discrete space

- SO(1) = $\{1\}$

- SO(2) is S^1

- SO(3) is $\mathbf{RP}^3$

- SO(4) is doubly covered by SU(2) $\times$ SU(2) = $S^3 \times S^3$.

21.3.4 Homotopy groups

In terms of algebraic topology, for $n > 2$ the fundamental group of SO(n, $\mathbf{R}$) is cyclic of order 2,[2] and the spin group Spin(n) is its universal cover. For $n = 2$ the fundamental group is infinite cyclic and the universal cover corresponds to the real line (the group Spin(2) is the unique connected 2-fold cover).

Generally, the homotopy groups $\pi k(O)$ of the real orthogonal group are related to homotopy groups of spheres, and thus are in general hard to compute. However, one can compute the homotopy groups of the stable orthogonal group (aka the infinite orthogonal group), defined as the direct limit of the sequence of inclusions:

$$\mathbf{O}(0) \subset \mathbf{O}(1) \subset \mathbf{O}(2) \subset \cdots \subset O = \bigcup_{k=0}^{\infty} \mathbf{O}(k)$$

Since the inclusions are all closed, hence cofibrations, this can also be interpreted as a union. On the other hand, S^n is a homogeneous space for O($n + 1$), and one has the following fiber bundle:

$$\mathbf{O}(n) \to \mathbf{O}(n+1) \to S^n,$$

which can be understood as "The orthogonal group O($n + 1$) acts transitively on the unit sphere S^n, and the stabilizer of a point (thought of as a unit vector) is the orthogonal group of the perpendicular complement, which is an orthogonal group one dimension lower. Thus the natural inclusion O(n) $\to$ O($n + 1$) is ($n - 1$)-connected, so the homotopy groups stabilize, and $\pi k(O(n+1)) = \pi k(O(n))$ for $n > k + 1$: thus the homotopy groups of the stable space equal the lower homotopy groups of the unstable spaces.

From Bott periodicity we obtain $\Omega^8 O \cong O$, therefore the homotopy groups of O are 8-fold periodic, meaning $\pi k + 8(O) = \pi k(O)$, and one needs only to list the lower 8 homotopy groups:

$\pi_0(O) = \mathbf{Z}/2$

$\pi_1(O) = \mathbf{Z}/2$

$\pi_2(O) = 0$

$\pi_3(O) = \mathbf{Z}$

$\pi_4(O) = 0$

$\pi_5(O) = 0$

$\pi_6(O) = 0$

$\pi_7(O) = \mathbf{Z}$

Relation to KO-theory

Via the clutching construction, homotopy groups of the stable space O are identified with stable vector bundles on spheres (up to isomorphism), with a dimension shift of 1: $\pi k(O) = \pi_{k+1}(BO)$. Setting $KO = BO \times \mathbf{Z} = \Omega^{-1}O \times \mathbf{Z}$ (to make π_0 fit into the periodicity), one obtains:

$\pi_0(KO) = \mathbf{Z}$

$\pi_1(KO) = \mathbf{Z}/2$

$\pi_2(KO) = \mathbf{Z}/2$

$\pi_3(KO) = 0$

$\pi_4(KO) = \mathbf{Z}$

$\pi_5(KO) = 0$

$\pi_6(KO) = 0$

$\pi_7(KO) = 0$

Computation and interpretation of homotopy groups

Low-dimensional groups The first few homotopy groups can be calculated by using the concrete descriptions of low-dimensional groups.

- $\pi_0(O) = \pi_0(O(1)) = \mathbf{Z}/2$, from orientation-preserving/reversing (this class survives to O(2) and hence stably)

- $\pi_1(O) = \pi_1(SO(3)) = \mathbf{Z}/2$, which is spin comes from $SO(3) = \mathbf{R}P^3 = S^3/(\mathbf{Z}/2)$.

- $\pi_2(O) = \pi_2(SO(3)) = 0$, which surjects onto $\pi_2(SO(4))$; this latter thus vanishes.

Lie groups From general facts about Lie groups, $\pi_2(G)$ always vanishes, and $\pi_3(G)$ is free (free abelian).

Vector bundles From the vector bundle point of view, $\pi_0(KO)$ is vector bundles over S^0, which is two points. Thus over each point, the bundle is trivial, and the non-triviality of the bundle is the difference between the dimensions of the vector spaces over the two points, so $\pi_0(KO) = \mathbf{Z}$ is dimension.

Loop spaces Using concrete descriptions of the loop spaces in Bott periodicity, one can interpret higher homotopy of O as lower homotopy of simple to analyze spaces. Using π_0, O and O/U have two components, $KO = BO \times \mathbf{Z}$ and $KSp = BSp \times \mathbf{Z}$ have countably many components, and the rest are connected.

Interpretation of homotopy groups

In a nutshell:[3]

- $\pi_0(KO) = \mathbf{Z}$ is about dimension

- $\pi_1(KO) = \mathbf{Z}/2$ is about orientation

- $\pi_2(KO) = \mathbf{Z}/2$ is about spin

- $\pi_4(KO) = \mathbf{Z}$ is about topological quantum field theory.

Let R be any of the four division algebras $\mathbf{R}, \mathbf{C}, \mathbf{H}, \mathbf{O}$, and let LR be the tautological line bundle over the projective line RP^1, and $[LR]$ its class in K-theory. Noting that $\mathbf{R}P^1 = S^1$, $\mathbf{C}P^1 = S^2$, $\mathbf{H}P^1 = S^4$, $\mathbf{O}P^1 = S^8$, these yield vector bundles over the corresponding spheres, and

- $\pi_1(KO)$ is generated by $[LR]$

- $\pi_2(KO)$ is generated by $[LC]$

- $\pi_4(KO)$ is generated by $[LH]$

- $\pi_8(KO)$ is generated by $[LO]$

From the point of view of symplectic geometry, $\pi_0(KO) \cong \pi_8(KO) = \mathbf{Z}$ can be interpreted as the Maslov index, thinking of it as the fundamental group $\pi_1(U/O)$ of the stable Lagrangian Grassmannian as $U/O \cong \Omega^7(KO)$, so $\pi_1(U/O) = \pi_{1+7}(KO)$.

21.4   Over the complex number field

Over the field $\mathbf{C}$ of complex numbers, $O(n, \mathbf{C})$ and $SO(n, \mathbf{C})$ are complex Lie groups of dimension $n(n-1)/2$ over $\mathbf{C}$ (it means the dimension over $\mathbf{R}$ is twice that). $O(n, \mathbf{C})$ has two connected components, and $SO(n, \mathbf{C})$ is the connected component containing the identity matrix. For $n \geq 2$ these groups are noncompact.

Just as in the real case SO(n, **C**) is not simply connected. For $n > 2$ the fundamental group of SO(n, **C**) is cyclic of order 2 whereas the fundamental group of SO(2, **C**) is infinite cyclic.

21.5 Over finite fields

Orthogonal groups can also be defined over finite fields **F**q, where q is a power of a prime p.

Over finite fields of characteristic not equal to 2, orthogonal groups come in two types in even dimension: O$^+$(2n, q) and O$^-$(2n, q); and one type in odd dimension: O(2$n + 1$, q).[4]

If V is the vector space on which the orthogonal group G acts, it can be written as a direct orthogonal sum as follows:

$$V = L_1 \oplus L_2 \oplus \cdots \oplus L_m \oplus W,$$

where L_i are hyperbolic lines and W contains no singular vectors. If W is the zero subspace, then G is of plus type. If W is one-dimensional then G has odd dimension. If W has dimension 2, G is of minus type.

In the special case where $n = 1$, O$^\epsilon$(2, q) is a dihedral group of order $2(q - \epsilon)$.

We have the following formulas for the order of O(n, q), when the characteristic is not two:

$$|\mathrm{O}(2n + 1, q)| = 2q^n \prod_{i=0}^{n-1} (q^{2n} - q^{2i}).$$

If -1 is a square in **F**q

$$|\mathrm{O}(2n, q)| = 2(q^n - 1) \prod_{i=1}^{n-1} (q^{2n} - q^{2i}).$$

If -1 is a non-square in **F**q

$$|\mathrm{O}(2n, q)| = 2(q^n + (-1)^{n+1}) \prod_{i=1}^{n-1} (q^{2n} - q^{2i}).$$

21.6 The Dickson invariant

For orthogonal groups, the **Dickson invariant** is a homomorphism from the orthogonal group to the quotient group **Z**/2**Z** (integers modulo 2), taking the value 0 in case the element is the product of an even number of reflections, and the value of 1 otherwise.[5]

Algebraically, the Dickson invariant can be defined as $D(f)$ = rank($I - f$) modulo 2, where I is the identity (Taylor 1992, Theorem 11.43). Over fields that are not of characteristic 2 it is equivalent to the determinant: the determinant is -1 to the power of the Dickson invariant. Over fields of characteristic 2, the determinant is always 1, so the Dickson invariant gives more information than the determinant.

The special orthogonal group is the kernel of the Dickson invariant[5] and usually has index 2 in O(n, F).[6] When the characteristic of F is not 2, the Dickson Invariant is 0 whenever the determinant is 1. Thus when the characteristic is not 2, SO(n, F) is commonly defined to be the elements of O(n, F) with determinant 1. Each element in O(n, F) has determinant ± 1. Thus in characteristic 2, the determinant is always 1.

The Dickson invariant can also be defined for Clifford groups and Pin groups in a similar way (in all dimensions).

21.7 Orthogonal groups of characteristic 2

Over fields of characteristic 2 orthogonal groups often exhibit special behaviors, some of which are listed in this section. (Formerly these groups were known as the **hypoabelian groups** but this term is no longer used.)

- Any orthogonal group over any field is generated by reflections, except for a unique example where the vector space is 4-dimensional over the field with 2 elements and the Witt index is 2.[7] Note that a reflection in characteristic two has a slightly different definition. In characteristic two, the reflection orthogonal to a vector **u** takes a vector **v** to **v** + B(**v**, **u**)/Q(**u**) · **u** where B is the bilinear form and Q is the quadratic form associated to the orthogonal geometry. Compare this to the Householder reflection of odd characteristic or characteristic zero, which takes **v** to **v** − 2·B(**v**, **u**)/Q(**u**) · **u**.

- The center of the orthogonal group usually has order 1 in characteristic 2, rather than 2, since $I = -I$.

- In odd dimensions $2n + 1$ in characteristic 2, orthogonal groups over perfect fields are the same as symplectic groups in dimension $2n$. In fact the symmetric form is alternating in characteristic 2, and as the dimension is odd it must have a kernel of dimension 1, and the quotient by this kernel is a symplectic space of dimension $2n$, acted upon by the orthogonal group.

- In even dimensions in characteristic 2 the orthogonal group is a subgroup of the symplectic group, because the symmetric bilinear form of the quadratic form is also an alternating form.

21.8 The spinor norm

The **spinor norm** is a homomorphism from an orthogonal group over a field F to the quotient group F^*/F^{*2} (the multiplicative group of the field F up to square elements), that takes reflection in a vector of norm n to the image of n in F^*/F^{*2}.[8]

For the usual orthogonal group over the reals it is trivial, but it is often non-trivial over other fields, or for the orthogonal group of a quadratic form over the reals that is not positive definite.

21.9 Galois cohomology and orthogonal groups

In the theory of Galois cohomology of algebraic groups, some further points of view are introduced. They have explanatory value, in particular in relation with the theory of quadratic forms; but were for the most part *post hoc*, as far as the discovery of the phenomena is concerned. The first point is that quadratic forms over a field can be identified as a Galois H^1, or twisted forms (torsors) of an orthogonal group. As an algebraic group, an orthogonal group is in general neither connected nor simply-connected; the latter point brings in the spin phenomena, while the former is related to the discriminant.

The 'spin' name of the spinor norm can be explained by a connection to the spin group (more accurately a pin group). This may now be explained quickly by Galois cohomology (which however postdates the introduction of the term by more direct use of Clifford algebras). The spin covering of the orthogonal group provides a short exact sequence of algebraic groups.

$$1 \to \mu_2 \to \mathrm{Pin}_V \to \mathrm{O}_V \to 1$$

Here μ_2 is the algebraic group of square roots of 1; over a field of characteristic not 2 it is roughly the same as a two-element group with trivial Galois action. The connecting homomorphism from $H^0(\mathrm{O}V)$, which is simply the group $\mathrm{O}V(F)$ of F-valued points, to $H^1(\mu_2)$ is essentially the spinor norm, because $H^1(\mu_2)$ is isomorphic to the multiplicative group of the field modulo squares.

There is also the connecting homomorphism from H^1 of the orthogonal group, to the H^2 of the kernel of the spin covering. The cohomology is non-abelian, so that this is as far as we can go, at least with the conventional definitions.

21.10 Lie algebra

The Lie algebra corresponding to Lie groups $\mathrm{O}(n, F)$ and $\mathrm{SO}(n, F)$ consists of the skew-symmetric $n \times n$ matrices, with the Lie bracket [,] given by the commutator. One Lie algebra corresponds to both groups. It is often denoted by $\mathfrak{o}(n, F)$ or $\mathfrak{so}(n, F)$, and called the **orthogonal Lie algebra** or **special orthogonal Lie algebra**. Over real numbers, these Lie algebras for different n are the compact real forms of two of the four families of semisimple Lie algebras: in odd dimension B_k, where $n = 2k + 1$, while in even dimension D_r, where $n = 2r$.

More intrinsically, given a vector space with an inner product, the special orthogonal Lie algebra is given by the bivectors on the space, which are sums of simple bivectors (2-blades) $\mathbf{v} \wedge \mathbf{w}$. The correspondence is given by the map $\mathbf{v} \wedge \mathbf{w} \mapsto \mathbf{v}^* \otimes \mathbf{w} - \mathbf{w}^* \otimes \mathbf{v}$, where $\mathbf{v}^*$ is the covector dual to the vector $\mathbf{v}$; in coordinates these are exactly the elementary skew-symmetric matrices.

Over real numbers, this characterization is used in interpreting the curl of a vector field (naturally a 2-vector) as an infinitesimal rotation or "curl", hence the name. Generalizing the inner product with a nondegenerate form yields the indefinite orthogonal Lie algebras $\mathfrak{so}(p, q)$.

The representation theory of the orthogonal Lie algebras includes both representations corresponding to linear representations of the orthogonal groups, and representations corresponding to projective representations of the orthogonal groups (linear representations of spin groups), the so-called spin representation, which are important in physics.

21.11 Related groups

The orthogonal groups and special orthogonal groups have a number of important subgroups, supergroups, quotient groups, and covering groups. These are listed below.

The inclusions $\mathrm{O}(n) \subset \mathrm{U}(n) \subset \mathrm{Sp}(n) = \mathrm{USp}(2n)$ and $\mathrm{USp}(n) \subset \mathrm{U}(n) \subset \mathrm{O}(2n)$ are part of a sequence of 8 inclusions used in a geometric proof of the Bott periodicity theorem, and the corresponding quotient spaces are symmetric spaces of independent interest – for example, $\mathrm{U}(n)/\mathrm{O}(n)$ is the Lagrangian Grassmannian.

21.11.1 Lie subgroups

In physics, particularly in the areas of Kaluza–Klein compactification, it is important to find out the subgroups of the orthogonal group. The main ones are:

$O(n) \supset O(n-1)$ – preserve an axis

$O(2n) \supset U(n) \supset SU(n)$ – $U(n)$ are those that preserve a compatible complex structure *or* a compatible symplectic structure – see 2-out-of-3 property; $SU(n)$ also preserves a complex orientation.

$O(2n) \supset USp(n)$

$O(7) \supset G_2$

21.11.2 Lie supergroups

The orthogonal group $O(n)$ is also an important subgroup of various Lie groups:

$U(n) \supset SU(n) \supset O(n)$

$USp(2n) \supset O(n)$

$G_2 \supset O(3)$

$F_4 \supset O(9)$

$E_6 \supset O(10)$

$E_7 \supset O(12)$

$E_8 \supset O(16)$

Conformal group

Main article: Conformal group

Being isometries, real orthogonal transforms preserve angles, and are thus conformal maps, though not all conformal linear transforms are orthogonal. In classical terms this is the difference between congruence and similarity, as exemplified by SSS (Side-Side-Side) congruence of triangles and AAA (Angle-Angle-Angle) similarity of triangles. The group of conformal linear maps of $\mathbf{R}^n$ is denoted $CO(n)$ for the **conformal orthogonal group**, and consists of the product of the orthogonal group with the group of dilations. If n is odd, these two subgroups do not intersect, and they are a direct product: $CO(2k+1) = O(2k+1) \times \mathbf{R}^*$, where $\mathbf{R}^* = \mathbf{R}\backslash\{0\}$ is the real multiplicative group, while if n is even, these subgroups intersect in ± 1, so this is not a direct product, but it is a direct product with the subgroup of dilation by a positive scalar: $CO(2k) = O(2k) \times \mathbf{R}^+$.

Similarly one can define $CSO(n)$; note that this is always: $CSO(n) = CO(n) \cap GL^+(n) = SO(n) \times \mathbf{R}^+$.

21.11.3 Discrete subgroups

As the orthogonal group is compact, discrete subgroups are equivalent to finite subgroups.[note 2] These subgroups are known as point group and can be realized as the symmetry groups of polytopes. A very important class of examples are the finite Coxeter groups, which include the symmetry groups of regular polytopes.

Dimension 3 is particularly studied – see point groups in three dimensions, polyhedral groups, and list of spherical symmetry groups. In 2 dimensions, the finite groups are either cyclic or dihedral – see point groups in two dimensions.

Other finite subgroups include:

- Permutation matrices (the Coxeter group An)

- Signed permutation matrices (the Coxeter group Bn); also equals the intersection of the orthogonal group with the integer matrices.[note 3]

21.11.4 Covering and quotient groups

The orthogonal group is neither simply connected nor centerless, and thus has both a covering group and a quotient group, respectively:

- Two covering Pin groups, $Pin_+(n) \to O(n)$ and $Pin_-(n) \to O(n)$,

- The quotient projective orthogonal group, $O(n) \to PO(n)$.

These are all 2-to-1 covers.

For the special orthogonal group, the corresponding groups are:

- Spin group, $Spin(n) \to SO(n)$,

- Projective special orthogonal group, $SO(n) \to PSO(n)$.

Spin is a 2-to-1 cover, while in even dimension, $PSO(2k)$ is a 2-to-1 cover, and in odd dimension $PSO(2k+1)$ is a 1-to-1 cover, i.e., isomorphic to $SO(2k+1)$. These groups, $Spin(n)$, $SO(n)$, and $PSO(n)$ are Lie group forms of the compact special orthogonal Lie algebra, $\mathfrak{so}(n, \mathbb{R})$ – Spin is the simply connected form, while PSO is the centerless form, and SO is in general neither.[note 4]

In dimension 3 and above these are the covers and quotients, while dimension 2 and below are somewhat degenerate; see specific articles for details.

21.12   Principal homogeneous space: Stiefel manifold

Main article: Stiefel manifold

The principal homogeneous space for the orthogonal group $O(n)$ is the Stiefel manifold $V_n(\mathbf{R}^n)$ of orthonormal bases (orthonormal n-frames).

In other words, the space of orthonormal bases is like the orthogonal group, but without a choice of base point: given an orthogonal space, there is no natural choice of orthonormal basis, but once one is given one, there is a one-to-one correspondence between bases and the orthogonal group. Concretely, a linear map is determined by where it sends a basis: just as an invertible map can take any basis to any other basis, an orthogonal map can take any *orthogonal* basis to any other *orthogonal* basis.

The other Stiefel manifolds $V_k(\mathbf{R}^n)$ for $k < n$ of *incomplete* orthonormal bases (orthonormal k-frames) are still homogeneous spaces for the orthogonal group, but not *principal* homogeneous spaces: any k-frame can be taken to any other k-frame by an orthogonal map, but this map is not uniquely determined.

21.13   See also

21.13.1   Specific transforms

- Coordinate rotations and reflections
- Reflection through the origin

21.13.2   Specific groups

- rotation group, SO(3, **R**)
- SO(8)

21.13.3   Related groups

- indefinite orthogonal group
- unitary group
- symplectic group

21.13.4   Lists of groups

- list of finite simple groups
- list of simple Lie groups

21.14   Notes

[1] The analogy is stronger: Weyl groups, a class of (representations of) Coxeter groups, can be considered as simple algebraic groups over the field with one element, and there are a number of analogies between algebraic groups and vector spaces on the one hand, and Weyl groups and sets on the other.

[2] Infinite subsets of a compact space have an accumulation point and are not discrete.

[3] $O(n) \cap GL(n, \mathbf{Z})$ equals the signed permutation matrices because an integer vector of norm 1 must have a single non-zero entry, which must be ±1 (if it has two non-zero entries or a larger entry, the norm will be larger than 1), and in an orthogonal matrix these entries must be in different coordinates, which is exactly the signed permutation matrices.

[4] In odd dimension, $SO(2k + 1) \cong PSO(2k + 1)$ is centerless (but not simply connected), while in even dimension $SO(2k)$ is neither centerless nor simply connected.

21.15   References

[1] For base fields of characteristic not 2, it is equivalent to use symmetric bilinear forms or quadratic forms. But in characteristic 2 these notions differ.

[2] Hall 2015 Proposition 13.10

[3] John Baez "This Week's Finds in Mathematical Physics" week 105

[4] Wilson, Robert A. (2009). *The finite simple groups*. Graduate Texts in Mathematics **251**. London: Springer. pp. 69–75. ISBN 978-1-84800-987-5. Zbl 1203.20012.

[5] Knus, Max-Albert (1991), *Quadratic and Hermitian forms over rings*, Grundlehren der Mathematischen Wissenschaften **294**, Berlin etc.: Springer-Verlag, p. 224, ISBN 3-540-52117-8, Zbl 0756.11008

[6] (Taylor 1992, page 160)

[7] (Grove 2002, Theorem 6.6 and 14.16)

[8] Cassels 1978, p. 178

- Cassels, J.W.S. (1978), *Rational Quadratic Forms*, London Mathematical Society Monographs **13**, Academic Press, ISBN 0-12-163260-1, Zbl 0395.10029

- Grove, Larry C. (2002), *Classical groups and geometric algebra*, Graduate Studies in Mathematics **39**, Providence, R.I.: American Mathematical Society, ISBN 978-0-8218-2019-3, MR 1859189

- Hall, Brian C. (2015), *Lie Groups, Lie Algebras, and Representations: An Elementary Introduction*, Graduate Texts in Mathematics **222** (2nd ed.), Springer

- Taylor, Donald E. (1992), *The Geometry of the Classical Groups*, Sigma Series in Pure Mathematics **9**, Berlin: Heldermann Verlag, ISBN 3-88538-009-9, MR 1189139, Zbl 0767.20001

21.16 External links

- Hazewinkel, Michiel, ed. (2001), "Orthogonal group", *Encyclopedia of Mathematics*, Springer, ISBN 978-1-55608-010-4

- John Baez "This Week's Finds in Mathematical Physics" week 105

- John Baez on Octonions

- (Italian) n-dimensional Special Orthogonal Group parametrization

Chapter 22

E8 (mathematics)

In mathematics, $\mathbf{E}_8$ is any of several closely related exceptional simple Lie groups, linear algebraic groups or Lie algebras of dimension 248; the same notation is used for the corresponding root lattice, which has rank 8. The designation E_8 comes from the Cartan–Killing classification of the complex simple Lie algebras, which fall into four infinite series labeled An, Bn, Cn, Dn, and five exceptional cases labeled E_6, E_7, E_8, F_4, and G_2. The E_8 algebra is the largest and most complicated of these exceptional cases.

Wilhelm Killing (1888a, 1888b, 1889, 1890) discovered the complex Lie algebra E_8 during his classification of simple compact Lie algebras, though he did not prove its existence, which was first shown by Élie Cartan. Cartan determined that a complex simple Lie algebra of type E_8 admits three real forms. Each of them gives rise to a simple Lie group of dimension 248, exactly one of which is compact. Chevalley (1955) introduced algebraic groups and Lie algebras of type E_8 over other fields: for example, in the case of finite fields they lead to an infinite family of finite simple groups of Lie type.

22.1 Basic description

The Lie group E_8 has dimension 248. Its rank, which is the dimension of its maximal torus, is 8.

Therefore, the vectors of the root system are in eight-dimensional Euclidean space: they are described explicitly later in this article. The Weyl group of E_8, which is the group of symmetries of the maximal torus which are induced by conjugations in the whole group, has order $2^{14}3^55^27 = 696729600$.

The compact group E_8 is unique among simple compact Lie groups in that its non-trivial representation of smallest dimension is the adjoint representation (of dimension 248) acting on the Lie algebra E_8 itself; it is also the unique one which has the following four properties: trivial center, compact, simply connected, and simply laced (all roots have the same length).

There is a Lie algebra Ek for every integer $k \geq 3$, which is infinite dimensional if k is greater than 8.

22.2 Real and complex forms

There is a unique complex Lie algebra of type E_8, corresponding to a complex group of complex dimension 248. The complex Lie group E_8 of complex dimension 248 can be considered as a simple real Lie group of real dimension 496. This is simply connected, has maximal compact subgroup the compact form (see below) of E_8, and has an outer automorphism group of order 2 generated by complex conjugation.

As well as the complex Lie group of type E_8, there are three real forms of the Lie algebra, three real forms of the group with trivial center (two of which have non-algebraic double covers, giving two further real forms), all of real dimension 248, as follows:

- The compact form (which is usually the one meant if no other information is given), which is simply connected and has trivial outer automorphism group.

- The split form, EVIII (or $E_{8(8)}$), which has maximal compact subgroup Spin(16)/($\mathbf{Z}/2\mathbf{Z}$), fundamental group of order 2 (implying that it has a double cover, which is a simply connected Lie real group but is not algebraic, see below) and has trivial outer automorphism group.

- EIX (or $E_{8(-24)}$), which has maximal compact subgroup $E_7 \times$SU(2)/($-1,-1$), fundamental group of order 2 (again implying a double cover, which is not algebraic) and has trivial outer automorphism group.

For a complete list of real forms of simple Lie algebras, see the list of simple Lie groups.

22.3 E_8 as an algebraic group

By means of a Chevalley basis for the Lie algebra, one can define E_8 as a linear algebraic group over the integers and, consequently, over any commutative ring and in particular over any field: this defines the so-called split (sometimes also known as "untwisted") form of E_8. Over an algebraically closed field, this is the only form; however, over other fields, there are often many other forms, or "twists" of E_8, which are classified in the general framework of Galois cohomology (over a perfect field k) by the set $H^1(k,\text{Aut}(E_8))$ which, because the Dynkin diagram of E_8 (see below) has no automorphisms, coincides with $H^1(k,E_8)$.[1]

Over **R**, the real connected component of the identity of these algebraically twisted forms of E_8 coincide with the three real Lie groups mentioned above, but with a subtlety concerning the fundamental group: all forms of E_8 are simply connected in the sense of algebraic geometry, meaning that they admit no non-trivial algebraic coverings; the non-compact and simply connected real Lie group forms of E_8 are therefore not algebraic and admit no faithful finite-dimensional representations.

Over finite fields, the Lang–Steinberg theorem implies that $H^1(k,E_8)=0$, meaning that E_8 has no twisted forms: see below.

22.4 Representation theory

The characters of finite dimensional representations of the real and complex Lie algebras and Lie groups are all given by the Weyl character formula. The dimensions of the smallest irreducible representations are (sequence A121732 in OEIS):

> 1,   248,   3875,   27000,   30380,   147250,
> 779247,   1763125,   2450240,   4096000,
> 4881384,   6696000,   26411008,   70680000,
> 76271625,   79143000,   146325270,
> 203205000,   281545875,   301694976,
> 344452500,   820260000,   1094951000,
> 2172667860,   2275896000,   2642777280,
> 2903770000,   3929713760,   4076399250,
> 4825673125,   6899079264,   8634368000
> (twice), 12692520960...

The 248-dimensional representation is the adjoint representation. There are two non-isomorphic irreducible representations of dimension 8634368000 (it is not unique; however, the next integer with this property is 175898504162692612600853299200000 (sequence A181746 in OEIS)). The fundamental representations are those with dimensions 3875, 6696000, 6899079264, 146325270, 2450240, 30380, 248 and 147250 (corresponding to the eight nodes in the Dynkin diagram in the order chosen for the Cartan matrix below, i.e., the nodes are read in the seven-node chain first, with the last node being connected to the third).

The coefficients of the character formulas for infinite dimensional irreducible representations of E_8 depend on some large square matrices consisting of polynomials, the Lusztig–Vogan polynomials, an analogue of Kazhdan–Lusztig polynomials introduced for reductive groups in general by George Lusztig and David Kazhdan (1983). The values at 1 of the Lusztig–Vogan polynomials give the coefficients of the matrices relating the standard representations (whose characters are easy to describe) with the irreducible representations.

These matrices were computed after four years of collaboration by a group of 18 mathematicians and computer scientists, led by Jeffrey Adams, with much of the programming done by Fokko du Cloux. The most difficult case (for exceptional groups) is the split real form of E_8 (see above), where the largest matrix is of size 453060×453060. The Lusztig–Vogan polynomials for all other exceptional simple groups have been known for some time; the calculation for the split form of E_8 is far longer than any other case. The announcement of the result in March 2007 received extraordinary attention from the media (see the external links), to the surprise of the mathematicians working on it.

The representations of the E_8 groups over finite fields are given by Deligne–Lusztig theory.

22.5 Constructions

One can construct the (compact form of the) E_8 group as the automorphism group of the corresponding $\mathbf{e}_8$ Lie algebra. This algebra has a 120-dimensional subalgebra $\mathbf{so}(16)$ generated by Jij as well as 128 new generators Qa that transform as a Weyl–Majorana spinor of $\mathbf{spin}(16)$. These statements determine the commutators

$$[J_{ij}, J_{k\ell}] = \delta_{jk}J_{i\ell} - \delta_{j\ell}J_{ik} - \delta_{ik}J_{j\ell} + \delta_{i\ell}J_{jk}$$

as well as

$$[J_{ij}, Q_a] = \frac{1}{4}(\gamma_i\gamma_j - \gamma_j\gamma_i)_{ab}Q_b,$$

while the remaining commutator (not anticommutator!) is defined as

$$[Q_a, Q_b] = \gamma^{[i}_{ac} \gamma^{j]}_{cb} J_{ij}.$$

It is then possible to check that the Jacobi identity is satisfied.

22.6   Geometry

The compact real form of E_8 is the isometry group of the 128-dimensional exceptional compact Riemannian symmetric space EVIII (in Cartan's classification). It is known informally as the "octooctonionic projective plane" because it can be built using an algebra that is the tensor product of the octonions with themselves, and is also known as a Rosenfeld projective plane, though it does not obey the usual axioms of a projective plane. This can be seen systematically using a construction known as the *magic square*, due to Hans Freudenthal and Jacques Tits (Landsberg & Manivel 2001).

22.7 E_8 root system

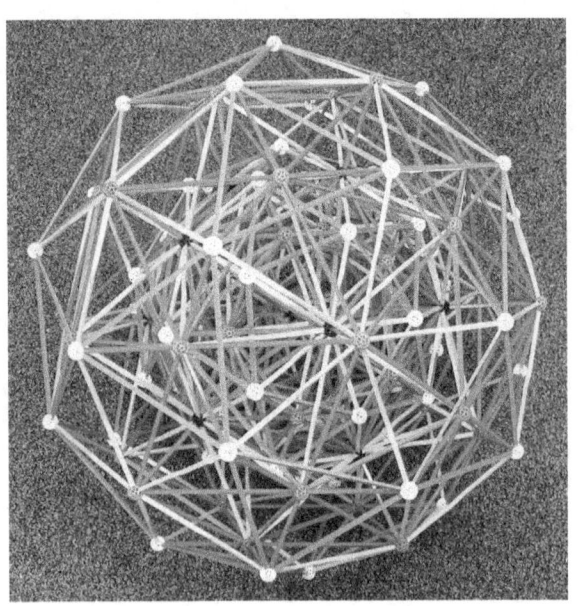

Zome model of the E_8 root system, projected into three-space, and represented by the vertices of the 4_{21} polytope,

A root system of rank r is a particular finite configuration of vectors, called *roots*, which span an r-dimensional Euclidean space and satisfy certain geometrical properties. In particular, the root system must be invariant under reflection through the hyperplane perpendicular to any root.

The **E_8 root system** is a rank 8 root system containing 240 root vectors spanning $\mathbf{R}^8$. It is irreducible in the sense that it cannot be built from root systems of smaller rank. All the root vectors in E_8 have the same length. It is convenient for a number of purposes to normalize them to have length $\sqrt{2}$. These 240 vectors are the vertices of a semi-regular polytope discovered by Thorold Gosset in 1900, sometimes known as the 4_{21} polytope.

22.7.1   Construction

In the so-called *even coordinate system*, E_8 is given as the set of all vectors in $\mathbf{R}^8$ with length squared equal to 2 such that coordinates are either all integers or all half-integers and the sum of the coordinates is even.

Explicitly, there are 112 roots with integer entries obtained from

$$(\pm 1, \pm 1, 0, 0, 0, 0, 0, 0)$$

by taking an arbitrary combination of signs and an arbitrary permutation of coordinates, and 128 roots with half-integer entries obtained from

$$\left(\pm \tfrac{1}{2}, \pm \tfrac{1}{2}, \pm \tfrac{1}{2}, \pm \tfrac{1}{2}, \pm \tfrac{1}{2}, \pm \tfrac{1}{2}, \pm \tfrac{1}{2}, \pm \tfrac{1}{2}\right)$$

by taking an even number of minus signs (or, equivalently, requiring that the sum of all the eight coordinates be even). There are 240 roots in all.

E8 with thread made by hand

The 112 roots with integer entries form a D_8 root system. The E_8 root system also contains a copy of A_8 (which has

72 roots) as well as E_6 and E_7 (in fact, the latter two are usually *defined* as subsets of E_8).

In the *odd coordinate system*, E_8 is given by taking the roots in the even coordinate system and changing the sign of any one coordinate. The roots with integer entries are the same while those with half-integer entries have an odd number of minus signs rather than an even number.

22.7.2 Dynkin diagram

The Dynkin diagram for E_8 is given by

This diagram gives a concise visual summary of the root structure. Each node of this diagram represents a simple root. A line joining two simple roots indicates that they are at an angle of 120° to each other. Two simple roots which are not joined by a line are orthogonal.

22.7.3 Cartan matrix

The Cartan matrix of a rank r root system is an $r \times r$ matrix whose entries are derived from the simple roots. Specifically, the entries of the Cartan matrix are given by

$$A_{ij} = 2\frac{(\alpha_i, \alpha_j)}{(\alpha_i, \alpha_i)}$$

where $(-,-)$ is the Euclidean inner product and αi are the simple roots. The entries are independent of the choice of simple roots (up to ordering).

The Cartan matrix for E_8 is given by

$$\begin{bmatrix} 2 & -1 & 0 & 0 & 0 & 0 & 0 & 0 \\ -1 & 2 & -1 & 0 & 0 & 0 & 0 & 0 \\ 0 & -1 & 2 & -1 & 0 & 0 & 0 & 0 \\ 0 & 0 & -1 & 2 & -1 & 0 & 0 & 0 \\ 0 & 0 & 0 & -1 & 2 & -1 & 0 & -1 \\ 0 & 0 & 0 & 0 & -1 & 2 & -1 & 0 \\ 0 & 0 & 0 & 0 & 0 & -1 & 2 & 0 \\ 0 & 0 & 0 & 0 & -1 & 0 & 0 & 2 \end{bmatrix}.$$

The determinant of this matrix is equal to 1.

22.7.4 Simple roots

A set of simple roots for a root system Φ is a set of roots that form a basis for the Euclidean space spanned by Φ with the special property that each root has components with respect to this basis that are either all nonnegative or all nonpositive.

Given the E_8 Cartan matrix (above) and a Dynkin diagram node ordering of:

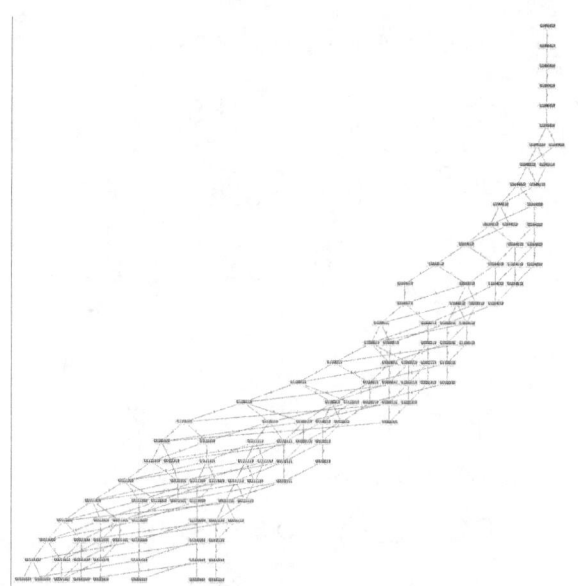

Hasse diagram of E8 root poset with edge labels identifying added simple root position.

One choice of simple roots is given by the rows of the following matrix:

$$\begin{bmatrix} 1 & -1 & 0 & 0 & 0 & 0 & 0 & 0 \\ 0 & 1 & -1 & 0 & 0 & 0 & 0 & 0 \\ 0 & 0 & 1 & -1 & 0 & 0 & 0 & 0 \\ 0 & 0 & 0 & 1 & -1 & 0 & 0 & 0 \\ 0 & 0 & 0 & 0 & 1 & -1 & 0 & 0 \\ 0 & 0 & 0 & 0 & 0 & 1 & 1 & 0 \\ -\frac{1}{2} & -\frac{1}{2} & -\frac{1}{2} & -\frac{1}{2} & -\frac{1}{2} & -\frac{1}{2} & -\frac{1}{2} & -\frac{1}{2} \\ 0 & 0 & 0 & 0 & 0 & 1 & -1 & 0 \end{bmatrix}.$$

22.7.5 Weyl group

The Weyl group of E_8 is of order 696729600, and can be described as O+
8(2): it is of the form 2.*G*.2 (that is, a stem extension by the cyclic group of order 2 of an extension of the cyclic group of order 2 by a group G) where G is the unique simple group of order 174182400 (which can be described as $PS\Omega_8{}^+(2)$).[2]

22.7.6 E_8 root lattice

Main article: E_8 lattice

The integral span of the E_8 root system forms a lattice in $\mathbf{R}^8$ naturally called the $\mathbf{E_8}$ **root lattice**. This lattice is rather remarkable in that it is the only (nontrivial) even, unimodular lattice with rank less than 16.

Subgroup Tree of E8

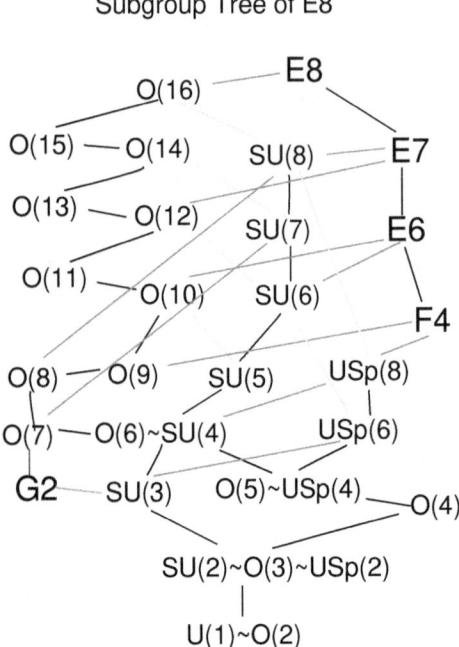

An incomplete simple subgroup tree of E_8

22.7.7 Simple subalgebras of E_8

The Lie algebra E8 contains as subalgebras all the exceptional Lie algebras as well as many other important Lie algebras in mathematics and physics. The height of the Lie algebra on the diagram approximately corresponds to the rank of the algebra. A line from an algebra down to a lower algebra indicates that the lower algebra is a subalgebra of the higher algebra.

22.8 Chevalley groups of type E_8

Chevalley (1955) showed that the points of the (split) algebraic group E_8 (see above) over a finite field with q elements form a finite Chevalley group, generally written $E_8(q)$, which is simple for any q,[3][4] and constitutes one of the infinite families addressed by the classification of finite simple groups. Its number of elements is given by the formula (sequence A008868 in OEIS):

$$q^{120}(q^{30}-1)(q^{24}-1)(q^{20}-1)(q^{18}-1)(q^{14}$$
$$-1)(q^{12}-1)(q^8-1)(q^2-1)$$

The first term in this sequence, the order of $E_8(2)$, namely 337804753143634480626138819061408559507991692242467651576160959909068800000 ≈ 3.38×10^{74}

, is already larger than the size of the Monster group. This group $E_8(2)$ is the last one described (but

without its character table) in the ATLAS of Finite Groups.[5]

The Schur multiplier of $E_8(q)$ is trivial, and its outer automorphism group is that of field automorphisms (i.e., cyclic of order f if $q=p^f$ where p is prime).

Lusztig (1979) described the unipotent representations of finite groups of type E_8.

22.9 Subgroups

The smaller exceptional groups E_7 and E_6 sit inside E_8. In the compact group, both $E_7×SU(2)/(-1,-1)$ and $E_6×SU(3)/(\mathbf{Z}/3\mathbf{Z})$ are maximal subgroups of E_8.

The 248-dimensional adjoint representation of E_8 may be considered in terms of its restricted representation to the first of these subgroups. It transforms under $E_7×SU(2)$ as a sum of tensor product representations, which may be labelled as a pair of dimensions as (3,1) + (1,133) + (2,56) (since there is a quotient in the product, these notations may strictly be taken as indicating the infinitesimal (Lie algebra) representations). Since the adjoint representation can be described by the roots together with the generators in the Cartan subalgebra, we may see that decomposition by looking at these. In this description,

- (3,1) consists of the roots (0,0,0,0,0,0,1,−1), (0,0,0,0,0,0,−1,1) and the Cartan generator corresponding to the last dimension;

- (1,133) consists of all roots with (1,1), (−1,−1), (0,0), (−½,−½) or (½,½) in the last two dimensions, together with the Cartan generators corresponding to the first seven dimensions;

- (2,56) consists of all roots with permutations of (1,0), (−1,0) or (½,−½) in the last two dimensions.

The 248-dimensional adjoint representation of E_8, when similarly restricted, transforms under $E_6×SU(3)$ as: (8,1) + (1,78) + (3,27) + (3,27). We may again see the decomposition by looking at the roots together with the generators in the Cartan subalgebra. In this description,

- (8,1) consists of the roots with permutations of (1,−1,0) in the last three dimensions, together with the Cartan generator corresponding to the last two dimensions;

- (1,78) consists of all roots with (0,0,0), (−½,−½,−½) or (½,½,½) in the last three dimensions, together with the Cartan generators corresponding to the first six dimensions;

- (3,27) consists of all roots with permutations of (1,0,0), (1,1,0) or (−½,½,½) in the last three dimensions.

- (3,27) consists of all roots with permutations of (−1,0,0), (−1,−1,0) or (½,−½,−½) in the last three dimensions.

The finite quasisimple groups that can embed in (the compact form of) E_8 were found by Griess & Ryba (1999).

The Dempwolff group is a subgroup of (the compact form of) E_8. It is contained in the Thompson sporadic group, which acts on the underlying vector space of the Lie group E_8 but does not preserve the Lie bracket. The Thompson group fixes a lattice and does preserve the Lie bracket of this lattice mod 3, giving an embedding of the Thompson group into $E_8(\mathbf{F}_3)$.

22.10   Applications

The E_8 Lie group has applications in theoretical physics and especially in string theory and supergravity. $E_8 \times E_8$ is the gauge group of one of the two types of heterotic string and is one of two anomaly-free gauge groups that can be coupled to the $N = 1$ supergravity in ten dimensions. E_8 is the U-duality group of supergravity on an eight-torus (in its split form).

One way to incorporate the standard model of particle physics into heterotic string theory is the symmetry breaking of E_8 to its maximal subalgebra SU(3)×E_6.

In 1982, Michael Freedman used the E_8 lattice to construct an example of a topological 4-manifold, the E_8 manifold, which has no smooth structure.

Antony Garrett Lisi's incomplete "An Exceptionally Simple Theory of Everything" attempts to describe all known fundamental interactions in physics as part of the E_8 Lie algebra.[6][7]

R. Coldea, D. A. Tennant, and E. M. Wheeler et al. (2010) reported an experiment where the electron spins of a cobalt-niobium crystal exhibited, under certain conditions, two of the eight peaks related to E_8 that were predicted by Zamolodchikov (1989).[8][9]

22.11   Notes

[1] Платонов, Владимир П.; Рапинчук, Андрей С. (1991), *Алгебраические группы и теория чисел*, Наука, ISBN 5-02-014191-7 (English translation: Platonov, Vladimir P.; Rapinchuk, Andrei S. (1994), *Algebraic groups and number theory*, Academic Press, ISBN 0-12-558180-7), §2.2.4

[2] Conway, John Horton; Curtis, Robert Turner; Norton, Simon Phillips; Parker, Richard A; Wilson, Robert Arnott (1985), *Atlas of Finite Groups: Maximal Subgroups and Ordinary Characters for Simple Groups*, Oxford University Press, p. 85, ISBN 0-19-853199-0

[3] Carter, Roger W. (1989), *Simple Groups of Lie Type*, Wiley Classics Library, John Wiley & Sons, ISBN 0-471-50683-4

[4] Wilson, Robert A. (2009), *The Finite Simple Groups*, Graduate Texts in Mathematics **251**, Springer-Verlag, ISBN 1-84800-987-9

[5] Conway &al, *op. cit.*, p. 235.

[6] A. G. Lisi; J. O. Weatherall (2010). "A Geometric Theory of Everything". *Scientific American* **303** (6): 54–61. doi:10.1038/scientificamerican1210-54. PMID 21141358.

[7] Greg Boustead (2008-11-17). "Garrett Lisi's Exceptional Approach to Everything". *SEED Magazine*.

[8] Most beautiful math structure appears in lab for first time, *New Scientist*, January 2010 (retrieved January 8, 2010).

[9] Did a 1-dimensional magnet detect a 248-dimensional Lie algebra?, *Notices of the American Mathematical Society*, September 2011.

22.12   References

- Adams, J. Frank (1996), *Lectures on exceptional Lie groups*, Chicago Lectures in Mathematics, University of Chicago Press, ISBN 978-0-226-00526-3, MR 1428422

- Baez, John C. (2002), "The octonions", *American Mathematical Society. Bulletin. New Series* **39** (2): 145–205, doi:10.1090/S0273-0979-01-00934-X, MR 1886087

- Chevalley, Claude (1955), "Sur certains groupes simples", *The Tohoku Mathematical Journal. Second Series* **7**: 14–66, doi:10.2748/tmj/1178245104, ISSN 0040-8735, MR 0073602

- Coldea, R.; Tennant, D. A.; Wheeler, E. M.; Wawrzynska, E.; Prabhakaran, D.; Telling, M.; Habicht, K.; Smeibidl, P.; Kiefer, K. (2010), "Quantum Criticality in an Ising Chain: Experimental Evidence for Emergent E_8 Symmetry", *Science* **327** (5962): 177–180, doi:10.1126/science.1180085

- Griess, Robert L.; Ryba, A. J. E. (1999), "Finite simple groups which projectively embed in an exceptional Lie group are classified!", *American Mathematical Society. Bulletin. New Series* **36** (1): 75–93, doi:10.1090/S0273-0979-99-00771-5, MR 1653177

- Killing, Wilhelm (1888a), "Die Zusammensetzung der stetigen endlichen Transformationsgruppen", *Mathematische Annalen* **31** (2): 252–290, doi:10.1007/BF01211904

- Killing, Wilhelm (1888b), "Die Zusammensetzung der stetigen endlichen Transformationsgruppen", *Mathematische Annalen* **33** (1): 1–48, doi:10.1007/BF01444109

- Killing, Wilhelm (1889), "Die Zusammensetzung der stetigen endlichen Transformationsgruppen", *Mathematische Annalen* **34** (1): 57–122, doi:10.1007/BF01446792

- Killing, Wilhelm (1890), "Die Zusammensetzung der stetigen endlichen Transformationsgruppen", *Mathematische Annalen* **36** (2): 161–189, doi:10.1007/BF01207837

- J.M. Landsberg and L. Manivel (2001), *The projective geometry of Freudenthal's magic square*, Journal of Algebra, Volume 239, Issue 2, pages 477–512, doi:10.1006/jabr.2000.8697, arXiv:math/9908039v1.

- Lusztig, George (1979), "Unipotent representations of a finite Chevalley group of type E8", *The Quarterly Journal of Mathematics. Oxford. Second Series* **30** (3): 315–338, doi:10.1093/qmath/30.3.301, ISSN 0033-5606, MR R545068

- Lusztig, George; Vogan, David (1983), "Singularities of closures of K-orbits on flag manifolds", *Inventiones Mathematicae* (Springer-Verlag) **71** (2): 365–379, doi:10.1007/BF01389103

- Zamolodchikov, A. B. (1989), "Integrals of motion and S-matrix of the (scaled) T=T$_c$ Ising model with magnetic field", *International Journal of Modern Physics A. Particles and Fields. Gravitation. Cosmology. Nuclear Physics* **4** (16): 4235–4248, doi:10.1142/S0217751X8900176X, MR 1017357

22.13   External links

Lusztig–Vogan polynomial calculation

- Atlas of Lie groups

- Kazhdan–Lusztig–Vogan Polynomials for E$_8$

- Narrative of the Project to compute Kazhdan–Lusztig Polynomials for E$_8$

- Slides for *The Character Table for E$_8$, or How We Wrote Down a 453,060 × 453,060 Matrix and Found Happiness* by D. Vogan.

- American Institute of Mathematics (March 2007), *Mathematicians Map E$_8$*

- The *n*-Category Café, a University of Texas blog posting by John Baez on E$_8$.

Other links

- Graphic representation of E$_8$ root system.

- The list of dimensions of irreducible representations of the complex form of E$_8$ is sequence A121732 in the OEIS.

Chapter 23

Spacetime

For other uses of this term, see Spacetime (disambiguation).

In physics, **spacetime** is any mathematical model that combines space and time into a single interwoven continuum. Since 300 BCE, the spacetime of our universe has historically been interpreted from a Euclidean space perspective, which regards space as consisting of three dimensions, and time as consisting of one dimension, the "fourth dimension". By combining space and time into a single manifold called Minkowski space in 1905, physicists have significantly simplified a large number of physical theories, as well as described in a more uniform way the workings of the universe at both the supergalactic and subatomic levels.

23.1   Explanation

In non-relativistic classical mechanics, the use of Euclidean space instead of spacetime is appropriate, because time is treated as universal with a constant rate of passage that is independent of the state of motion of an observer. In relativistic contexts, time cannot be separated from the three dimensions of space, because the observed rate at which time passes for an object depends on the object's velocity relative to the observer and also on the strength of gravitational fields, which can slow the passage of time for an object as seen by an observer outside the field.

In cosmology, the concept of spacetime combines space and time to a single abstract universe. Mathematically it is a manifold consisting of "events" which are described by some type of coordinate system. Typically **three spatial dimensions** (length, width, height), and one **temporal dimension** (time) are required. Dimensions are independent components of a coordinate grid needed to locate a point in a certain defined "space". For example, on the globe the latitude and longitude are two independent coordinates which together uniquely determine a location. In spacetime, a coordinate grid that spans the 3+1 dimensions locates events (rather than just points in space), i.e., time is added as another dimension to the coordinate grid. This way the coordinates specify *where* and *when* events occur. However, the unified nature of spacetime and the freedom of coordinate choice it allows imply that to express the temporal coordinate in one coordinate system requires both temporal and spatial coordinates in another coordinate system. Unlike in normal spatial coordinates, there are still restrictions for how measurements can be made spatially and temporally (see Spacetime intervals). These restrictions correspond roughly to a particular mathematical model which differs from Euclidean space in its manifest symmetry.

Until the beginning of the 20th century, time was believed to be independent of motion, progressing at a fixed rate in all reference frames; however, following its prediction by special relativity, later experiments confirmed that time slows at higher speeds of the reference frame relative to another reference frame. Such slowing, called time dilation, is explained in special relativity theory. Many experiments have confirmed time dilation, such as the relativistic decay of muons from cosmic ray showers and the slowing of atomic clocks aboard a Space Shuttle relative to synchronized Earth-bound inertial clocks.[1] The duration of time can therefore vary according to events and reference frames.

When dimensions are understood as mere components of the grid system, rather than physical attributes of space, it is easier to understand the alternate dimensional views as being simply the result of coordinate transformations.

The term *spacetime* has taken on a generalized meaning beyond treating spacetime events with the normal 3+1 dimensions. It is really the combination of space and time. Other proposed spacetime theories include additional dimensions—normally spatial but there exist some speculative theories that include additional temporal dimensions and even some that include dimensions that are neither temporal nor spatial (e.g., superspace). How many dimensions are needed to describe the universe is still an open question. Speculative theories such as string theory predict 10 or 26 dimensions (with M-theory predicting 11 dimensions: 10

spatial and 1 temporal), but the existence of more than four dimensions would only appear to make a difference at the subatomic level.[2]

23.2   Spacetime in literature

Incas regarded space and time as a single concept, referred to as **pacha** (Quechua: *pacha*, Aymara: *pacha*).[3][4] The peoples of the Andes maintain a similar understanding.[5]

The idea of a unified spacetime is stated by Edgar Allan Poe in his essay on cosmology titled *Eureka* (1848) that "Space and duration are one". In 1895, in his novel *The Time Machine*, H. G. Wells wrote, "There is no difference between time and any of the three dimensions of space except that our consciousness moves along it", and that "any real body must have extension in four directions: it must have Length, Breadth, Thickness, and Duration".

Marcel Proust, in his novel *Swann's Way* (published 1913), describes the village church of his childhood's Combray as "a building which occupied, so to speak, four dimensions of space—the name of the fourth being Time".

23.2.1   Mathematical concept

In Encyclopedie, published in 1754, under the term *dimension* Jean le Rond d'Alembert speculated that duration (time) might be considered a fourth dimension if the idea was not too novel.[6]

Another early venture was by Joseph Louis Lagrange in his *Theory of Analytic Functions* (1797, 1813). He said, "One may view mechanics as a geometry of four dimensions, and mechanical analysis as an extension of geometric analysis".[7]

The ancient idea of the cosmos gradually was described mathematically with differential equations, differential geometry, and abstract algebra. These mathematical articulations blossomed in the nineteenth century as electrical technology stimulated men like Michael Faraday and James Clerk Maxwell to describe the reciprocal relations of electric and magnetic fields. Daniel Siegel phrased Maxwell's role in relativity as follows:

> [...] the idea of the propagation of forces at the velocity of light through the electromagnetic field as described by Maxwell's equations—rather than instantaneously at a distance—formed the necessary basis for relativity theory.[8]

Maxwell used vortex models in his papers on On Physical

Lines of Force, but ultimately gave up on any substance but the electromagnetic field. Pierre Duhem wrote:

> [Maxwell] was not able to create the theory that he envisaged except by giving up the use of any model, and by extending by means of analogy the abstract system of electrodynamics to displacement currents.[9]

In Siegel's estimation, "this very abstract view of the electromagnetic fields, involving no visualizable picture of what is going on out there in the field, is Maxwell's legacy."[10] Describing the behaviour of electric fields and magnetic fields led Maxwell to view the combination as an electromagnetic field. These fields have a value at every point of spacetime. It is the intermingling of electric and magnetic manifestations, described by Maxwell's equations, that give spacetime its structure. In particular, the rate of motion of an observer determines the electric and magnetic profiles of the electromagnetic field. The propagation of the field is determined by the electromagnetic wave equation, which requires spacetime for description.

Spacetime was described as an affine space with quadratic form in Minkowski space of 1908.[11] In his 1914 textbook *The Theory of Relativity*, Ludwik Silberstein used biquaternions to represent events in Minkowski space. He also exhibited the Lorentz transformations between observers of differing velocities as biquaternion mappings. Biquaternions were described in 1853 by W. R. Hamilton, so while the physical interpretation was new, the mathematics was well known in English literature, making relativity an instance of applied mathematics.

The first inkling of general relativity in spacetime was articulated by W. K. Clifford. Description of the effect of gravitation on space and time was found to be most easily visualized as a "warp" or stretching in the geometrical fabric of space and time, in a smooth and continuous way that changed smoothly from point-to-point along the spacetime fabric. In 1947 James Jeans provided a concise summary of the development of spacetime theory in his book *The Growth of Physical Science*.[12]

23.3   Basic concepts

The basic elements of spacetime are events. In any given spacetime, an event is a unique position at a unique time. Because events are spacetime points, an example of an event in classical relativistic physics is (x, y, z, t), the location of an elementary (point-like) particle at a particular time. A spacetime itself can be viewed as the union of all events in the same way that a line is the union of all of its points, formally organized into a manifold, a space which can be described at small scales using coordinate systems.

A spacetime is independent of any observer.[13] However, in describing physical phenomena (which occur at certain moments of time in a given region of space), each observer chooses a convenient metrical coordinate system. Events are specified by four real numbers in any such coordinate system. The trajectories of elementary (point-like) particles through space and time are thus a continuum of events called the world line of the particle. Extended or composite objects (consisting of many elementary particles) are thus a union of many world lines twisted together by virtue of their interactions through spacetime into a "world-braid".

However, in physics, it is common to treat an extended object as a "particle" or "field" with its own unique (e.g., center of mass) position at any given time, so that the world line of a particle or light beam is the path that this particle or beam takes in the spacetime and represents the history of the particle or beam. The world line of the orbit of the Earth (in such a description) is depicted in two spatial dimensions *x* and *y* (the plane of the Earth's orbit) and a time dimension orthogonal to *x* and *y*. The orbit of the Earth is an ellipse in space alone, but its world line is a helix in spacetime.[14]

The unification of space and time is exemplified by the common practice of selecting a metric (the measure that specifies the interval between two events in spacetime) such that all four dimensions are measured in terms of units of distance: representing an event as $(x_0, x_1, x_2, x_3) = (ct, x, y, z)$ (in the Lorentz metric) or $(x_1, x_2, x_3, x_4) = (x, y, z, ict)$ (in the original Minkowski metric) where *c* is the speed of light.[15] The metrical descriptions of Minkowski Space and spacelike, lightlike, and timelike intervals given below follow this convention, as do the conventional formulations of the Lorentz transformation.

23.3.1   Spacetime intervals in flat space

In a Euclidean space, the separation between two points is measured by the distance between the two points. The distance is purely spatial, and is always positive. In spacetime, the displacement four-vector ΔR is given by the space displacement vector Δr and the time difference Δt between the events. The *spacetime interval*, also called *invariant interval*, between the two events, s^2,[16] is defined as:

$$s^2 = \Delta r^2 - c^2 \Delta t^2 \text{ (spacetime interval)},$$

where *c* is the speed of light. The choice of signs for s^2 above follows the space-like convention (−+++).[17] Spacetime intervals may be classified into three distinct types, based on whether the temporal separation ($c^2 \Delta t^2$) or the spatial separation (Δr^2) of the two events is greater: time-like, light-like or space-like.

Certain types of world lines are called geodesics of the spacetime – straight lines in the case of Minkowski space and their closest equivalent in the curved spacetime of general relativity. In the case of purely time-like paths, geodesics are (locally) the paths of greatest separation (spacetime interval) as measured along the path between two events, whereas in Euclidean space and Riemannian manifolds, geodesics are paths of shortest distance between two points.[18][19] The concept of geodesics becomes central in general relativity, since geodesic motion may be thought of as "pure motion" (inertial motion) in spacetime, that is, free from any external influences.

Time-like interval

$$c^2 \Delta t^2 > \Delta r^2$$
$$s^2 < 0$$

For two events separated by a time-like interval, enough time passes between them that there could be a cause–effect relationship between the two events. For a particle traveling through space at less than the speed of light, any two events which occur to or by the particle must be separated by a time-like interval. Event pairs with time-like separation define a negative spacetime interval ($s^2 < 0$) and may be said to occur in each other's future or past. There exists a reference frame such that the two events are observed to occur in the same spatial location, but there is no reference frame in which the two events can occur at the same time.

The measure of a time-like spacetime interval is described by the proper time interval, $\Delta \tau$:

$$\Delta \tau = \sqrt{\Delta t^2 - \frac{\Delta r^2}{c^2}} \text{ (proper time interval).}$$

The proper time interval would be measured by an observer with a clock traveling between the two events in an inertial reference frame, when the observer's path intersects each event as that event occurs. (The proper time interval defines a real number, since the interior of the square root is positive.)

Light-like interval

$$c^2 \Delta t^2 = \Delta r^2$$
$$s^2 = 0$$

In a light-like interval, the spatial distance between two events is exactly balanced by the time between the two events. The events define a spacetime interval of zero (

$s^2 = 0$). Light-like intervals are also known as "null" intervals.

Events which occur to or are initiated by a photon along its path (i.e., while traveling at c, the speed of light) all have light-like separation. Given one event, all those events which follow at light-like intervals define the propagation of a light cone, and all the events which preceded from a light-like interval define a second (graphically inverted, which is to say "*pastward*") light cone.

Space-like interval

$c^2 \Delta t^2 < \Delta r^2$

$s^2 > 0$

When a space-like interval separates two events, not enough time passes between their occurrences for there to exist a causal relationship crossing the spatial distance between the two events at the speed of light or slower. Generally, the events are considered not to occur in each other's future or past. There exists a reference frame such that the two events are observed to occur at the same time, but there is no reference frame in which the two events can occur in the same spatial location.

For these space-like event pairs with a positive spacetime interval ($s^2 > 0$), the measurement of space-like separation is the proper distance, $\Delta \sigma$:

$\Delta \sigma = \sqrt{s^2} = \sqrt{\Delta r^2 - c^2 \Delta t^2}$ (proper distance).

Like the proper time of time-like intervals, the proper distance of space-like spacetime intervals is a real number value.

23.3.2   Interval as area

The interval has been presented as the area of an oriented rectangle formed by two events and isotropic lines through them. Time-like or space-like separations correspond to oppositely oriented rectangles, one type considered to have rectangles of negative area. The case of two events separated by light corresponds to the rectangle degenerating to the segment between the events and zero area.[20] The transformations leaving interval-length invariant are the area-preserving squeeze mappings.

The parameters traditionally used rely on quadrature of the hyperbola, which is the natural logarithm. This transcendental function is essential in mathematical analysis as its inverse unites circular functions and hyperbolic functions:

The exponential function, e^t, t a real number, used in the hyperbola (e^t, e^{-t}), generates hyperbolic sectors and the hyperbolic angle parameter. The functions cosh and sinh, used with rapidity as hyperbolic angle, provide the common representation of squeeze in the form $\begin{pmatrix} \cosh \phi & \sinh \phi \\ \sinh \phi & \cosh \phi \end{pmatrix}$, or as the split-complex unit $e^{j\phi} = \cosh \phi + j \sinh \phi$.

23.4   Mathematics of spacetimes

For physical reasons, a spacetime continuum is mathematically defined as a four-dimensional, smooth, connected Lorentzian manifold (M, g) . This means the smooth Lorentz metric g has signature $(3, 1)$. The metric determines the geometry of spacetime, as well as determining the geodesics of particles and light beams. About each point (event) on this manifold, coordinate charts are used to represent observers in reference frames. Usually, Cartesian coordinates (x, y, z, t) are used. Moreover, for simplicity's sake, units of measurement are usually chosen such that the speed of light c is equal to 1.

A reference frame (observer) can be identified with one of these coordinate charts; any such observer can describe any event p . Another reference frame may be identified by a second coordinate chart about p . Two observers (one in each reference frame) may describe the same event p but obtain different descriptions.

Usually, many overlapping coordinate charts are needed to cover a manifold. Given two coordinate charts, one containing p (representing an observer) and another containing q (representing another observer), the intersection of the charts represents the region of spacetime in which both observers can measure physical quantities and hence compare results. The relation between the two sets of measurements is given by a non-singular coordinate transformation on this intersection. The idea of coordinate charts as local observers who can perform measurements in their vicinity also makes good physical sense, as this is how one actually collects physical data—locally.

For example, two observers, one of whom is on Earth, but the other one who is on a fast rocket to Jupiter, may observe a comet crashing into Jupiter (this is the event p). In general, they will disagree about the exact location and timing of this impact, i.e., they will have different 4-tuples (x, y, z, t) (as they are using different coordinate systems). Although their kinematic descriptions will differ, dynamical (physical) laws, such as momentum conservation and the first law of thermodynamics, will still hold. In fact, relativity theory requires more than this in the sense that it stipulates these (and all other physical) laws must take the same form in all coordinate systems. This introduces tensors into relativity, by which all physical quantities are represented.

Geodesics are said to be time-like, null, or space-like if the tangent vector to one point of the geodesic is of this nature. Paths of particles and light beams in spacetime are represented by time-like and null (light-like) geodesics, respectively.

23.4.1 Topology

Main article: Spacetime topology

The assumptions contained in the definition of a spacetime are usually justified by the following considerations.

The connectedness assumption serves two main purposes. First, different observers making measurements (represented by coordinate charts) should be able to compare their observations on the non-empty intersection of the charts. If the connectedness assumption were dropped, this would not be possible. Second, for a manifold, the properties of connectedness and path-connectedness are equivalent, and one requires the existence of paths (in particular, geodesics) in the spacetime to represent the motion of particles and radiation.

Every spacetime is paracompact. This property, allied with the smoothness of the spacetime, gives rise to a smooth linear connection, an important structure in general relativity. Some important theorems on constructing spacetimes from compact and non-compact manifolds include the following:

- A compact manifold can be turned into a spacetime if, and only if, its Euler characteristic is 0. (Proof idea: the existence of a Lorentzian metric is shown to be equivalent to the existence of a nonvanishing vector field.)

- Any non-compact 4-manifold can be turned into a spacetime.[21]

23.4.2 Spacetime symmetries

Main article: Spacetime symmetries

Often in relativity, spacetimes that have some form of symmetry are studied. As well as helping to classify spacetimes, these symmetries usually serve as a simplifying assumption in specialized work. Some of the most popular ones include:

- Axisymmetric spacetimes

- Spherically symmetric spacetimes

- Static spacetimes

- Stationary spacetimes

23.4.3 Causal structure

Main article: Causal structure
See also: Causality (physics) and Causality

The causal structure of a spacetime describes causal relationships between pairs of points in the spacetime based on the existence of certain types of curves joining the points.

23.5 Spacetime in special relativity

Main article: Minkowski space

The geometry of spacetime in special relativity is described by the Minkowski metric on R^4. This spacetime is called Minkowski space. The Minkowski metric is usually denoted by η and can be written as a four-by-four matrix:

$$\eta_{ab} = \mathrm{diag}(1, -1, -1, -1)$$

where the Landau–Lifshitz time-like convention is being used. A basic assumption of relativity is that coordinate transformations must leave spacetime intervals invariant. Intervals are invariant under Lorentz transformations. This invariance property leads to the use of four-vectors (and other tensors) in describing physics.

Strictly speaking, one can also consider events in Newtonian physics as a single spacetime. This is Galilean–Newtonian relativity, and the coordinate systems are related by Galilean transformations. However, since these preserve spatial and temporal distances independently, such a spacetime can be decomposed into spatial coordinates plus temporal coordinates, which is not possible in the general case.

23.6 Spacetime in general relativity

In general relativity, it is assumed that spacetime is curved by the presence of matter (energy), this curvature being represented by the Riemann tensor. In special relativity, the Riemann tensor is identically zero, and so this concept of "non-curvedness" is sometimes expressed by the statement *Minkowski spacetime is flat.*

The earlier discussed notions of time-like, light-like and space-like intervals in special relativity can similarly be used to classify one-dimensional curves through curved spacetime. A time-like curve can be understood as one where the interval between any two infinitesimally close events on the curve is time-like, and likewise for light-like and space-like curves. Technically the three types of curves are usually defined in terms of whether the tangent vector at each point on the curve is time-like, light-like or space-like. The world line of a slower-than-light object will always be a time-like curve, the world line of a massless particle such as a photon will be a light-like curve, and a space-like curve could be the world line of a hypothetical tachyon. In the local neighborhood of any event, time-like curves that pass through the event will remain inside that event's past and future light cones, light-like curves that pass through the event will be on the surface of the light cones, and space-like curves that pass through the event will be outside the light cones. One can also define the notion of a three-dimensional "space-like hypersurface", a continuous three-dimensional "slice" through the four-dimensional property with the property that every curve that is contained entirely within this hypersurface is a space-like curve.[22]

Many spacetime continua have physical interpretations which most physicists would consider bizarre or unsettling. For example, a compact spacetime has closed timelike curves, which violate our usual ideas of causality (that is, future events could affect past ones). For this reason, mathematical physicists usually consider only restricted subsets of all the possible spacetimes. One way to do this is to study "realistic" solutions of the equations of general relativity. Another way is to add some additional "physically reasonable" but still fairly general geometric restrictions and try to prove interesting things about the resulting spacetimes. The latter approach has led to some important results, most notably the Penrose–Hawking singularity theorems.

23.7   Quantized spacetime

Main article: Quantum spacetime

In general relativity, spacetime is assumed to be smooth and continuous—and not just in the mathematical sense. In the theory of quantum mechanics, there is an inherent discreteness present in physics. In attempting to reconcile these two theories, it is sometimes postulated that spacetime should be quantized at the very smallest scales. Current theory is focused on the nature of spacetime at the Planck scale. Causal sets, loop quantum gravity, string theory, causal dynamical triangulation, and black hole thermodynamics all predict a quantized spacetime with agreement on the order of magnitude. Loop quantum gravity makes precise predictions about the geometry of spacetime at the Planck scale.

Spin networks provide a language to describe quantum geometry of space. Spin foam does the same job on spacetime. A spin network is a one-dimensional graph, together with labels on its vertices and edges which encodes aspects of a spatial geometry.

23.8   See also

- Anthropic_principle § Applications of the principle §§ Spacetime

- Basic introduction to the mathematics of curved spacetime

- Four-vector

- Frame-dragging

- Global spacetime structure

- Hole argument

- List of mathematical topics in relativity

- Local spacetime structure

- Lorentz invariance

- Manifold

- Mathematics of general relativity

- Metric space

- Philosophy of space and time

- Relativity of simultaneity

- Strip photography

- World manifold

23.9   References

[1] Ashby, Neil (2003). "Relativity in the Global Positioning System" (PDF). *Living Reviews in Relativity* **6**: 16. Bibcode:2003LRR.....6....1A. doi:10.12942/lrr-2003-1.

[2] Kopeikin, Sergei; Efroimsky, Michael; Kaplan, George (2011). *Relativistic Celestial Mechanics of the Solar System*. John Wiley & Sons. p. 157. ISBN 3527634576. Retrieved 2016-02-28. Extract of page 157

[3] Atuq Eusebio Manga Qespi, Instituto de lingüística y Cultura Amerindia de la Universidad de Valencia. *Pacha: un concepto andino de espacio y tiempo*. Revísta española de Antropología Americana, 24, p. 155–189. Edit. Complutense, Madrid. 1994

[4] Paul Richard Steele, Catherine J. Allen, *Handbook of Inca mythology*, p. 86, (ISBN 1-57607-354-8)

[5] Shirley Ardener, University of Oxford, *Women and space: ground rules and social maps*, p. 36 (ISBN 0-85496-728-1)

[6] Jean d'Alembert (1754) Dimension from ARTFL Encyclopedie project

[7] R.C. Archibald (1914) *Time as a fourth dimension Bulletin of the American Mathematical Society 20:409.*

[8] Daniel M. Siegel (2014) "Maxwell's contributions to electricity and magnetism", chapter 10 in *James Clerk Maxwell: Perspectives on his Life and Work*, Raymond Flood, Mark McCartney, Andrew Whitaker, editors, Oxford University Press ISBN 978-0-19-966437-5

[9] Pierre Duhem (1954) *The Aim and Structure of Physical Theory*, page 98, Princeton University Press

[10] Siegel 2014 p 191

[11] Minkowski, Hermann (1909), "Raum und Zeit", *Physikalische Zeitschrift* **10**: 75–88

- Various English translations on Wikisource: Space and Time.

[12] James Jeans (1947) The Growth of Physical Science, "Space-time", pp. 205–301, link from Internet Archive

[13] Matolcsi, Tamás (1994). *Spacetime Without Reference Frames*. Budapest: Akadémiai Kiadó.

[14] Ellis, G. F. R.; Williams, Ruth M. (2000). *Flat and curved space–times* (2nd ed.). Oxford University Press. p. 9. ISBN 0-19-850657-0.

[15] Petkov, Vesselin (2010). *Minkowski Spacetime: A Hundred Years Later*. Springer. p. 70. ISBN 90-481-3474-9. Retrieved 2016-02-28., Section 3.4, p. 70

[16] Note that the term *spacetime interval* is applied by several authors to the quantity s^2 and not to s. The reason that the quantity s^2 is used and not s is that s^2 can be positive, zero or negative, and is a more generally convenient and useful quantity than the Minkowski norm with a timelike/null/spacelike distinguisher: the pair $(\sqrt{|s^2|}, \text{sgn}(s^2))$. Despite the notation, it should not be regarded as the square of a number, but as a symbol. The cost for this convenience is that this "interval" is quadratic in linear separation along a straight line.

[17] More generally the spacetime interval in flat space can be written as $s^2 = g_{\alpha\beta}\Delta x^\alpha \Delta x^\beta$ with metric tensor g independent of spacetime position.

[18] This characterization is not universal: both the arcs between two points of a great circle on a sphere are geodesics.

[19] Berry, Michael V. (1989). *Principles of Cosmology and Gravitation*. CRC Press. p. 58. ISBN 0-85274-037-9. Retrieved 2016-02-28. Extract of page 58, caption of Fig. 25

[20] I. M. Yaglom (1979) *A Simple Non-Euclidean Geometry and its Physical Basis*, page 178, Springer, ISBN 0387-90332-1, MR 520230

[21] Geroch, Robert; Horowitz, Gary T. (1979). "Chapter 5. Global structure of spacetimes". In Hawking, S.W.; Israel, W. *General Relativity: An Einstein Centenary Survey*. Cambridge University Press. p. 219. ISBN 0521299284.

[22] See "Quantum Spacetime and the Problem of Time in Quantum Gravity" by Leszek M. Sokolowski, where on this page he writes "Each of these hypersurfaces is spacelike, in the sense that every curve, which entirely lies on one of such hypersurfaces, is a spacelike curve." More commonly a spacelike hypersurface is defined technically as a surface such that the normal vector at every point is time-like, but the definition above may be somewhat more intuitive.

23.10 Further reading

- Albert Einstein on Space-Time 13th edition Encyclopedia Britannica Historical: Albert Einstein's 1926 article

- Ehrenfest, Paul (1920) "How do the fundamental laws of physics make manifest that Space has 3 dimensions?" *Annalen der Physik 366*: 440.

- George F. Ellis and Ruth M. Williams (1992) *Flat and curved space–times*. Oxford Univ. Press. ISBN 0-19-851164-7

- Encyclopedia of Space-time and gravitation Scholarpedia Expert articles

23.11 External links

- http://universaltheory.org

- Barrow, John D.; Tipler, Frank J. (1988). *The Anthropic Cosmological Principle*. Oxford University Press. ISBN 978-0-19-282147-8. LCCN 87028148.

- Isenberg, J. A. (1981). "Wheeler–Einstein–Mach spacetimes". *Phys. Rev. D* **24** (2): 251–256. Bibcode:1981PhRvD..24..251I. doi:10.1103/PhysRevD.24.251.

- Kant, Immanuel (1929) "Thoughts on the true estimation of living forces" in J. Handyside, trans., *Kant's Inaugural Dissertation and Early Writings on Space*. Univ. of Chicago Press.

- Lorentz, H. A., Einstein, Albert, Minkowski, Hermann, and Weyl, Hermann (1952) *The Principle of Relativity: A Collection of Original Memoirs*. Dover.

- Lucas, John Randolph (1973) *A Treatise on Time and Space*. London: Methuen.

- Penrose, Roger (2004). *The Road to Reality*. Oxford: Oxford University Press. ISBN 0-679-45443-8. Chpts. 17–18.

- Poe, Edgar A. (1848). *Eureka; An Essay on the Material and Spiritual Universe*. Hesperus Press Limited. ISBN 1-84391-009-8.

- Robb, A. A. (1936). *Geometry of Time and Space*. University Press.

- Erwin Schrödinger (1950) *Space–time structure*. Cambridge Univ. Press.

- Schutz, J. W. (1997). *Independent axioms for Minkowski Space–time*. Addison-Wesley Longman. ISBN 0-582-31760-6.

- Tangherlini, F. R. (1963). "Schwarzschild Field in n Dimensions and the Dimensionality of Space Problem". *Nuovo Cimento* **14** (27): 636.

- Taylor, E. F.; Wheeler, John A. (1963). *Spacetime Physics*. W. H. Freeman. ISBN 0-7167-2327-1.

- Wells, H.G. (2004). *The Time Machine*. New York: Pocket Books. ISBN 0-671-57554-6. (pp. 5–6)

- Stanford Encyclopedia of Philosophy: "Space and Time: Inertial Frames" by Robert DiSalle.

Chapter 24

Winding number

The term winding number *may also refer to the rotation number of an iterated map.*

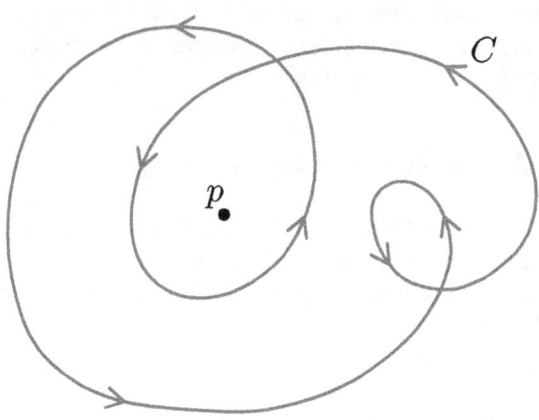

This curve has winding number two around the point p.

In mathematics, the **winding number** of a closed curve in the plane around a given point is an integer representing the total number of times that curve travels counterclockwise around the point. The winding number depends on the orientation of the curve, and is negative if the curve travels around the point clockwise.

Winding numbers are fundamental objects of study in algebraic topology, and they play an important role in vector calculus, complex analysis, geometric topology, differential geometry, and physics, including string theory.

24.1 Intuitive description

Suppose we are given a closed, oriented curve in the xy plane. We can imagine the curve as the path of motion of some object, with the orientation indicating the direction in which the object moves. Then the **winding number** of the curve is equal to the total number of counterclockwise turns that the object makes around the origin.

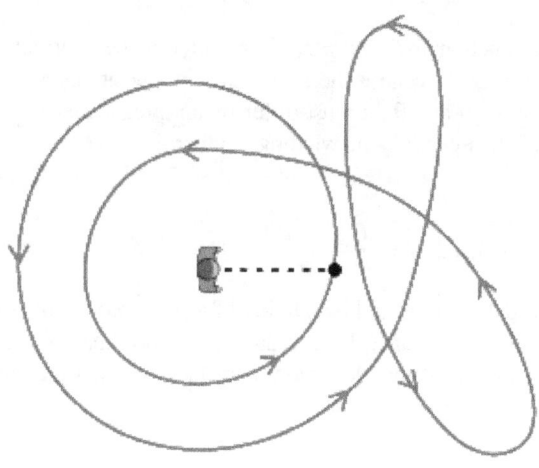

An object traveling along the red curve makes two counterclockwise turns around the person at the origin.

When counting the total number of turns, counterclockwise motion counts as positive, while clockwise motion counts as negative. For example, if the object first circles the origin four times counterclockwise, and then circles the origin once clockwise, then the total winding number of the curve is three.

Using this scheme, a curve that does not travel around the origin at all has winding number zero, while a curve that travels clockwise around the origin has negative winding number. Therefore, the winding number of a curve may be any integer. The following pictures show curves with winding numbers between −2 and 3:

24.2 Formal definition

A curve in the xy plane can be defined by parametric equations:

$$x = x(t) \quad \text{and} \quad y = y(t) \qquad \text{for} 0 \le t \le 1.$$

If we think of the parameter t as time, then these equations specify the motion of an object in the plane between $t = 0$ and $t = 1$. The path of this motion is a curve as long as the functions $x(t)$ and $y(t)$ are continuous. This curve is closed as long as the position of the object is the same at $t = 0$ and $t = 1$.

We can define the **winding number** of such a curve using the polar coordinate system. Assuming the curve does not pass through the origin, we can rewrite the parametric equations in polar form:

$$r = r(t) \quad \text{and} \quad \theta = \theta(t) \qquad \text{for} 0 \le t \le 1.$$

The functions $r(t)$ and $\theta(t)$ are required to be continuous, with $r > 0$. Because the initial and final positions are the same, $\theta(0)$ and $\theta(1)$ must differ by an integer multiple of 2π. This integer is the winding number:

$$\text{number winding} = \frac{\theta(1) - \theta(0)}{2\pi}.$$

This defines the winding number of a curve around the origin in the xy plane. By translating the coordinate system, we can extend this definition to include winding numbers around any point p.

24.3   Alternative definitions

Winding number is often defined in different ways in various parts of mathematics. All of the definitions below are equivalent to the one given above:

24.3.1   Alexander numbering

A simple combinatorial rule for defining the winding number was proposed by August Ferdinand Möbius in 1865[1] and again independently by James Waddell Alexander II in 1928.[2] Any curve partitions the plane into several connected regions, one of which is unbounded. The winding numbers of the curve around two points in the same region are equal. The winding number around (any point in) the unbounded region is zero. Finally, the winding numbers for any two adjacent regions differ by exactly 1; the region with the larger winding number appears on the left side of the curve.

24.3.2   Differential geometry

In differential geometry, parametric equations are usually assumed to be differentiable (or at least piecewise differen-

tiable). In this case, the polar coordinate θ is related to the rectangular coordinates x and y by the equation:

$$d\theta = \frac{1}{r^2}\left(x\,dy - y\,dx\right) \quad \text{where} r^2 = x^2 + y^2.$$

By the fundamental theorem of calculus, the total change in θ is equal to the integral of $d\theta$. We can therefore express the winding number of a differentiable curve as a line integral:

$$\text{number winding} = \frac{1}{2\pi} \oint_C \frac{x}{r^2}\,dy - \frac{y}{r^2}\,dx.$$

The one-form $d\theta$ (defined on the complement of the origin) is closed but not exact, and it generates the first de Rham cohomology group of the punctured plane. In particular, if ω is any closed differentiable one-form defined on the complement of the origin, then the integral of ω along closed loops gives a multiple of the winding number.

24.3.3   Complex analysis

In complex analysis, the winding number of a closed curve C in the complex plane can be expressed in terms of the complex coordinate $z = x + iy$. Specifically, if we write $z = re^{i\theta}$, then

$$dz = e^{i\theta}dr + ire^{i\theta}d\theta$$

and therefore

$$\frac{dz}{z} = \frac{dr}{r} + i\,d\theta = d[\ln r] + i\,d\theta.$$

The total change in $\ln(r)$ is zero, and thus the integral of dz/z is equal to i multiplied by the total change in θ. Therefore:

$$\text{number winding} = \frac{1}{2\pi i} \oint_C \frac{dz}{z}.$$

More generally, the winding number of C around any complex number a is given by

$$\frac{1}{2\pi i} \oint_C \frac{dz}{z - a}.$$

This is a special case of the famous Cauchy integral formula. Winding numbers play a very important role throughout complex analysis (c.f. the statement of the residue theorem).

24.3.4 Topology

In topology, the winding number is an alternate term for the degree of a continuous mapping. In physics, winding numbers are frequently called topological quantum numbers. In both cases, the same concept applies.

The above example of a curve winding around a point has a simple topological interpretation. The complement of a point in the plane is homotopy equivalent to the circle, such that maps from the circle to itself are really all that need to be considered. It can be shown that each such map can be continuously deformed to (is homotopic to) one of the standard maps $S^1 \to S^1 : s \mapsto s^n$, where multiplication in the circle is defined by identifying it with the complex unit circle. The set of homotopy classes of maps from a circle to a topological space form a group, which is called the first homotopy group or fundamental group of that space. The fundamental group of the circle is the group of the integers, **Z**; and the winding number of a complex curve is just its homotopy class.

Maps from the 3-sphere to itself are also classified by an integer which is also called the winding number or sometimes Pontryagin index.

24.3.5 Polygons

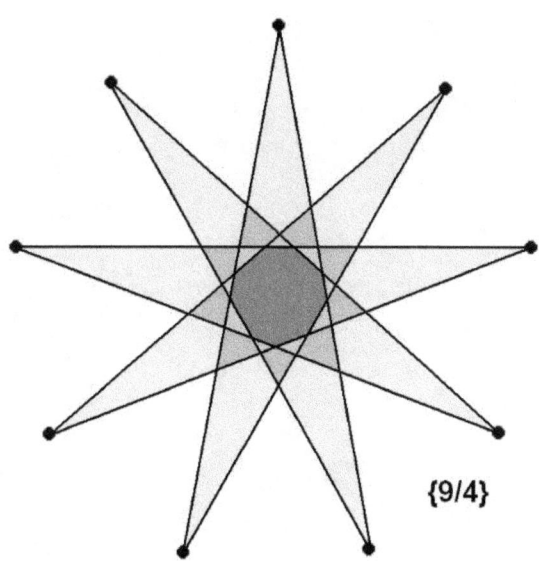

The boundary of the regular Enneagram {9/4} winds around its centre 4 times, so it has a density of 4.

In polygons, the winding number is referred to as the polygon density. For convex polygons, and more generally simple polygons (not self-intersecting), the density is 1, by the Jordan curve theorem. By contrast, for a regular star polygon $\{p/q\}$, the density is q.

24.4 Turning number

One can also consider the winding number of the path with respect to the tangent of the path itself. As a path followed through time, this would be the winding number with respect to the origin of the velocity vector. In this case the example illustrated at the beginning of this article has a winding number of 3, because the small loop *is* counted.

This is only defined for immersed paths (i.e., for differentiable paths with nowhere vanishing derivatives), and is the degree of the tangential Gauss map.

This is called the **turning number**, and can be computed as the total curvature divided by 2π.

24.5 Winding number and Heisenberg ferromagnet equations

Finally, note that the winding number is closely related with the $(2 + 1)$-dimensional continuous Heisenberg ferromagnet equations and its integrable extensions: the Ishimori equation etc. Solutions of the last equations are classified by the winding number or topological charge (topological invariant and/or topological quantum number).

24.6 See also

- Argument principle
- Linking coefficient
- Writhe
- Polygon density
- Residue theorem
- Topological degree theory
- Topological quantum number
- Wilson loop
- Nonzero-rule

24.7 References

[1] Möbius, August (1865). "Über der Bestimmung des In-
    haltes eines Polyëders". *Berichte über die Verhandlungen
    der Königlich Sächsischen Gesellschaft der Wissenschaften,
    Mathematisch-Physische Klasse* **17**: 31–68.

[2] Alexander, J. W. (April 1928). "Topological Invariants of
    Knots and Links". *Transactions of the American Mathemat-
    ical Society* **30** (2): 275–306. doi:10.2307/1989123.

24.8 External links

- Winding number at PlanetMath.org.

Chapter 25

Boson

For other uses, see Boson (disambiguation).

In quantum mechanics, a **boson** (/ˈboʊsɒn/,[1] /ˈboʊzɒn/[2])

Satyendra Nath Bose

is a particle that follows Bose–Einstein statistics. Bosons make up one of the two classes of particles, the other being fermions.[3] The name boson was coined by Paul Dirac[4] to commemorate the contribution of the Indian physicist Satyendra Nath Bose[5][6] in developing, with Einstein, Bose–Einstein statistics—which theorizes the characteristics of elementary particles.[7] Examples of bosons include fundamental particles such as photons, gluons, and W and Z bosons (the four force-carrying gauge bosons of the Standard Model), the recently discovered Higgs boson, and the hypothetical graviton of quantum gravity; composite particles (e.g. mesons and stable nuclei of even mass number such as deuterium (with one proton and one neutron, mass number = 2), helium-4, or lead-208[Note 1]); and some quasiparticles (e.g. Cooper pairs, plasmons, and phonons).[8]:130

An important characteristic of bosons is that their statistics do not restrict the number of them that occupy the same quantum state. This property is exemplified by helium-4 when it is cooled to become a superfluid.[9] Unlike bosons, two identical fermions cannot occupy the same quantum space. Whereas the elementary particles that make up matter (i.e. leptons and quarks) are fermions, the elementary bosons are force carriers that function as the 'glue' holding matter together.[10] This property holds for all particles with integer spin (s = 0, 1, 2 etc.) as a consequence of the spin–statistics theorem. When a gas of Bose particles is cooled down to temperatures very close to absolute zero then the kinetic energy of the particles decreases to a negligible amount and they condense into a lowest energy level state. This state is called Bose-Einstein condensation. It is believed that this property is the explanation of superfluidity.

25.1 Types

Bosons may be either elementary, like photons, or composite, like mesons.

While most bosons are composite particles, in the Standard Model there are five bosons which are elementary:

- the four gauge bosons ($\gamma \cdot g \cdot Z \cdot W\pm$)

- the only scalar boson (the Higgs boson (H0))

Additionally, the graviton (G) is a hypothetical elementary particle not incorporated in the Standard Model. If it exists, a graviton must be a boson, and could conceivably be a gauge boson.

Composite bosons are important in superfluidity and other applications of Bose–Einstein condensates. When a gas of Bose particles is cooled to temperatures very close to absolute zero its kinetic energy decreases up to a negligible amount then the particles would condense into the lowest energy state. This phenomenon is known as Bose-Einstein condensation and it is believed that this phenomenon is the secret behind superfluidity of liquids.

25.2 Properties

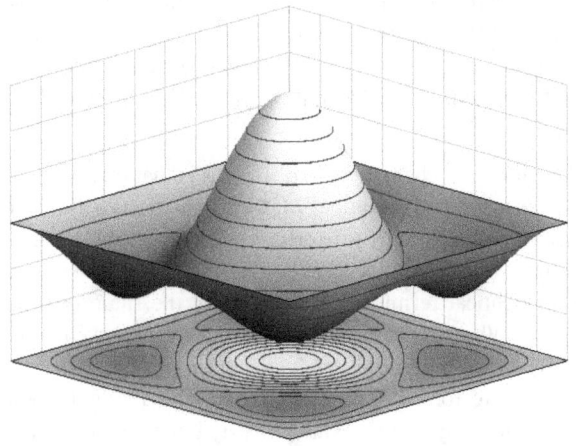

Symmetric wavefunction for a (bosonic) 2-particle state in an infinite square well potential.

Bosons differ from fermions, which obey Fermi–Dirac statistics. Two or more identical fermions cannot occupy the same quantum state (see Pauli exclusion principle).

Since bosons with the same energy can occupy the same place in space, bosons are often force carrier particles. Fermions are usually associated with matter (although in quantum physics the distinction between the two concepts is not clear cut).

Bosons are particles which obey Bose–Einstein statistics: when one swaps two bosons (of the same species), the wavefunction of the system is unchanged.[11] Fermions, on the other hand, obey Fermi–Dirac statistics and the Pauli exclusion principle: two fermions cannot occupy the same quantum state, resulting in a "rigidity" or "stiffness" of matter which includes fermions. Thus fermions are sometimes said to be the constituents of matter, while bosons are said to be the particles that transmit interactions (force carriers), or the constituents of radiation. The quantum fields of bosons are bosonic fields, obeying canonical commutation relations.

The properties of lasers and masers, superfluid helium-4 and Bose–Einstein condensates are all consequences of

statistics of bosons. Another result is that the spectrum of a photon gas in thermal equilibrium is a Planck spectrum, one example of which is black-body radiation; another is the thermal radiation of the opaque early Universe seen today as microwave background radiation. Interactions between elementary particles are called fundamental interactions. The fundamental interactions of virtual bosons with real particles result in all forces we know.

All known elementary and composite particles are bosons or fermions, depending on their spin: particles with half-integer spin are fermions; particles with integer spin are bosons. In the framework of nonrelativistic quantum mechanics, this is a purely empirical observation. However, in relativistic quantum field theory, the spin–statistics theorem shows that half-integer spin particles cannot be bosons and integer spin particles cannot be fermions.[12]

In large systems, the difference between bosonic and fermionic statistics is only apparent at large densities—when their wave functions overlap. At low densities, both types of statistics are well approximated by Maxwell–Boltzmann statistics, which is described by classical mechanics.

25.3 Elementary bosons

See also: List of particles § Bosons

All observed elementary particles are either fermions or bosons. The observed elementary bosons are all gauge bosons: photons, W and Z bosons, gluons, except the Higgs boson which is a scalar boson.

- Photons are the force carriers of the electromagnetic field.

- W and Z bosons are the force carriers which mediate the weak force.

- Gluons are the fundamental force carriers underlying the strong force.

- Higgs bosons give W and Z bosons mass via the Higgs mechanism. Their existence was confirmed by CERN on 14 March 2013.

Finally, many approaches to quantum gravity postulate a force carrier for gravity, the graviton, which is a boson of spin plus or minus two.

25.4 Composite bosons

See also: List of particles: Composite particles

Composite particles (such as hadrons, nuclei, and atoms) can be bosons or fermions depending on their constituents. More precisely, because of the relation between spin and statistics, a particle containing an even number of fermions is a boson, since it has integer spin.

Examples include the following:

- Any meson, since mesons contain one quark and one antiquark.

- The nucleus of a carbon-12 atom, which contains 6 protons and 6 neutrons.

- The helium-4 atom, consisting of 2 protons, 2 neutrons and 2 electrons.

The number of bosons within a composite particle made up of simple particles bound with a potential has no effect on whether it is a boson or a fermion.

25.5 To which states can bosons crowd?

Bose–Einstein statistics encourages identical bosons to crowd into one quantum state, but not any state is necessarily convenient for it. Aside of statistics, bosons can interact – for example, helium-4 atoms are repulsed by intermolecular force on a very close approach, and if one hypothesizes their condensation in a spatially-localized state, then gains from the statistics cannot overcome a prohibitive force potential. A spatially-delocalized state (i.e. with low $| \psi(x) |$) is preferable: if the number density of the condensate is about the same as in ordinary liquid or solid state, then the repulsive potential for the N-particle condensate in such state can be no higher than for a liquid or a crystalline lattice of the same N particles described without quantum statistics. Thus, Bose–Einstein statistics for a material particle is not a mechanism to bypass physical restrictions on the density of the corresponding substance, and superfluid liquid helium has the density comparable to the density of ordinary liquid matter. Spatially-delocalized states also permit for a low momentum according to uncertainty principle, hence for low kinetic energy; this is why superfluidity and superconductivity are usually observed in low temperatures.

Photons do not interact with themselves and hence do not experience this difference in states where to crowd (see squeezed coherent state).

25.6 See also

- Anyon

- Bose gas

- Identical particles

- Parastatistics

- Fermion

25.7 Notes

[1] Even-mass-number nuclides, which comprise $152/255 = \sim$ 60% of all stable nuclides, are bosons, i.e. they have integer spin. Almost all (148 of the 152) are even-proton, even-neutron (EE) nuclides, which necessarily have spin 0 because of pairing. The remainder of the stable bosonic nuclides are 5 odd-proton, odd-neutron stable nuclides (see even and odd atomic nuclei#Odd proton, odd neutron); these odd–odd bosons are: 2
1H, 6
3Li,10
5B, 14
7N and 180m
73Ta). All have nonzero integer spin.

25.8 References

[1] Wells, John C. (1990). *Longman pronunciation dictionary.* Harlow, England: Longman. ISBN 0582053838. entry "Boson"

[2] "boson". *Collins Dictionary.*

[3] Carroll, Sean (2007) *Dark Matter, Dark Energy: The Dark Side of the Universe*, Guidebook Part 2 p. 43, The Teaching Company, ISBN 1598033506 "...boson: A force-carrying particle, as opposed to a matter particle (fermion). Bosons can be piled on top of each other without limit. Examples include photons, gluons, gravitons, weak bosons, and the Higgs boson. The spin of a boson is always an integer, such as 0, 1, 2, and so on..."

[4] Notes on Dirac's lecture *Developments in Atomic Theory* at Le Palais de la Découverte, 6 December 1945, UK-NATARCHI Dirac Papers BW83/2/257889. See note 64 to p. 331 in "The Strangest Man" by Graham Farmelo

[5] Daigle, Katy (10 July 2012). "India: Enough about Higgs, let's discuss the boson". *AP News*. Retrieved 10 July 2012.

[6] Bal, Hartosh Singh (19 September 2012). "The Bose in the Boson". *New York Times blog*. Retrieved 21 September 2012.

[7] "Higgs boson: The poetry of subatomic particles". *BBC News*. 4 July 2012. Retrieved 6 July 2012.

[8] Charles P. Poole, Jr. (11 March 2004). *Encyclopedic Dictionary of Condensed Matter Physics*. Academic Press. ISBN 978-0-08-054523-3.

[9] "boson". *Merriam-Webster Online Dictionary*. Retrieved 21 March 2010.

[10] Carroll, Sean. "Explain it in 60 seconds: Bosons". *Symmetry Magazine*. Fermilab/SLAC. Retrieved 15 February 2013.

[11] Srednicki, Mark (2007). *Quantum Field Theory*, Cambridge University Press, pp. 28–29, ISBN 978-0-521-86449-7.

[12] Sakurai, J.J. (1994). *Modern Quantum Mechanics* (Revised Edition), p. 362. Addison-Wesley, ISBN 0-201-53929-2.

Chapter 26

Fermion

Enrico Fermi

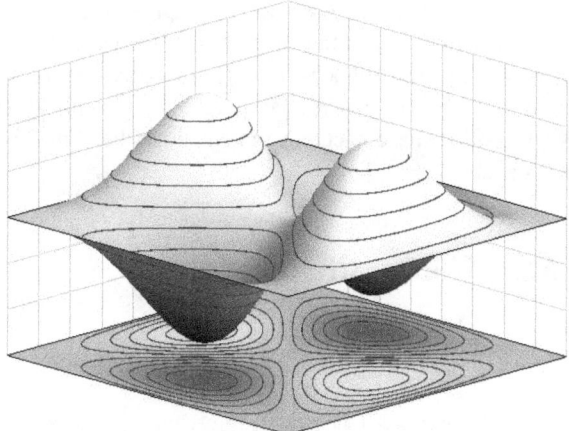

Antisymmetric wavefunction for a (fermionic) 2-particle state in an infinite square well potential.

In particle physics, a **fermion** (a name coined by Paul Dirac[1] from the surname of Enrico Fermi) is any particle characterized by Fermi–Dirac statistics. These particles obey the Pauli exclusion principle. Fermions include all quarks and leptons, as well as any composite particle made of an odd number of these, such as all baryons and many atoms and nuclei. Fermions differ from bosons, which obey Bose–Einstein statistics.

A fermion can be an elementary particle, such as the electron, or it can be a composite particle, such as the proton. According to the spin-statistics theorem in any reasonable relativistic quantum field theory, particles with integer spin are bosons, while particles with half-integer spin are fermions.

Besides this spin characteristic, fermions have another spe-

cific property: they possess conserved baryon or lepton quantum numbers. Therefore, what is usually referred as the spin statistics relation is in fact a spin statistics-quantum number relation.[2]

As a consequence of the Pauli exclusion principle, only one fermion can occupy a particular quantum state at any given time. If multiple fermions have the same spatial probability distribution, then at least one property of each fermion, such as its spin, must be different. Fermions are usually associated with matter, whereas bosons are generally force carrier particles, although in the current state of particle physics the distinction between the two concepts is unclear. Weakly interacting fermions can also display bosonic behavior under extreme conditions. At low temperature fermions show superfluidity for uncharged particles and superconductivity for charged particles.

Composite fermions, such as protons and neutrons, are the key building blocks of everyday matter.

26.1 Elementary fermions

The Standard Model recognizes two types of elementary fermions: quarks and leptons. In all, the model distinguishes 24 different fermions. There are six quarks (up, down, strange, charm, bottom and top quarks), and six leptons (electron, electron neutrino, muon, muon neutrino, tau particle and tau neutrino), along with the corresponding antiparticle of each of these.

Mathematically, fermions come in three types - Weyl fermions (massless), Dirac fermions (massive), and Majorana fermions (each its own antiparticle). Most Standard Model fermions are believed to be Dirac fermions, although it is unknown at this time whether the neutrinos are Dirac or Majorana fermions. Dirac fermions can be treated as a combination of two Weyl fermions.[3]:106 In July 2015, Weyl fermions have been experimentally realized in Weyl semimetals.

26.2 Composite fermions

See also: List of particles § Composite particles

Composite particles (such as hadrons, nuclei, and atoms) can be bosons or fermions depending on their constituents. More precisely, because of the relation between spin and statistics, a particle containing an odd number of fermions is itself a fermion. It will have half-integer spin.

Examples include the following:

- A baryon, such as the proton or neutron, contains three fermionic quarks and thus it is a fermion.

- The nucleus of a carbon-13 atom contains six protons and seven neutrons and is therefore a fermion.

- The atom helium-3 (^{3}He) is made of two protons, one neutron, and two electrons, and therefore it is a fermion.

The number of bosons within a composite particle made up of simple particles bound with a potential has no effect on whether it is a boson or a fermion.

Fermionic or bosonic behavior of a composite particle (or system) is only seen at large (compared to size of the system) distances. At proximity, where spatial structure begins to be important, a composite particle (or system) behaves according to its constituent makeup.

Fermions can exhibit bosonic behavior when they become loosely bound in pairs. This is the origin of superconductivity and the superfluidity of helium-3: in superconducting materials, electrons interact through the exchange of phonons, forming Cooper pairs, while in helium-3, Cooper pairs are formed via spin fluctuations.

The quasiparticles of the fractional quantum Hall effect are also known as composite fermions, which are electrons with an even number of quantized vortices attached to them.

26.2.1 Skyrmions

Main article: Skyrmion

In a quantum field theory, there can be field configurations of bosons which are topologically twisted. These are coherent states (or solitons) which behave like a particle, and they can be fermionic even if all the constituent particles are bosons. This was discovered by Tony Skyrme in the early 1960s, so fermions made of bosons are named skyrmions after him.

Skyrme's original example involved fields which take values on a three-dimensional sphere, the original nonlinear sigma model which describes the large distance behavior of pions. In Skyrme's model, reproduced in the large N or string approximation to quantum chromodynamics (QCD), the proton and neutron are fermionic topological solitons of the pion field.

Whereas Skyrme's example involved pion physics, there is a much more familiar example in quantum electrodynamics with a magnetic monopole. A bosonic monopole with the smallest possible magnetic charge and a bosonic version of the electron will form a fermionic dyon.

The analogy between the Skyrme field and the Higgs field of the electroweak sector has been used[4] to postulate that all fermions are skyrmions. This could explain why all known fermions have baryon or lepton quantum numbers and provide a physical mechanism for the Pauli exclusion principle.

26.3 See also

26.4 Notes

[1] Notes on Dirac's lecture *Developments in Atomic Theory* at Le Palais de la Découverte, 6 December 1945, UK-NATARCHI Dirac Papers BW83/2/257889. See note 64 on page 331 in "The Strangest Man: The Hidden Life of Paul Dirac, Mystic of the Atom" by Graham Farmelo

[2] Physical Review D volume 87, page 0550003, year 2013, author Weiner, Richard M., title "Spin-statistics-quantum number connection and supersymmetry" arxiv:1302.0969

[3] T. Morii; C. S. Lim; S. N. Mukherjee (1 January 2004). *The Physics of the Standard Model and Beyond*. World Scientific. ISBN 978-981-279-560-1.

[4] Weiner, Richard M. (2010). "The Mysteries of Fermions". *International Journal of Theoretical Physics* **49** (5): 1174–1180.   arXiv:0901.3816.   Bibcode:2010IJTP...49.1174W. doi:10.1007/s10773-010-0292-7.

Chapter 27

Kaluza–Klein theory

This article is about gravitation and electromagnetism. For the mathematical generalization of K theory, see KK-theory.

In physics, **Kaluza–Klein theory** (**KK theory**) is a unified field theory of gravitation and electromagnetism built around the idea of a fifth dimension beyond the usual four of space and time. It is considered to be an important precursor to string theory.

The five-dimensional theory was developed in three steps. The original hypothesis came from Theodor Kaluza, who sent his results to Einstein in 1919,[1] and published them in 1921.[2] Kaluza's theory was a purely classical extension of general relativity to five dimensions. The five-dimensional metric has 15 components. Ten components are identified with the four-dimensional spacetime metric, four components with the electromagnetic vector potential, and one component with an unidentified scalar field sometimes called the "radion" or the "dilaton". Correspondingly, the five-dimensional Einstein equations yield the four-dimensional Einstein field equations, the Maxwell equations for the electromagnetic field, and an equation for the scalar field. Kaluza also introduced the hypothesis known as the "cylinder condition", that no component of the five-dimensional metric depends on the fifth dimension. Without this assumption, the field equations of five-dimensional relativity are enormously more complex. Standard four-dimensional physics seems to manifest the cylinder condition. Kaluza also set the scalar field equal to a constant, in which case standard general relativity and electrodynamics are recovered identically.

In 1926, Oskar Klein gave Kaluza's classical five-dimensional theory a quantum interpretation,[3][4] to accord with the then-recent discoveries of Heisenberg and Schrödinger. Klein introduced the hypothesis that the fifth dimension was curled up and microscopic, to explain the cylinder condition. Klein also calculated a scale for the fifth dimension based on the quantum of charge.

It wasn't until the 1940s that the classical theory was completed, and the full field equations including the scalar field were obtained by three independent research groups:[5] Thiry,[6][7][8] working in France on his dissertation under Lichnerowicz; Jordan, Ludwig, and Müller in Germany,[9][10][11][12][13] with critical input from Pauli and Fierz; and Scherrer [14][15][16] working alone in Switzerland. Jordan's work led to the scalar-tensor theory of Brans & Dicke;[17] Brans and Dicke were apparently unaware of Thiry or Scherrer. The full Kaluza equations under the cylinder condition are quite complex, and most English-language reviews as well as the English translations of Thiry contain some errors. The complete Kaluza equations were evaluated using tensor algebra software in 2015.[18]

27.1 Kaluza hypothesis

In his 1921 paper,[2] Kaluza established all the elements of the classical five-dimensional theory: the metric, the field equations, the equations of motion, the stress-energy tensor, and the cylinder condition. The theory has no free parameters; it merely extends general relativity to five dimensions. One starts by hypothesizing a form of the five-dimensional metric $\widetilde{g}_{ab}$, where Roman indices span five dimensions. Let one also introduce the four-dimensional spacetime metric $g_{\mu\nu}$, where Greek indices span the usual four dimensions of space and time; a 4-vector A^{μ} which will be identified with the electromagnetic vector potential; and a scalar field ϕ. Then decompose the 5D metric so that the 4D metric is framed by the electromagnetic vector potential, with the scalar field at the fifth diagonal. This can be visualized as:

$$\widetilde{g}_{ab} \equiv \begin{bmatrix} g_{\mu\nu} + \phi^2 A_\mu A_\nu & \phi^2 A_\mu \\ \phi^2 A_\nu & \phi^2 \end{bmatrix}.$$

More precisely, one can write

$$\widetilde{g}_{\mu\nu} \equiv g_{\mu\nu} + \phi^2 A_\mu A_\nu, \qquad \widetilde{g}_{5\nu} \equiv \widetilde{g}_{\nu 5}$$

$$\equiv \phi^2 A_\nu, \qquad \widetilde{g}_{55} \equiv \phi^2$$

where the index 5 indicates the fifth coordinate by convention even though the first four coordinates are indexed with 0, 1, 2, and 3. The associated inverse metric is

$$\widetilde{g}^{ab} \equiv \begin{bmatrix} g^{\mu\nu} & -A^{\mu} \\ -A^{\nu} & g_{\alpha\beta}A^{\alpha}A^{\beta} + \frac{1}{\phi^2} \end{bmatrix}.$$

So far, this decomposition is quite general and all terms are dimensionless. Kaluza then applies the machinery of standard general relativity to this metric. The field equations are obtained from five-dimensional Einstein equations, and the equations of motion are obtained from the five-dimensional geodesic hypothesis. The resulting field equations provide both the equations of general relativity and of electrodynamics; the equations of motion provide the four-dimensional geodesic equation and the Lorentz force law, and one finds that electric charge is identified with motion in the fifth dimension.

The hypothesis for the metric implies an invariant five-dimensional length element ds :

$$ds^2 \equiv \widetilde{g}_{ab}dx^a dx^b = g_{\mu\nu}dx^\mu dx^\nu + \phi^2 (A_\nu dx^\nu + dx^5)^2$$

27.2 Field equations from the Kaluza hypothesis

The field equations of the 5-dimensional theory were never adequately provided by Kaluza or Klein, mainly regarding the scalar field. The full Kaluza field equations are generally attributed to Thiry,[7] who most famously obtained vacuum field equations, although Kaluza [2] originally provided a stress-energy tensor for his theory and Thiry included a stress-energy tensor in his thesis. But as described by Gonner,[5] several independent groups worked on the field equations in the 1940s and earlier. Thiry is perhaps best known only because an English translation was provided by Applequist, Chodos, & Freund in their review book.[19] Applequist et al. also provided an English translation of Kaluza's paper. There are no English translations of the Jordan papers.[9][10][12]

To obtain the 5D field equations, the 5D connections $\widetilde{\Gamma}^a_{bc}$ are calculated from the 5D metric $\widetilde{g}_{ab}$, and the 5D Ricci tensor $\widetilde{R}_{ab}$ is calculated from the 5D connections.

The classic results of Thiry and other authors presume the cylinder condition:

$$\frac{\partial \widetilde{g}_{ab}}{\partial x^5} = 0$$

Without this assumption, the field equations become much more complex, providing many more degrees of freedom that can be identified with various new fields. Paul Wesson and colleagues have pursued relaxation of the cylinder condition to gain extra terms that can be identified with the matter fields,[20] for which Kaluza [2] otherwise inserted a stress-energy tensor by hand.

It has been an objection to the original Kaluza hypothesis to invoke the fifth dimension only to negate its dynamics. But Thiry argued [5] that the interpretation of the Lorentz force law in terms of a 5-dimensional geodesic mitigates strongly for a fifth dimension irrespective of the cylinder condition. Most authors have therefore employed the cylinder condition in deriving the field equations. Furthermore, vacuum equations are typically assumed for which

$$\widetilde{R}_{ab} = 0$$

where

$$\widetilde{R}_{ab} \equiv \partial_c \widetilde{\Gamma}^c_{ab} - \partial_b \widetilde{\Gamma}^c_{ca} + \widetilde{\Gamma}^c_{cd}\widetilde{\Gamma}^d_{ab} - \widetilde{\Gamma}^c_{bd}\widetilde{\Gamma}^d_{ac}$$

and

$$\widetilde{\Gamma}^a_{bc} \equiv \frac{1}{2}\widetilde{g}^{ad}(\partial_b \widetilde{g}_{dc} + \partial_c \widetilde{g}_{db} - \partial_d \widetilde{g}_{bc})$$

The vacuum field equations obtained in this way by Thiry [7] and Jordan's group [9][10][12] are as follows.

The field equation for ϕ is obtained from

$$\widetilde{R}_{55} = 0 \Rightarrow \Box\phi = \frac{1}{4}\phi^3 F^{\alpha\beta}F_{\alpha\beta}$$

where $F_{\alpha\beta} \equiv \partial_\alpha A_\beta - \partial_\beta A_\alpha$, where $\Box \equiv g^{\mu\nu}\nabla_\mu \nabla_\nu$, and where ∇_μ is a standard, 4D covariant derivative. It shows that the electromagnetic field is a source for the scalar field. Note that the scalar field cannot be set to a constant without constraining the electromagnetic field. The earlier treatments by Kaluza and Klein did not have an adequate description of the scalar field, and did not realize the implied constraint on the electromagnetic field by assuming the scalar field to be constant.

The field equation for A^ν is obtained from

$$\widetilde{R}_{5\alpha} = 0 = \frac{1}{2}g^{\beta\mu}\nabla_\mu(\phi^3 F_{\alpha\beta})$$

It has the form of the vacuum Maxwell equations if the scalar field is constant.

The field equation for the 4D Ricci tensor $R_{\mu\nu}$ is obtained from

$$\widetilde{R}_{\mu\nu} - \frac{1}{2}\widetilde{g}_{\mu\nu}\widetilde{R} = 0 \Rightarrow R_{\mu\nu} - \frac{1}{2}g_{\mu\nu}R = \frac{1}{2}\phi^2 \left(g^{\alpha\beta} \right.$$

$$\left. F_{\mu\alpha}F_{\nu\beta} - \frac{1}{4}g_{\mu\nu}F_{\alpha\beta}F^{\alpha\beta} \right) + \frac{1}{\phi}\left(\nabla_\mu\nabla_\nu\phi - g_{\mu\nu}\Box\phi \right)$$

where R is the standard 4D Ricci scalar.

This equation shows the remarkable result, called the "Kaluza miracle", that the precise form for the electromagnetic stress-energy tensor emerges from the 5D vacuum equations as a source in the 4D equations: field from the vacuum. This relation allows the definitive identification of A^μ with the electromagnetic vector potential. Therefore, the field needs to be rescaled with a conversion constant k such that $A^\mu \to kA^\mu$:

The relation above shows that we must have

$$\frac{k^2}{2} = \frac{8\pi G}{c^4}\frac{1}{\mu_0} = \frac{2G}{c^2}4\pi\epsilon_0$$

where G is the gravitational constant and μ_0 is the permeability of free space. In the Kaluza theory, the gravitational constant can be understood as an electromagnetic coupling constant in the metric. There is also a stress-energy tensor for the scalar field. The scalar field behaves like a variable gravitational constant, in terms of modulating the coupling of electromagnetic stress energy to spacetime curvature. The sign of ϕ^2 in the metric is fixed by correspondence with 4D theory so that electromagnetic energy densities are positive. This turns out to imply that the 5th coordinate is spacelike in its signature in the metric.

In the presence of matter, the 5D vacuum condition can not be assumed. Indeed, Kaluza did not assume it. The full field equations require evaluation of the 5D Einstein tensor

$$\widetilde{G}_{ab} \equiv \widetilde{R}_{ab} - \frac{1}{2}\widetilde{g}_{ab}\widetilde{R}$$

as seen in the recovery of the electromagnetic stress-energy tensor above. The 5D curvature tensors are complex, and most English-language reviews contain errors in either $\widetilde{G}_{ab}$ or $\widetilde{R}_{ab}$, as does the English translation of.[7] See [18] for a complete set of 5D curvature tensors under the cylinder condition, evaluated using tensor algebra software.

27.3 Equations of motion from the Kaluza hypothesis

The equations of motion are obtained from the five-dimensional geodesic hypothesis [2] in terms of a 5-velocity $\widetilde{U}^a \equiv dx^a/ds$:

$$\widetilde{U}^b\widetilde{\nabla}_b\widetilde{U}^a = \frac{d\widetilde{U}^a}{ds} + \widetilde{\Gamma}^a_{bc}\widetilde{U}^b\widetilde{U}^c = 0$$

This equation can be recast in several ways, and it has been studied in various forms by authors including Kaluza,[2] Pauli,[21] Gross & Perry,[22] Gegenberg & Kunstatter,[23] and Wesson & Ponce de Leon,[24] but it is instructive to convert it back to the usual 4-dimensional length element $c^2 d\tau^2 \equiv g_{\mu\nu}dx^\mu dx^\nu$, which is related to the 5-dimensional length element ds as given above:

$$ds^2 = c^2 d\tau^2 + \phi^2(kA_\nu dx^\nu + dx^5)^2$$

Then the 5D geodesic equation can be written [25] for the spacetime components of the 4velocity, $U^\nu \equiv dx^\nu/d\tau$: $\frac{dU^\nu}{d\tau} + \widetilde{\Gamma}^\mu_{\alpha\beta}U^\alpha U^\beta + 2\widetilde{\Gamma}^\mu_{5\alpha}U^\alpha U^5 + \widetilde{\Gamma}^\mu_{55}(U^5)^2 + U^\mu\frac{d}{d\tau}\ln\left(\frac{cd\tau}{ds}\right) = 0$

The term quadratic in U^ν provides the 4D geodesic equation plus some electromagnetic terms:

$$\widetilde{\Gamma}^\mu_{\alpha\beta} = \Gamma^\mu_{\alpha\beta} + \frac{1}{2}g^{\mu\nu}k^2\phi^2(A_\alpha F_{\beta\nu} + A_\beta F_{\alpha\nu} + A_\alpha A_\beta \partial_\nu \ln\phi^2)$$

The term linear in U^ν provides the Lorentz force law:

$$\widetilde{\Gamma}^\mu_{5\alpha} = \frac{1}{2}g^{\mu\nu}k\phi^2(F_{\alpha\nu} - A_\alpha \partial_\nu \ln\phi^2)$$

This is another expression of the "Kaluza miracle". The same hypothesis for the 5D metric that provides electromagnetic stress-energy in the Einstein equations, also provides the Lorentz force law in the equation of motions along with the 4D geodesic equation. Yet correspondence with the Lorentz force law requires that we identify the component of 5-velocity along the 5th dimension with electric charge:

$$kU^5 = k\frac{dx^5}{d\tau} \to \frac{q}{mc}$$

where m is particle mass and q is particle electric charge. Thus, electric charge is understood as motion along the 5th dimension. The fact that the Lorentz force law could be understood as a geodesic in 5 dimensions was to Kaluza a primary motivation for considering the 5-dimensional hypothesis, even in the presence of the aesthetically-unpleasing cylinder condition.

Yet there is a problem: the term quadratic in U^5.

$$\widetilde{\Gamma}^\mu_{55} = -\frac{1}{2}g^{\mu\alpha}\partial_\alpha\phi^2$$

If there is no gradient in the scalar field, the term quadratic in U^5 vanishes. But otherwise the expression above implies

$$U^5 \sim c \frac{q/m}{G^{1/2}}$$

For elementary particles, $U^5 > 10^{20} c$. The term quadratic in U^5 should dominate the equation, perhaps in contradiction to experience. This was the main shortfall of the 5-dimensional theory as Kaluza saw it,[2] and he gives it some discussion in his original article.

The equation of motion for U^5 is particularly simple under the cylinder condition. Start with the alternate form of the geodesic equation, written for the covariant 5-velocity:

$$\frac{d\widetilde{U}_a}{ds} = \frac{1}{2}\widetilde{U}^b \widetilde{U}^c \frac{\partial \widetilde{g}_{bc}}{\partial x^a}$$

This means that under the cylinder condition, $\widetilde{U}_5$ is a constant of the 5-dimensional motion:

$$\widetilde{U}_5 = \widetilde{g}_{5a}\widetilde{U}^a = \phi^2 \frac{cd\tau}{ds}(kA_\nu U^\nu + U^5) = \text{constant}$$

27.4 Kaluza's hypothesis for the matter stress-energy tensor

Kaluza [2] proposed a 5D matter stress tensor $\widetilde{T}_M^{ab}$ of the form

$$\widetilde{T}_M^{ab} = \rho \frac{dx^a}{ds}\frac{dx^b}{ds}$$

where ρ is a density and the length element ds is as defined above.

Then, the spacetime component gives a typical "dust" stress energy tensor:

$$\widetilde{T}_M^{\mu\nu} = \rho \frac{dx^\mu}{ds}\frac{dx^\nu}{ds}$$

The mixed component provides a 4-current source for the Maxwell equations:

$$\widetilde{T}_M^{5\mu} = \rho \frac{dx^\mu}{ds}\frac{dx^5}{ds} = \rho U^\mu \frac{q}{kmc}$$

Just as the five-dimensional metric comprises the 4-D metric framed by the electromagnetic vector potential, the 5-dimensional stress-energy tensor comprises the 4-D stress-energy tensor framed by the vector 4-current.

27.5 Quantum interpretation of Klein

Kaluza's original hypothesis was purely classical and extended discoveries of general relativity. By the time of Klein's contribution, the discoveries of Heisenberg, Schroedinger, and de Broglie were receiving a lot of attention. Klein's *Nature* paper [4] suggested that the fifth dimension is closed and periodic, and that the identification of electric charge with motion in the fifth dimension be interpreted as standing waves of wavelength λ^5, much like the electrons around a nucleus in the Bohr model of the atom. The quantization of electric charge could then be nicely understood in terms of integer multiples of fifth-dimensional momentum. Combining the previous Kaluza result for U^5 in terms of electric charge, and a de Broglie relation for momentum $p^5 = h/\lambda^5$, Klein [4] obtained an expression for the 0th mode of such waves:

$$mU^5 = \frac{cq}{G^{1/2}} = \frac{h}{\lambda^5} \to \lambda^5 \sim \frac{hG^{1/2}}{cq}$$

where h is the Planck constant. Klein found $\lambda^5 \sim 10^{-30}$ cm, and thereby an explanation for the cylinder condition in this small value.

Klein's *Zeitschrift für Physik* paper of the same year,[3] gave a more-detailed treatment that explicitly invoked the techniques of Schroedinger and de Broglie. It recapitulated much of the classical theory of Kaluza described above, and then departed into Klein's quantum interpretation. Klein solved a Schroedinger-like wave equation using an expansion in terms of fifth-dimensional waves resonating in the closed, compact fifth dimension.

27.6 Quantum field theory interpretation

27.7 Group theory interpretation

A splitting of five-dimensional spacetime into the Einstein equations and Maxwell equations in four dimensions was first discovered by Gunnar Nordström in 1914, in the context of his theory of gravity, but subsequently forgotten. Kaluza published his derivation in 1921 as an attempt to unify electromagnetism with Einstein's general relativity.

In 1926, Oskar Klein proposed that the fourth spatial dimension is curled up in a circle of a very small radius, so that a particle moving a short distance along that axis would return to where it began. The distance a particle can travel

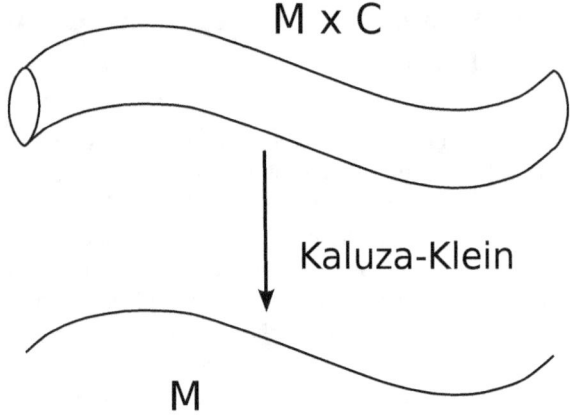

The space M × C *is compactified over the compact set* C, *and after Kaluza–Klein decomposition one has an effective field theory over* M.

before reaching its initial position is said to be the size of the dimension. This extra dimension is a compact set, and the phenomenon of having a space-time with compact dimensions is referred to as compactification.

In modern geometry, the extra fifth dimension can be understood to be the circle group $U(1)$, as electromagnetism can essentially be formulated as a gauge theory on a fiber bundle, the circle bundle, with gauge group $U(1)$. In Kaluza–Klein theory this group suggests that gauge symmetry is the symmetry of circular compact dimensions. Once this geometrical interpretation is understood, it is relatively straightforward to replace $U(1)$ by a general Lie group. Such generalizations are often called Yang–Mills theories. If a distinction is drawn, then it is that Yang–Mills theories occur on a flat space-time, whereas Kaluza–Klein treats the more general case of curved spacetime. The base space of Kaluza–Klein theory need not be four-dimensional space-time; it can be any (pseudo-)Riemannian manifold, or even a supersymmetric manifold or orbifold or even a noncommutative space.

The construction can be outlined, roughly, as follows.[26] One starts by considering a principle fiber bundle P with gauge group G over a manifold M. Given a connection on the bundle, and a metric on the base manifold, and a gauge invariant metric on the tangent of each fiber, one can construct a bundle metric defined on the entire bundle. Computing the scalar curvature of this bundle metric, one finds that it constant on each fiber: this is the "Kaluza miracle". One did not have to explicitly impose a cylinder condition, or to compactify: by assumption, the gauge group is already compact. Next, one takes this scalar curvature as the Lagrangian density, and, from this, constructs the Einstein–Hilbert action for the bundle, as a whole. The equations of motion, the Euler–Lagrange equations, can be then obtained by considering where the action is stationary with respect to variations of either the metric on the base manifold, or of the gauge connection. Variations with respect to the base metric gives the Einstein field equations on the base manifold, with the energy-momentum tensor given by the curvature (field strength) of the gauge connection. On the flip side, the action is stationary against variations of the gauge connection precisely when the gauge connection solves the Yang-Mills equations. Thus, by applying a single idea: the principle of least action, to a single quantity: the scalar curvature on the bundle (as a whole), one obtains simultaneously all of the needed field equations, for both the space-time and the gauge field.

As an approach to the unification of the forces, it is straightforward to apply the Kaluza–Klein theory in an attempt to unify gravity with the strong and electroweak forces by using the symmetry group of the Standard Model, $SU(3) \times SU(2) \times U(1)$. However, an attempt to convert this interesting geometrical construction into a bona-fide model of reality flounders on a number of issues, including the fact that the fermions must be introduced in an artificial way (in nonsupersymmetric models). Nonetheless, KK remains an important touchstone in theoretical physics and is often embedded in more sophisticated theories. It is studied in its own right as an object of geometric interest in K-theory.

Even in the absence of a completely satisfying theoretical physics framework, the idea of exploring extra, compactified, dimensions is of considerable interest in the experimental physics and astrophysics communities. A variety of predictions, with real experimental consequences, can be made (in the case of large extra dimensions and warped models). For example, on the simplest of principles, one might expect to have standing waves in the extra compactified dimension(s). If a spatial extra dimension is of radius R, the invariant mass of such standing waves would be $Mn = nh/Rc$ with n an integer, h being Planck's constant and c the speed of light. This set of possible mass values is often called the **Kaluza–Klein tower**. Similarly, in Thermal quantum field theory a compactification of the euclidean time dimension leads to the Matsubara frequencies and thus to a discretized thermal energy spectrum.

However, Klein's approach to a quantum theory is flawed and, for example, leads to a calculated electron mass of $3 \times (10^{30})$ MeV instead of the measured value 0.511 MeV.

Examples of experimental pursuits include work by the CDF collaboration, which has re-analyzed particle collider data for the signature of effects associated with large extra dimensions/warped models.

Brandenberger and Vafa have speculated that in the early universe, cosmic inflation causes three of the space dimensions to expand to cosmological size while the remaining dimensions of space remained microscopic.

27.8   Space-time-matter theory

One particular variant of Kaluza–Klein theory is **space-time-matter theory** or **induced matter theory**, chiefly promulgated by Paul Wesson and other members of the so-called Space-Time-Matter Consortium.[27] In this version of the theory, it is noted that solutions to the equation

$$\widetilde{R}_{ab} = 0$$

may be re-expressed so that in four dimensions, these solutions satisfy Einstein's equations

$$G_{\mu\nu} = 8\pi T_{\mu\nu}$$

with the precise form of the $T\mu\nu$ following from the Ricci-flat condition on the five-dimensional space. In other words, the cylinder condition of the previous development is dropped, and the stress-energy now comes from the derivatives of the 5D metric with respect to the fifth coordinate. Since the energy–momentum tensor is normally understood to be due to concentrations of matter in four-dimensional space, the above result is interpreted as saying that four-dimensional matter is induced from geometry in five-dimensional space.

In particular, the soliton solutions of $\widetilde{R}_{ab} = 0$ can be shown to contain the Friedmann–Lemaître–Robertson–Walker metric in both radiation-dominated (early universe) and matter-dominated (later universe) forms. The general equations can be shown to be sufficiently consistent with classical tests of general relativity to be acceptable on physical principles, while still leaving considerable freedom to also provide interesting cosmological models.

27.9   Geometric interpretation

The Kaluza–Klein theory has a particularly elegant presentation in terms of geometry. In a certain sense, it looks just like ordinary gravity in free space, except that it is phrased in five dimensions instead of four.

27.9.1   Einstein equations

The equations governing ordinary gravity in free space can be obtained from an action, by applying the variational principle to a certain action. Let M be a (pseudo-)Riemannian manifold, which may be taken as the spacetime of general relativity. If g is the metric on this manifold, one defines the action $S(g)$ as

$$S(g) = \int_M R(g)\mathrm{vol}(g)$$

where $R(g)$ is the scalar curvature and vol(g) is the volume element. By applying the variational principle to the action

$$\frac{\delta S(g)}{\delta g} = 0$$

one obtains precisely the Einstein equations for free space:

$$R_{ij} - \frac{1}{2}g_{ij}R = 0$$

Here, *Rij* is the Ricci tensor.

27.9.2   Maxwell equations

By contrast, the Maxwell equations describing electromagnetism can be understood to be the Hodge equations of a principal U(1)-bundle or circle bundle π: $P \to M$ with fiber U(1). That is, the electromagnetic field F is a harmonic 2-form in the space $\Omega^2(M)$ of differentiable 2-forms on the manifold M. In the absence of charges and currents, the free-field Maxwell equations are

$$\mathrm{d}F = 0 \text{ and } \mathrm{d}{*}F = 0.$$

where * is the Hodge star.

27.9.3   Kaluza–Klein geometry

To build the Kaluza–Klein theory, one picks an invariant metric on the circle $\mathbf{S}^1$ that is the fiber of the U(1)-bundle of electromagnetism. In this discussion, an *invariant metric* is simply one that is invariant under rotations of the circle. Suppose this metric gives the circle a total length of Λ. One then considers metrics $\widehat{g}$ on the bundle P that are consistent with both the fiber metric, and the metric on the underlying manifold M. The consistency conditions are:

- The projection of $\widehat{g}$ to the vertical subspace $\mathrm{Vert}_p P \subset T_p P$ needs to agree with metric on the fiber over a point in the manifold M.

- The projection of $\widehat{g}$ to the horizontal subspace $\mathrm{Hor}_p P \subset T_p P$ of the tangent space at point $p \in P$ must be isomorphic to the metric g on M at $\pi(p)$.

The Kaluza–Klein action for such a metric is given by

$$S(\widehat{g}) = \int_P R(\widehat{g}) \, \mathrm{vol}(\widehat{g})$$

The scalar curvature, written in components, then expands to

$$R(\widehat{g}) = \pi^* \left(R(g) - \frac{\Lambda^2}{2} |F|^2 \right)$$

where π^* is the pullback of the fiber bundle projection π: $P \to M$. The connection A on the fiber bundle is related to the electromagnetic field strength as

$$\pi^* F = \mathrm{d}A$$

That there always exists such a connection, even for fiber bundles of arbitrarily complex topology, is a result from homology and specifically, K-theory. Applying Fubini's theorem and integrating on the fiber, one gets

$$S(\widehat{g}) = \Lambda \int_M \left(R(g) - \frac{1}{\Lambda^2} |F|^2 \right) \mathrm{vol}(g)$$

Varying the action with respect to the component A, one regains the Maxwell equations. Applying the variational principle to the base metric g, one gets the Einstein equations

$$R_{ij} - \frac{1}{2} g_{ij} R = \frac{1}{\Lambda^2} T_{ij}$$

with the stress–energy tensor being given by

$$T^{ij} = F^{ik} F^{jl} g_{kl} - \frac{1}{4} g^{ij} |F|^2,$$

sometimes called the **Maxwell stress tensor**.

The original theory identifies Λ with the fiber metric g_{55}, and allows Λ to vary from fiber to fiber. In this case, the coupling between gravity and the electromagnetic field is not constant, but has its own dynamical field, the radion.

27.9.4 Generalizations

In the above, the size of the loop Λ acts as a coupling constant between the gravitational field and the electromagnetic field. If the base manifold is four-dimensional, the Kaluza–Klein manifold P is five-dimensional. The fifth dimension is a compact space, and is called the **compact dimension**. The technique of introducing compact dimensions to obtain a higher-dimensional manifold is referred to as compactification. Compactification does not produce group actions on chiral fermions except in very specific cases: the dimension of the total space must be 2 mod 8 and the G-index of the Dirac operator of the compact space must be nonzero.[28]

The above development generalizes in a more-or-less straightforward fashion to general principal G-bundles for some arbitrary Lie group G taking the place of U(1). In such a case, the theory is often referred to as a Yang–Mills theory, and is sometimes taken to be synonymous. If the underlying manifold is supersymmetric, the resulting theory is a super-symmetric Yang–Mills theory.

27.10 Empirical tests

Up to now, no experimental or observational signs of extra dimensions have been officially reported. Many theoretical search techniques for detecting Kaluza–Klein resonances have been proposed using the mass couplings of such resonances with the top quark, however until the Large Hadron Collider (LHC) reaches full operational power observation of such resonances are unlikely. An analysis of results from the LHC in December 2010 severely constrains theories with large extra dimensions.[29]

The observation of a Higgs-like boson at the LHC puts a brand new empirical test in the search for Kaluza–Klein resonances and supersymmetric particles. The loop Feynman diagrams that exist in the Higgs interactions allow any particle with electric charge and mass to run in such a loop. Standard Model particles besides the top quark and W boson do not make big contributions to the cross-section observed in the H → γγ decay, but if there are new particles beyond the Standard Model, they could potentially change the ratio of the predicted Standard Model H → γγ cross-section to the experimentally observed cross-section. Hence a measurement of any dramatic change to the H → γγ cross section predicted by the Standard Model is crucial in probing the physics beyond it.

27.11 See also

- Classical theories of gravitation

- DGP model

- Quantum gravity

- Randall–Sundrum model

- String theory

- Supergravity

- Superstring theory

27.12 Notes

[1] Pais, Abraham (1982). *Subtle is the Lord ...: The Science and the Life of Albert Einstein*. Oxford: Oxford University Press. pp. 329–330.

[2] Kaluza, Theodor (1921). "Zum Unitätsproblem in der Physik". *Sitzungsber. Preuss. Akad. Wiss. Berlin. (Math. Phys.)*: 966–972.

[3] Klein, Oskar (1926). "Quantentheorie und fünfdimensionale Relativitätstheorie". *Zeitschrift für Physik A* **37** (12): 895–906. Bibcode:1926ZPhy...37..895K. doi:10.1007/BF01397481.

[4] Klein, Oskar (1926). "The Atomicity of Electricity as a Quantum Theory Law". *Nature* **118**: 516. Bibcode:1926Natur.118..516K. doi:10.1038/118516a0.

[5] Goenner, H. (2012). "Some remarks on the genesis of scalar-tensor theories". *General Relativity and Gravitation* **44**: 2077–2097. arXiv:1204.3455. Bibcode:2012GReGr..44.2077G. doi:10.1007/s10714-012-1378-8.

[6] Lichnerowicz, A.; Thiry, M.Y. (1947). *Compt. Rend. Acad. Sci. Paris* **224**: 529–531. Missing or empty |title= (help)

[7] Thiry, M.Y. (1948). *Compt. Rend. Acad. Sci. Paris* **226**: 216–218. Missing or empty |title= (help)

[8] Thiry, M.Y. (1948). *Compt. Rend. Acad. Sci. Paris* **226**: 1881–1882. Missing or empty |title= (help)

[9] Jordan, P. (1946). *Naturwiss.* **11**: 250–251. Missing or empty |title= (help)

[10] Jordan, P.; Müller, C. (1947). *Z. Naturforsch.* **2a**: 1–2. Missing or empty |title= (help)

[11] Ludwig, G. (1947). *Z. Naturforsch.* **2a**: 3–5. Missing or empty |title= (help)

[12] Jordan, P. (1948). *Astron. Nachr.* **276**: 193–208. Missing or empty |title= (help)

[13] Ludwig, G.; Müller, C. (1948). *Annalen der Physik* **2** (6): 76–84. Missing or empty |title= (help)

[14] Scherrer, W. (1941). *Helv. Phys. Acta* **14** (2): 130. Missing or empty |title= (help)

[15] Scherrer, W. (1949). *Helv. Phys. Acta* **22**: 537–551. Missing or empty |title= (help)

[16] Scherrer, W. (1949). *Helv. Phys. Acta* **23**: 547–555. Missing or empty |title= (help)

[17] Brans, C. H.; Dicke, R. H. (November 1, 1961). "Mach's Principle and a Relativistic Theory of Gravitation". *Physical Review* **124** (3): 925–935. Bibcode:1961PhRv..124..925B. doi:10.1103/PhysRev.124.925.

[18] Williams, L.L. (2015). "Field Equations and Lagrangian for the Kaluza Metric Evaluated with Tensor Algebra Software". *Journal of Gravitation* **2015**: 901870. doi:10.1155/2015/901870.

[19] Appelquist, Thomas; Chodos, Alan; Freund, Peter G. O. (1987). *Modern Kaluza–Klein Theories*. Menlo Park, Cal.: Addison–Wesley. ISBN 0-201-09829-6.

[20] Wesson, Paul S. (1999). *Space-Time-Matter, Modern Kaluza-Klein Theory*. Singapore: World Scientific. ISBN 981-02-3588-7.

[21] Pauli, Wolfgang (1958). *Theory of Relativity* (translated by George Field ed.). New York: Pergamon Press. pp. Supplement 23.

[22] Gross, D.J.; Perry, M.J. (1983). "Magnetic monopoles in Kaluza-Klein theories". *Nucl. Phys. B* **226**: 29–48. Bibcode:1983NuPhB.226...29G. doi:10.1016/0550-3213(83)90462-5.

[23] Gegenberg, J.; Kunstatter, G. (1984). *Phys. Lett.* **106A**: 410. Missing or empty |title= (help)

[24] Wesson, P.S.; Ponce de Leon, J. (1995). *Astronomy and Astrophysics* **294**: 1. Bibcode:1995A&A...294....1W. Missing or empty |title= (help)

[25] Williams, L.L. (2012). "Physics of the Electromagnetic Control of Spacetime and Gravity". *Proceedings of 48th AIAA Joint Propulsion Conference*. AIAA 2012-3916. doi:10.2514/6.2012-3916.

[26] David Bleecker, "Gauge Theory and Variational Principles" (1982) D. Reidel Publishing (*See chapter 9*)

[27] 5Dstm.org

[28] L. Castellani et al., Supergravity and superstrings, Vol 2, chapter V.11

[29] CMS Collaboration, "Search for Microscopic Black Hole Signatures at the Large Hadron Collider", http://arxiv.org/abs/1012.3375

27.13 References

- Nordström, Gunnar (1914). "Über die Möglichkeit, das elektromagnetische Feld und das Gravitationsfeld zu vereinigen". *Physikalische Zeitschrift* **15**: 504–506. OCLC 1762351.

- Kaluza, Theodor (1921). "Zum Unitätsproblem in der Physik". *Sitzungsber. Preuss. Akad. Wiss. Berlin. (Math. Phys.)*: 966–972. https://archive.org/details/sitzungsberichte1921preussi

- Klein, Oskar (1926). "Quantentheorie und fünfdimensionale Relativitätstheorie". *Zeitschrift für Physik A* **37** (12): 895–906. Bibcode:1926ZPhy...37..895K. doi:10.1007/BF01397481.

- Witten, Edward (1981). "Search for a realistic Kaluza–Klein theory". *Nuclear Physics B* **186** (3): 412–428. Bibcode:1981NuPhB.186..412W. doi:10.1016/0550-3213(81)90021-3.

- Appelquist, Thomas; Chodos, Alan; Freund, Peter G. O. (1987). *Modern Kaluza–Klein Theories*. Menlo Park, Cal.: Addison–Wesley. ISBN 0-201-09829-6. *(Includes reprints of the above articles as well as those of other important papers relating to Kaluza–Klein theory.)*

- Brandenberger, Robert; Vafa, Cumrun (1989). "Superstrings in the early universe". *Nuclear Physics B* **316** (2): 391–410. Bibcode:1989NuPhB.316..391B. doi:10.1016/0550-3213(89)90037-0.

- Duff, M. J. (1994). "Kaluza–Klein Theory in Perspective". In Lindström, Ulf (ed.). *Proceedings of the Symposium 'The Oskar Klein Centenary'*. Singapore: World Scientific. pp. 22–35. ISBN 981-02-2332-3.

- Overduin, J. M.; Wesson, P. S. (1997). "Kaluza–Klein Gravity". *Physics Reports* **283** (5): 303–378. arXiv:gr-qc/9805018. Bibcode:1997PhR...283..303O. doi:10.1016/S0370-1573(96)00046-4.

- Wesson, Paul S. (1999). *Space-Time-Matter, Modern Kaluza-Klein Theory*. Singapore: World Scientific. ISBN 981-02-3588-7.

- Wesson, Paul S. (2006). *Five-Dimensional Physics: Classical and Quantum Consequences of Kaluza-Klein Cosmology*. Singapore: World Scientific. ISBN 981-256-661-9.

- Coquereaux, R.; Esposito-Farese, G. (1990). "The Theory of Kaluza-Klein-Jordan-Thiry revisited". *Annales de l'I.H.P., Section A* **52**: 113–150.

- Kaku, Michio and Robert O'Keefe. *Hyperspace: A Scientific Odyssey Through Parallel Universes, Time Warps, and the Tenth Dimension*. New York: Oxford University Press, 1994. ISBN 0-19-286189-1

- The CDF Collaboration, *Search for Extra Dimensions using Missing Energy at CDF*, (2004) *(A simplified presentation of the search made for extra dimensions at the Collider Detector at Fermilab (CDF) particle physics facility.)*

- John M. Pierre, *SUPERSTRINGS! Extra Dimensions*, (2003).

- TeV scale gravity, mirror universe, and ... dinosaurs Article from Acta Physica Polonica B by Z.K. Silagadze.

- Chris Pope, *Lectures on Kaluza–Klein Theory*.

- Edward Witten (2014). "A Note On Einstein, Bergmann, and the Fifth Dimension", arXiv:1401.8048; pdf

27.14   Further reading

- Grøn, Øyvind; Hervik, Sigbjørn (2007). *Einstein's General Theory of Relativity*. New York: Springer. ISBN 978-0-387-69199-2.

Chapter 28

N = 4 supersymmetric Yang–Mills theory

$N = 4$ supersymmetric Yang–Mills theory is a mathematical and physical model created to study particles through a simple system, similar to string theory, with conformal symmetry. It is a simplified toy theory based on Yang–Mills theory that does not describe the real world, but is useful because it can act as a proving ground for approaches for attacking problems in more complex theories.[1] It describes a universe containing boson fields and fermion fields which are related by 4 supersymmetries (this means that swapping boson, fermion and scalar fields in a certain way leaves the predictions of the theory invariant). It is one of the simplest (because it has no free parameters except for the gauge group) and one of the few finite quantum field theories in 4 dimensions. It can be thought of as the most symmetric field theory that does not involve gravity.

$$L = \text{tr}\left\{ \frac{1}{g^2} F_{IJ} F^{IJ} - i\bar{\lambda}\Gamma^I D_I \lambda \right\}$$

where I and J are now run from 0 through 9 and Γ^I are the 32 by 32 gamma matrices. ($32 = 2^{10/2}$) followed by adding the term with θ_I which is a topological term.

The components A_i of the gauge field for $i = 4$ to 9 become scalars upon eliminating the extra dimensions. This also gives an interpretation of the SO(6) R-symmetry as rotations in the extra compact dimensions.

By compactification on a T^6, all the supercharges are preserved, giving $N = 4$ in the 4-dimensional theory.

A Type IIB string theory interpretation of the theory is the worldvolume theory of a stack of D3-branes.

28.1 Lagrangian

The Lagrangian for the theory is:

$$L = \text{tr}\left\{ -\frac{1}{2g^2} F_{\mu\nu} F^{\mu\nu} + \frac{\theta_I}{8\pi^2} F_{\mu\nu} \bar{F}^{\mu\nu} - i\bar{\lambda}^a \bar{\sigma}^\mu D_\mu \lambda_a - D_\mu X^i D^\mu X^i + g C_i^{ab} \lambda_a [X^i, \lambda_b] + g \bar{C}_{iab} \bar{\lambda}^a [X^i, \bar{\lambda}^b] + \frac{g^2}{2} [X^i, X^j]^2 \right\}$$

where $F_{\mu\nu}^k = \partial_\mu A_\nu^k - \partial_\nu A_\mu^k + f^{klm} A_\mu^l A_\nu^m$ and the indices $i,j = 1, ..., 6$ and the indices $a, b = 1, ..., 4$. f are the group constants of the particular gauge group. C_i^{ab} are the structure constants of the R-symmetry group $SU(4)$ which rotates the 4 supersymmetries. As a consequence of the nonrenormalization theorems, this supersymmetric field theory is in fact a superconformal field theory.

28.2 Ten-dimensional Lagrangian

The above Lagrangian can be found by beginning with the simpler ten-dimensional Lagrangian

28.3 S-duality

Main article: S-duality
Main article: Montonen–Olive duality

The coupling constants θ_I and g naturally pair together in the form:

$$\tau = \frac{\theta}{2\pi} + \frac{4\pi i}{g^2}.$$

The theory has symmetries that shift τ by integers. The S-duality conjecture says there is also a symmetry which sends : $\tau \mapsto \frac{-1}{n_G \tau}$ as well as switching the group G to its Langlands dual group.

28.4 AdS/CFT correspondence

Main article: AdS/CFT correspondence

145

This theory is important also in the context of the holographic principle. There is a duality between Type IIB string theory on $AdS_5 \times S^5$ space (a product of 5-dimensional AdS space with a 5-dimensional sphere) and $N = 4$ Super Yang–Mills on the 4-dimensional boundary of AdS_5. It is the most successful realization of the holographic principle, a speculative idea about quantum gravity originally proposed by Gerard 't Hooft and improved and promoted by Leonard Susskind.

28.5   Integrability

As the number of colors goes to infinity, the amplitudes scale like N^{2-2g}, so that only the genus 0 contribution survives. For more details, see 1/N expansion.

Beisert et al. give a review article demonstrating how in this situation local operators can be expressed via certain states in "spin" chains, but based on a larger Lie superalgebras rather than su(2) for ordinary spin. These are amenable to Bethe ansatz techniques. They also construct an action of the associated Yangian on scattering amplitudes.[3]

Nima Arkani-Hamed et al. have also researched this subject. Using twistor theory, they find a description in terms of the positive Grassmannian.[4] See Amplituhedron.

28.6   Relation to 11 dimensional M-Theory

$N = 4$ Super Yang–Mills can be derived from a simpler 10 dimensional theory, and yet supergravity and M-Theory exist in 11 dimensions. The connection is that if the gauge group $U(N)$ of SYM becomes infinite as $N \to \infty$ it becomes equivalent to an 11-dimensional theory known as matrix theory.

28.7   See also

- 6D (2,0) superconformal field theory

- Extended supersymmetry

28.8   References

[1] Matt von Hippel. "Earning a PhD by studying a theory that we know is wrong". Ars Technica.

[2] Luke Wassink (2009). "$N = 4$ Super Yang–Mills theory" (PDF). Retrieved 2013-05-22.

[3] Beisert, Niklas (January 2012). "Review of AdS/CFT Integrability: An Overview". *Letters In Mathematical Physics* **99**: 425. arXiv:1012.4000. Bibcode:2012LMaPh..99..425K. doi:10.1007/s11005-011-0516-7.

[4] Nima Arkani-Hamed; Bourjaily, Jacob L.; Freddy Cachazo; Goncharov, Alexander B.; Alexander Postnikov; Jaroslav Trnka (2012). "Scattering Amplitudes and the Positive Grassmannian". arXiv:1212.5605 [hep-th].

- * Kapustin, Anton; Witten, Edward (2007). "Electric-magnetic duality and the geometric Langlands program". *Communications in Number Theory and Physics* **1** (1): 1–236. arXiv:hep-th/0604151v3. Bibcode:2007CNTP....1....1K. doi:10.4310/cntp.2007.v1.n1.a1.

Chapter 29

Coupling constant

For the Murray–von Neumann coupling constant, see von Neumann algebra. For the coupling constant in NMR spectroscopy, see Nuclear magnetic resonance spectroscopy and Proton NMR. For coupling strength in bibliometrics, see Bibliographic coupling.

In physics, a **coupling constant** or **gauge coupling parameter** is a number that determines the strength of the force exerted in an interaction. Usually, the Lagrangian or the Hamiltonian of a system describing an interaction can be separated into a *kinetic part* and an *interaction part*. The coupling constant determines the strength of the interaction part with respect to the kinetic part, or between two sectors of the interaction part. For example, the electric charge of a particle is a coupling constant that sits beside an interaction with two matter fields and one photon field (hence the common Feynman diagram with two arrows and one wavy line). Since photons carry electromagnetism, this coupling constant determines how strongly electrons feel such a force and has its value fixed by experiment.

A coupling constant plays an important role in dynamics. For example, one often sets up hierarchies of approximation based on the importance of various coupling constants. In the motion of a large lump of magnetized iron, the magnetic forces may be more important than the gravitational forces because of the relative magnitudes of the coupling constants. However, in classical mechanics one usually makes these decisions directly by comparing forces.

29.1 Fine-structure constant

The coupling constant arises naturally in a quantum field theory. A special role is played in relativistic quantum theories by coupling constants which are dimensionless, i.e., are pure numbers. For example, the fine-structure constant,

$$\alpha = \frac{e^2}{4\pi\varepsilon_0\hbar c}$$

(where e is the charge of an electron, ε_0 is the permittivity of free space, $\hbar$ is the reduced Planck constant and c is the speed of light) is such a dimensionless coupling constant that determines the strength of the electromagnetic force on an electron.

29.2 Gauge coupling

In a non-Abelian Gauge theory, the **gauge coupling parameter**, g, appears in the Lagrangian as

$$\frac{1}{4g^2}\mathrm{Tr}\,G_{\mu\nu}G^{\mu\nu}$$

(where G is the gauge field tensor) in some conventions. In another widely used convention, G is rescaled so that the coefficient of the kinetic term is 1/4 and g appears in the covariant derivative. This should be understood to be similar to a dimensionless version of the elementary charge defined as

$$\sqrt{4\pi\alpha} \approx 0.30282212 \, .$$

29.3 Weak and strong coupling

In a quantum field theory with a dimensionless coupling constant g, if g is much less than 1 then the theory is said to be *weakly coupled*. In this case it is well described by an expansion in powers of g, called perturbation theory. If the coupling constant is of order one or larger, the theory is said to be *strongly coupled*. An example of the latter is the hadronic theory of strong interactions (which is why it is called strong in the first place). In such a case non-perturbative methods have to be used to investigate the theory.

29.4   Running coupling

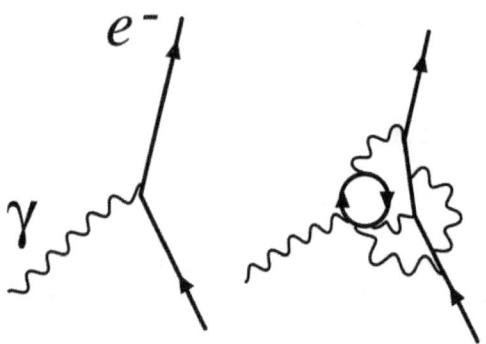

Virtual particles renormalize the coupling

One can probe a quantum field theory at short times or distances by changing the wavelength or momentum, **k**, of the probe one uses. With a high frequency (i.e., short time) probe, one sees virtual particles taking part in every process. This apparent violation of the conservation of energy can be understood heuristically by examining the uncertainty relation

$$\Delta E \Delta t \geq \hbar,$$

which allows such violations at short times. The previous remark only applies to some formulations of quantum field theory, in particular, canonical quantization in the interaction picture. In other formulations, the same event is described by "virtual" particles going off the mass shell. Such processes renormalize the coupling and make it dependent on the energy scale, μ at which one observes the coupling. The dependence of a coupling $g(\mu)$ on the energy-scale is known as running of the coupling. The theory of the running of couplings is known as the renormalization group.

29.4.1   Beta functions

Main article: Beta function (physics)

In quantum field theory, a *beta function* $\beta(g)$ encodes the running of a coupling parameter, g. It is defined by the relation

$$\beta(g) = \mu \frac{\partial g}{\partial \mu} = \frac{\partial g}{\partial \ln \mu},$$

where μ is the energy scale of the given physical process. If the beta functions of a quantum field theory vanish, then the theory is scale-invariant.

The coupling parameters of a quantum field theory can flow even if the corresponding classical field theory is scale-invariant. In this case, the non-zero beta function tells us that the classical scale-invariance is anomalous.

29.4.2   QED and the Landau pole

If a beta function is positive, the corresponding coupling increases with increasing energy. An example is quantum electrodynamics (QED), where one finds by using perturbation theory that the beta function is positive. In particular, at low energies, $\alpha \approx 1/137$, whereas at the scale of the Z boson, about 90 GeV, one measures $\alpha \approx 1/127$.

Moreover, the perturbative beta function tells us that the coupling continues to increase, and QED becomes *strongly coupled* at high energy. In fact the coupling apparently becomes infinite at some finite energy. This phenomenon was first noted by Lev Landau, and is called the Landau pole. However, one cannot expect the perturbative beta function to give accurate results at strong coupling, and so it is likely that the Landau pole is an artifact of applying perturbation theory in a situation where it is no longer valid. The true scaling behaviour of α at large energies is not known.

29.4.3   QCD and asymptotic freedom

In non-Abelian gauge theories, the beta function can be negative, as first found by Frank Wilczek, David Politzer and David Gross. An example of this is the beta function for Quantum Chromodynamics (QCD), and as a result the QCD coupling decreases at high energies.

Furthermore, the coupling decreases logarithmically, a phenomenon known as asymptotic freedom (the discovery of which was awarded with the Nobel Prize in Physics in 2004). The coupling decreases approximately as

$$\alpha_s(k^2) \stackrel{\text{def}}{=} \frac{g_s^2(k^2)}{4\pi} \approx \frac{1}{\beta_0 \ln(k^2/\Lambda^2)},$$

where β_0 is a constant computed by Wilczek, Gross and Politzer.

Conversely, the coupling increases with decreasing energy. This means that the coupling becomes large at low energies, and one can no longer rely on perturbation theory.

29.4.4 QCD scale

In quantum chromodynamics (QCD), the quantity Λ is called the **QCD scale**. The value is

$$\Lambda_{MS} = 217^{+25}_{-23} \text{ MeV}.$$

This value is to be used at a scale above the bottom quark mass of about 5 GeV. The meaning of ΛMS is given in the article on dimensional transmutation.

The proton-to-electron mass ratio is primarily determined by the QCD scale.

29.5 String theory

A remarkably different situation exists in string theory since it includes a dilaton. An analysis of the string spectrum shows that this field must be present, either in the bosonic string or the NS-NS sector of the superstring. Using vertex operators, it can be seen that exciting this field is equivalent to adding a term to the action where a scalar field couples to the Ricci scalar. This field is therefore an entire function worth of coupling constants. These coupling constants are not pre-determined, adjustable, or universal parameters; they depend on space and time in a way that is determined dynamically. Sources that describe the string coupling as if it were fixed are usually referring to the vacuum expectation value. This is free to have any value in the bosonic theory where there is no superpotential.

29.6 See also

- Canonical quantization, renormalization and dimensional regularization

- Fine structure constant

- Gravitational coupling constant

- Quantum field theory, especially quantum electrodynamics and quantum chromodynamics

- Gluon field, Gluon field strength tensor

29.7 References

- *An introduction to quantum field theory*, by M.E.Peskin and H.D.Schroeder, ISBN 0-201-50397-2

29.8 External links

- The Nobel Prize in Physics 2004 – Information for the Public

- Department of Physics and Astronomy of the Georgia State University - Coupling Constants for the Fundamental Forces

Chapter 30

Montonen–Olive duality

In theoretical physics, **Montonen–Olive duality** is the oldest known example of S-duality or a strong-weak duality. It generalizes the electro-magnetic symmetry of Maxwell's equations. It is named after Finnish Claus Montonen and British David Olive.

while the gauge group G is simultaneously replaced by its Langlands dual group ^LG and n_G is an integer depending on the choice of gauge group. In the case the theta-angle is 0, this reduces to the simple form of Montonen–Olive duality stated above.

30.1 Overview

In a four-dimensional Yang-Mills theory with $N=4$ supersymmetry, which is the case where the Montonen–Olive duality applies, one obtains a physically equivalent theory if one replaces the gauge coupling constant g by $1/g$. This also involves an interchange of the electrically charged particles and magnetic monopoles. See also Seiberg duality.

In fact, there exists a larger SL(2,$\mathbf{Z}$) symmetry where both g as well as theta-angle are transformed non-trivially.

30.2 Mathematical formalism

The gauge coupling and theta-angle can be combined together to form one complex coupling

$$\tau = \frac{\theta}{2\pi} + \frac{4\pi i}{g^2}.$$

Since the theta-angle is periodic, there is a symmetry

$$\tau \mapsto \tau + 1.$$

The quantum mechanical theory with gauge group G (but not the classical theory, except in the case when the G is abelian) is also invariant under the symmetry

$$\tau \mapsto \frac{-1}{n_G \tau}$$

30.3 References

- Edward Witten, *Notes from the 2006 Bowen Lectures*, an overview of electric–magnetic duality in gauge theory and its relation to the Langlands program

Chapter 31

Supermembranes

Supermembranes are hypothesized objects that live in the 11 dimensional theory called M-Theory and should also exist in 11 dimensional supergravity. Supermembranes are a generalisation of superstrings to another dimension. Supermembranes are 2-dimensional surfaces. For example they can be spherical or shaped like a torus. As in superstring theory the vibrations of the supermembranes correspond to different particles. Supermembranes also exhibit a symmetry called supersymmetry without which the vibrations would only correspond to bosons and not fermions.

31.1   Energy

The energy of a classical supermembrane is given by its surface area. One consequence of this is that there is no difference between one or two membranes since two membranes can be connected by a long 1 dimensional string of zero area. Hence, the idea of 'membrane-number' has no meaning. A second consequence is that unlike strings a supermembrane's vibrations can represent several particles at once. In technical terms this means it is already 'second-quantized'. All the particles in the Universe can be thought to arise as vibrations of a single membrane.

31.2   Spectrum

When going from the classical theory to the quantum theory of supermembranes it is found that they can only exist in 11 dimensions, just as superstrings can only exist in 10 dimensions. When examining the energy spectrum (the allowed frequencies that a string can vibrate in) it was found that they can only be in discrete values corresponding to the masses of different particles.

It has been shown:

- The energy spectrum for the classical bosonic membrane is continuous.

- The energy spectrum for the quantum bosonic membrane is discrete.

- The energy spectrum for the quantum supermembrane is continuous.

At first the discovery that the spectrum was continuous was thought to mean the theory didn't make sense. But it was realised that it meant that supermembranes actually correspond to multiple particles. (The continuous degrees of freedom corresponding to the coordinates/momenta of the additional particles).

31.3   Action

The action for a classical membrane is simply the surface area of the world sheet. The quantum version is harder to write down, is non-linear and very difficult to solve. Unlike the superstring action which is quadratic, the supermembrane action is quartic which makes it exponentially harder. Adding to this the fact that a membrane can represent many particles at once not much progress has been made on supermembranes.

31.4   Low energy sector

It has been proven that the low energy vibrations of the supermembrane correspond to the particles in 11 dimensional supergravity.

31.5   Topology

A supermembrane can have multiple thing tubes or strings coming out of it with little or no extra energy cost since strings, for example, have no area. This means that all orientable topologies of membranes are physically the same.

Also, joined and disjointed supermembanes are physically the same. Thus the topology of a supermembrane has no physical meaning.

31.6   Mathematics

The infinite supermembrane can be described in terms of an infinite number of patches. The coordinates of (each patch of) a supermembrane at any casual slice of time are 11 dimensional and depend on two continuous parameters (σ, θ) and a third integer parameter (k) denoting the patch number:

$$X_\mu^k(\sigma, \theta) = x_\mu^k + O(\sigma, \theta)$$

Therefor the super membrane can describe an infinite number of particles if we associate somehow the coordinate of each particle with some topological property of the patches - perhaps holes in the membrane or closed loops.

31.7   Supermembrane Field Theory

Since supermembranes correspond to multiple particles the field theory of membranes correspond to a Fock space. Informally, let $\mathbf{a}(x)$ denote the continuous degrees of freedom in the energy spectrum:

$$\Phi[X] = \Phi[x, \mathbf{a}, ..] = \phi(x) + \int \mathbf{a}(y)\phi(x, y)dy + \int \mathbf{a}(y)\mathbf{a}(z)\phi(x, y, z)dydz + ...$$

The action can be written as

$$S = \int \Phi[X]\mathbf{Q}\Phi[X]D[X]$$

where $\mathbf{Q}$ is the kinetic operator. No interaction terms are needed since there is no concept of membrane number. Everything is the same membrane. The action is not quite the same type as the one for superstrings or particles since it involves terms with multiple particles. The terms relating to single fields must recover the classical field equations of Dirac, Maxwell and Einstein. The propagator to get from a state with membrane $\mathbf{X}$ to one at another conformal slice with membrane $\mathbf{Y}$ is:

$$G(X, Y) = \mathbf{Q}^{-1}(X, Y)$$

And since each membrane corresponds to any number of identical particles this is equivalent to all the Green's functions for many particle collisions at once!

Although it looks like a lot of things simplify in the supermembrane picture, the actual form of the kinetic operator $\mathbf{Q}$ is yet unknown and must be a very complicated operator acting on an infinite Fock-like space. Hence the seeming simplicity of the theory is hidden in this operator.

31.8   Cosmology

Since the vibrations of a supermembrane of infinite energy can correspond to every particle in the Universe at once it is possible to interpret the supermembrane as equivalent to the Universe. i.e. all that exists is the supermembrane. It makes no difference to say we live on this supermembrane or that we in the 11 dimensional space-time. Every state of the Universe corresponds to a supermembrane and every history of the Universe corresponds to a supermembrane world volume. What we think of as space-time coordinates can equally be thought of as vector fields on the 2+1 dimensional supermembrane.

For a supermembrane moving at the speed of light, its world volume can be zero due to the metric (+++-). Thus the Big Bang can be thought of as a spherical membrane expanding at the speed of light. This has interesting interpretations in terms of the holographic principle.

31.9   Geometry

Because the supermembrane(s) correspond to all particles at a particular causal time slice, it also corresponds to all the gravitons particles (which are particular vibrational modes). Thus the geometry of the 2+1D supermembrane contains within it the description of the geometry of the (macroscopic) 10+1D space-time. But as it is a quantum theory it gives probabilities for different space-times consistent with observation. The different space-times may only differ microscopically whereas the macroscopic space-time is smooth. In other words the geometry of the membrane determines the geometry of (macroscopic) space-time. This is different from string theory where only condensates of many separate strings can macroscopically determine the space-time.

31.10   Super-5-branes

M-Theory and 11D supergravity also predict 5+1D objects called super-5-branes. An alternative cosmological theory is that we live on one of these branes.

31.11 Compactification

Compactifying one space-time dimension on a circle and wrapping the membrane around this circle gives us superstring theory. To get back to our 3+1 dimensional universe the space-time coordinates need to be compactified on a 7 dimensional manifold (of G2 holonomy). Not much is known about these types of shapes.

31.12 Matrix Theory

Matrix theory is a particular way of formulating supermembrane theory. It is still in development. The diagonal entries of an infinite dimensional matrix can be thought of as different supermembranes (parts) connected by 1 dimensional strings.

31.13 References

- Jansson, Ronnie (2003). *The Membrane Vacuum State*. The Membrane Vacuum State

- Howe,Sezgin (2005). *The supermembrane revisited*. The supermembrane revisited

Chapter 32

Matrix theory (physics)

In theoretical physics, the **BFSS matrix model** or **matrix theory** is a quantum mechanical model proposed by Tom Banks, Willy Fischler, Stephen Shenker, and Leonard Susskind in 1997.[1] This theory describes the behavior of a set of nine large matrices. In their original paper, these authors showed, among other things, that the low energy limit of this matrix model is described by eleven-dimensional supergravity. These calculations led them to propose that the BFSS matrix model is exactly equivalent to M-theory. The BFSS matrix model can therefore be used as a prototype for a correct formulation of M-theory and a tool for investigating the properties of M-theory in a relatively simple setting.

32.1 Noncommutative geometry

Main articles: Noncommutative geometry and Noncommutative quantum field theory

In geometry, it is often useful to introduce coordinates. For example, in order to study the geometry of the Euclidean plane, one defines the coordinates x and y as the distances between any point in the plane and a pair of axes. In ordinary geometry, the coordinates of a point are numbers, so they can be multiplied, and the product of two coordinates does not depend on the order of multiplication. That is, $xy = yx$. This property of multiplication is known as the commutative law, and this relationship between geometry and the commutative algebra of coordinates is the starting point for much of modern geometry.[2]

Noncommutative geometry is a branch of mathematics that attempts to generalize this situation. Rather than working with ordinary numbers, one considers some similar objects, such as matrices, whose multiplication does not satisfy the commutative law (that is, objects for which xy is not necessarily equal to yx). One imagines that these noncommuting objects are coordinates on some more general notion of "space" and proves theorems about these generalized spaces by exploiting the analogy with ordinary geometry.[3]

In a paper from 1998, Alain Connes, Michael R. Douglas, and Albert Schwarz showed that some aspects of matrix models and M-theory are described by a noncommutative quantum field theory, a special kind of physical theory in which the coordinates on spacetime do not satisfy the commutativity property.[4] This established a link between matrix models and M-theory on the one hand, and noncommutative geometry on the other hand. It quickly led to the discovery of other important links between noncommutative geometry and various physical theories.[5][6]

32.2 See also

- Matrix string theory

32.3 Notes

[1] Banks et al. 1997

[2] Connes 1994, p. 1

[3] Connes 1994

[4] Connes, Douglas, and Schwarz 1998

[5] Nekrasov and Schwarz 1998

[6] Seiberg and Witten 1999

32.4 References

- Banks, Tom; Fischler, Willy; Schenker, Stephen; Susskind, Leonard (1997). "M theory as a matrix model: A conjecture". *Physical Review D* **55** (8): 5112. arXiv:hep-th/9610043. Bibcode:1997PhRvD..55.5112B. doi:10.1103/physrevd.55.5112.

- Connes, Alain (1994). *Noncommutative Geometry*. Academic Press. ISBN 978-0-12-185860-5.

- Connes, Alain; Douglas, Michael; Schwarz, Albert (1998). "Noncommutative geometry and matrix theory". *Journal of High Energy Physics*. 19981 (2): 003. arXiv:hep-th/9711162. Bibcode:1998JHEP...02..003C. doi:10.1088/1126-6708/1998/02/003.

- Nekrasov, Nikita; Schwarz, Albert (1998). "Instantons on noncommutative $\mathbf{R}^4$ and (2,0) superconformal six dimensional theory". *Communications in Mathematical Physics* **198** (3): 689–703. arXiv:hep-th/9802068. Bibcode:1998CMaPh.198..689N. doi:10.1007/s002200050490.

- Seiberg, Nathan; Witten, Edward (1999). "String Theory and Noncommutative Geometry". *Journal of High Energy Physics* **1999** (9): 032. arXiv:hep-th/9908142. Bibcode:1999JHEP...09..032S. doi:10.1088/1126-6708/1999/09/032.

Chapter 33

Noncommutative geometry

Noncommutative geometry (**NCG**) is a branch of mathematics concerned with a geometric approach to noncommutative algebras, and with the construction of *spaces* that are locally presented by noncommutative algebras of functions (possibly in some generalized sense). A noncommutative algebra is an associative algebra in which the multiplication is not commutative, that is, for which xy does not always equal yx ; or more generally an algebraic structure in which one of the principal binary operations is not commutative; one also allows additional structures, e.g. topology or norm, to be possibly carried by the noncommutative algebra of functions.

33.1 Motivation

The main motivation is to extend the commutative duality between spaces and functions to the noncommutative setting. In mathematics, *spaces*, which are geometric in nature, can be related to numerical functions on them. In general, such functions will form a commutative ring. For instance, one may take the ring $C(X)$ of continuous complex-valued functions on a topological space X. In many cases (*e.g.*, if X is a compact Hausdorff space), we can recover X from $C(X)$, and therefore it makes some sense to say that X has *commutative topology*.

More specifically, in topology, compact Hausdorff topological spaces can be reconstructed from the Banach algebra of functions on the space (Gel'fand-Neimark). In commutative algebraic geometry, algebraic schemes are locally prime spectra of commutative unital rings (A. Grothendieck), and schemes can be reconstructed from the categories of quasi-coherent sheaves of modules on them (P. Gabriel-A. Rosenberg). For Grothendieck topologies, the cohomological properties of a site are invariant of the corresponding category of sheaves of sets viewed abstractly as a topos (A. Grothendieck). In all these cases, a space is reconstructed from the algebra of functions or its categorified version— some category of sheaves on that space.

Functions on a topological space can be multiplied and added pointwise hence they form a commutative algebra; in fact these operations are local in the topology of the base space, hence the functions form a sheaf of commutative rings over the base space.

The dream of noncommutative geometry is to generalize this duality to the duality between

- noncommutative algebras, or sheaves of noncommutative algebras, or sheaf-like noncommutative algebraic or operator-algebraic structures

- and geometric entities of certain kind,

and interact between the algebraic and geometric description of those via this duality.

Regarding that the commutative rings correspond to usual affine schemes, and commutative C*-algebras to usual topological spaces, the extension to noncommutative rings and algebras requires non-trivial generalization of topological spaces, as "non-commutative spaces". For this reason, some talk about non-commutative topology, though the term also has other meanings.

33.1.1 Applications in mathematical physics

Some applications in particle physics are described on the entries Noncommutative standard model and Noncommutative quantum field theory. Sudden rise in interest in noncommutative geometry in physics, follows after the speculations of its role in M-theory made in 1997.[1]

33.1.2 Motivation from ergodic theory

Some of the theory developed by Alain Connes to handle noncommutative geometry at a technical level has roots in older attempts, in particular in ergodic theory. The proposal of George Mackey to create a *virtual subgroup* theory, with respect to which ergodic group actions would become homogeneous spaces of an extended kind, has by now been subsumed.

33.2 Noncommutative C*-algebras, von Neumann algebras

(The formal duals of) non-commutative C*-algebras are often now called non-commutative spaces. This is by analogy with the Gelfand representation, which shows that commutative C*-algebras are dual to locally compact Hausdorff spaces. In general, one can associate to any C*-algebra S a topological space $\hat{S}$; see spectrum of a C*-algebra.

For the duality between σ-finite measure spaces and commutative von Neumann algebras, noncommutative von Neumann algebras are called *non-commutative measure spaces*.

33.3 Noncommutative differentiable manifolds

A smooth Riemannian manifold M is a topological space with a lot of extra structure. From its algebra of continuous functions $C(M)$ we only recover M topologically. The algebraic invariant that recovers the Riemannian structure is a spectral triple. It is constructed from a smooth vector bundle E over M, e.g. the exterior algebra bundle. The Hilbert space $L^2(M,E)$ of square integrable sections of E carries a representation of $C(M)$ by multiplication operators, and we consider an unbounded operator D in $L^2(M,E)$ with compact resolvent (e.g. the signature operator), such that the commutators $[D,f]$ are bounded whenever f is smooth. A recent deep theorem[2] states that M as a Riemannian manifold can be recovered from this data.

This suggests that one might define a noncommutative Riemannian manifold as a spectral triple (A,H,D), consisting of a representation of a C*-algebra A on a Hilbert space H, together with an unbounded operator D on H, with compact resolvent, such that $[D,a]$ is bounded for all a in some dense subalgebra of A. Research in spectral triples is very active, and many examples of noncommutative manifolds have been constructed.

33.4 Noncommutative affine and projective schemes

In analogy to the duality between affine schemes and commutative rings, we define a category of **noncommutative affine schemes** as the dual of the category of associative unital rings. There are certain analogues of Zariski topology in that context so that one can glue such affine schemes to more general objects.

There are also generalizations of the Cone and of the Proj of a commutative graded ring, mimicking a Serre's theorem on Proj. Namely the category of quasicoherent sheaves of O-modules on a Proj of a commutative graded algebra is equivalent to the category of graded modules over the ring localized on Serre's subcategory of graded modules of finite length; there is also analogous theorem for coherent sheaves when the algebra is Noetherian. This theorem is extended as a definition of **noncommutative projective geometry** by Michael Artin and J. J. Zhang,[3] who add also some general ring-theoretic conditions (e.g. Artin-Schelter regularity).

Many properties of projective schemes extend to this context. For example, there exist an analog of the celebrated Serre duality for noncommutative projective schemes of Artin and Zhang.[4]

A. L. Rosenberg has created a rather general relative concept of **noncommutative quasicompact scheme** (over a base category), abstracting the Grothendieck's study of morphisms of schemes and covers in terms of categories of quasicoherent sheaves and flat localization functors.[5] There is also another interesting approach via localization theory, due to Fred Van Oystaeyen, Luc Willaert and Alain Verschoren, where the main concept is that of a **schematic algebra**.[6]

33.5 Invariants for noncommutative spaces

Some of the motivating questions of the theory are concerned with extending known topological invariants to formal duals of noncommutative (operator) algebras and other replacements and candidates for noncommutative spaces. One of the main starting points of the Alain Connes' direction in noncommutative geometry is his discovery of a new homology theory associated to noncommutative associative algebras and noncommutative operator algebras, namely the cyclic homology and its relations to the algebraic K-theory (primarily via Connes-Chern character map).

The theory of characteristic classes of smooth manifolds has been extended to spectral triples, employing the tools of operator K-theory and cyclic cohomology. Several generalizations of now classical index theorems allow for effective extraction of numerical invariants from spectral triples. The fundamental characteristic class in cyclic cohomology, the JLO cocycle, generalizes the classical Chern character.

33.6 Examples of noncommutative spaces

- In the phase space formulation of quantum mechanics, the symplectic phase space of classical mechanics is deformed into a non-commutative phase space generated by the position and momentum operators.

- The standard model of particle physics is another example of a noncommutative geometry, cf noncommutative standard model.

- The noncommutative torus, deformation of the function algebra of the ordinary torus, can be given the structure of a spectral triple. This class of examples has been studied intensively and still functions as a test case for more complicated situations.

- **Snyder space**[7]

- Noncommutative algebras arising from foliations.

- Examples related to dynamical systems arising from number theory, such as the Gauss shift on continued fractions, give rise to noncommutative algebras that appear to have interesting noncommutative geometries.

33.7 See also

- Commutativity
- Phase space formulation
- Moyal product
- Fuzzy sphere
- Noncommutative algebraic geometry
- Noncommutative topology

33.8 Notes

[1] Alain Connes, Michael R. Douglas, Albert Schwarz, Non-commutative geometry and matrix theory: compactification on tori. J. High Energy Phys. 1998, no. 2, Paper 3, 35 pp. doi, hep-th/9711162

[2] Connes, Alain, On the spectral characterization of manifolds, arXiv:0810.2088v1

[3] M. Artin, J. J. Zhang, Noncommutative projective schemes, Adv. Math. 109 (1994), no. 2, 228-–287, doi

[4] Amnon Yekutieli, James J. Zhang, Serre duality for noncommutative projective schemes, Proc. Amer. Math. Soc. 125, n. 3, 1997, 697-707, pdf

[5] A. L. Rosenberg, Noncommutative schemes, Compositio Mathematica 112 (1998) 93-–125, doi; Underlying spaces of noncommutative schemes, preprint MPIM2003-111, dvi, ps; MSRI lecture *Noncommutative schemes and spaces* (Feb 2000): video

[6] Freddy van Oystaeyen, Algebraic geometry for associative algebras, ISBN 0-8247-0424-X - New York: Dekker, 2000.- 287 p. - (Monographs and textbooks in pure and applied mathematics , 232); F. van Oystaeyen, L. Willaert, Grothendieck topology, coherent sheaves and Serre's theorem for schematic algebras, J. Pure Appl. Alg. 104 (1995), p. 109-–122

[7] H. S. Snyder, Quantized Space-Time, Phys. Rev. 71 (1947) 38

33.9 References

- Connes, Alain (1994), *Non-commutative geometry* (PDF), Boston, MA: Academic Press, ISBN 978-0-12-185860-5

- Connes, Alain; Marcolli, Matilde (2008), "A walk in the noncommutative garden", *An invitation to noncommutative geometry*, World Sci. Publ., Hackensack, NJ, pp. 1–128, arXiv:math/0601054, Bibcode:2006math......1054C, MR 2408150

- Connes, Alain; Marcolli, Matilde (2008), *Noncommutative geometry, quantum fields and motives* (PDF), American Mathematical Society Colloquium Publications **55**, Providence, R.I.: American Mathematical Society, ISBN 978-0-8218-4210-2, MR 2371808

- Gracia-Bondia, Jose M; Figueroa, Hector; Varilly, Joseph C (2000), *Elements of Non-commutative geometry*, Birkhauser, ISBN 978-0-8176-4124-5

- Landi, Giovanni (1997), *An introduction to noncommutative spaces and their geometries*, Lecture Notes in Physics. New Series m: Monographs **51**, Berlin, New York: Springer-Verlag, arXiv:hep-th/9701078, Bibcode:1997hep.th....1078L, ISBN 978-3-540-63509-3, MR 1482228

- Van Oystaeyen, Fred; Verschoren, Alain (1981), *Noncommutative algebraic geometry*, Lecture Notes in Mathematics **887**, Springer-Verlag, ISBN 978-3-540-11153-5

33.10   Further reading

- Consani, Caterina; Connes, Alain, eds. (2011), *Noncommutative geometry, arithmetic, and related topics. Proceedings of the 21st meeting of the Japan-U.S. Mathematics Institute (JAMI) held at Johns Hopkins University, Baltimore, MD, USA, March 23–26, 2009*, Baltimore, MD: Johns Hopkins University Press, ISBN 1-4214-0352-8, Zbl 1245.00040

- Grensing, Gerhard (2013). *Structural aspects of quantum field theory and noncommutative geometry*. Hackensack New Jersey: World Scientific. ISBN 978-981-4472-69-2.

33.11   External links

- Introduction to Quantum Geometry by Micho Đurđevich

- Lectures on Noncommutative Geometry by Victor Ginzburg

- Very Basic Noncommutative Geometry by Masoud Khalkhali

- Lectures on Arithmetic Noncommutative Geometry by Matilde Marcolli

- Noncommutative Geometry for Pedestrians by J. Madore

- An informal introduction to the ideas and concepts of noncommutative geometry by Thierry Masson (an easier introduction that is still rather technical)

- Noncommutative geometry on arxiv.org

- MathOverflow, Theories of Noncommutative Geometry

- S. Mahanta, On some approaches towards noncommutative algebraic geometry, math.QA/0501166

- G. Sardanashvily, *Lectures on Differential Geometry of Modules and Rings* (Lambert Academic Publishing, Saarbrücken, 2012); arXiv: 0910.1515

- Noncommutative geometry and particle physics

Chapter 34

AdS/CFT correspondence

In theoretical physics, the **anti-de Sitter/conformal field theory correspondence**, sometimes called **Maldacena duality** or **gauge/gravity duality**, is a conjectured relationship between two kinds of physical theories. On one side are anti-de Sitter spaces (AdS) which are used in theories of quantum gravity, formulated in terms of string theory or M-theory. On the other side of the correspondence are conformal field theories (CFT) which are quantum field theories, including theories similar to the Yang–Mills theories that describe elementary particles.

The duality represents a major advance in our understanding of string theory and quantum gravity.[1] This is because it provides a non-perturbative formulation of string theory with certain boundary conditions and because it is the most successful realization of the holographic principle, an idea in quantum gravity originally proposed by Gerard 't Hooft and promoted by Leonard Susskind.

It also provides a powerful toolkit for studying strongly coupled quantum field theories.[2] Much of the usefulness of the duality results from the fact that it is a strong-weak duality: when the fields of the quantum field theory are strongly interacting, the ones in the gravitational theory are weakly interacting and thus more mathematically tractable. This fact has been used to study many aspects of nuclear and condensed matter physics by translating problems in those subjects into more mathematically tractable problems in string theory.

The AdS/CFT correspondence was first proposed by Juan Maldacena in late 1997. Important aspects of the correspondence were elaborated in articles by Steven Gubser, Igor Klebanov, and Alexander Markovich Polyakov, and by Edward Witten. By 2015, Maldacena's article had over 10,000 citations, becoming the most highly cited article in the field of high energy physics.[3]

34.1 Background

34.1.1 Quantum gravity and strings

Main articles: Quantum gravity and String theory

Our current understanding of gravity is based on Albert Einstein's general theory of relativity.[4] Formulated in 1915, general relativity explains gravity in terms of the geometry of space and time, or spacetime. It is formulated in the language of classical physics[5] developed by physicists such as Isaac Newton and James Clerk Maxwell. The other nongravitational forces are explained in the framework of quantum mechanics. Developed in the first half of the twentieth century by a number of different physicists, quantum mechanics provides a radically different way of describing physical phenomena based on probability.[6]

Quantum gravity is the branch of physics that seeks to describe gravity using the principles of quantum mechanics. Currently, the most popular approach to quantum gravity is string theory,[7] which models elementary particles not as zero-dimensional points but as one-dimensional objects called strings. In the AdS/CFT correspondence, one typically considers theories of quantum gravity derived from string theory or its modern extension, M-theory.[8]

In everyday life, there are three familiar dimensions of space (up/down, left/right, and forward/backward), and there is one dimension of time. Thus, in the language of modern physics, one says that spacetime is four-dimensional.[9] One peculiar feature of string theory and M-theory is that these theories require extra dimensions of spacetime for their mathematical consistency: in string theory spacetime is ten-dimensional, while in M-theory it is eleven-dimensional.[10] The quantum gravity theories appearing in the AdS/CFT correspondence are typically obtained from string and M-theory by a process known as compactification. This produces a theory in which spacetime has effectively a lower number of dimensions and the extra dimensions are "curled up" into circles.[11]

A standard analogy for compactification is to consider a multidimensional object such as a garden hose. If the hose is viewed from a sufficient distance, it appears to have only one dimension, its length, but as one approaches the hose, one discovers that it contains a second dimension, its circumference. Thus, an ant crawling inside it would move in two dimensions.[12]

34.1.2 Quantum field theory

Main articles: Quantum field theory and Conformal field theory

The application of quantum mechanics to physical objects such as the electromagnetic field, which are extended in space and time, is known as quantum field theory.[13] In particle physics, quantum field theories form the basis for our understanding of elementary particles, which are modeled as excitations in the fundamental fields. Quantum field theories are also used throughout condensed matter physics to model particle-like objects called quasiparticles.[14]

In the AdS/CFT correspondence, one considers, in addition to a theory of quantum gravity, a certain kind of quantum field theory called a conformal field theory. This is a particularly symmetric and mathematically well behaved type of quantum field theory.[15] Such theories are often studied in the context of string theory, where they are associated with the surface swept out by a string propagating through spacetime, and in statistical mechanics, where they model systems at a thermodynamic critical point.[16]

34.2 Overview of the correspondence

34.2.1 The geometry of anti-de Sitter space

For more details on the mathematics described here, see Anti-de Sitter space.

In the AdS/CFT correspondence, one considers string theory or M-theory on an anti-de Sitter background. This means that the geometry of spacetime is described in terms of a certain vacuum solution of Einstein's equation called anti-de Sitter space.[17]

In very elementary terms, anti-de Sitter space is a mathematical model of spacetime in which the notion of distance between points (the metric) is different from the notion of distance in ordinary Euclidean geometry. It is closely related to hyperbolic space, which can be viewed as a disk as illustrated on the right.[18] This image shows a tessellation of a disk by triangles and squares. One can define the dis-

A tessellation of the hyperbolic plane by triangles and squares.

tance between points of this disk in such a way that all the triangles and squares are the same size and the circular outer boundary is infinitely far from any point in the interior.[19]

Now imagine a stack of hyperbolic disks where each disk represents the state of the universe at a given time. The resulting geometric object is three-dimensional anti-de Sitter space.[18] It looks like a solid cylinder in which any cross section is a copy of the hyperbolic disk. Time runs along the vertical direction in this picture. The surface of this cylinder plays an important role in the AdS/CFT correspondence. As with the hyperbolic plane, anti-de Sitter space is curved in such a way that any point in the interior is actually infinitely far from this boundary surface.[20]

This construction describes a hypothetical universe with only two space and one time dimension, but it can be generalized to any number of dimensions. Indeed, hyperbolic space can have more than two dimensions and one can "stack up" copies of hyperbolic space to get higher-dimensional models of anti-de Sitter space.[18]

34.2.2 The idea of AdS/CFT

An important feature of anti-de Sitter space is its boundary (which looks like a cylinder in the case of three-dimensional anti-de Sitter space). One property of this boundary is that, locally around any point, it looks just like Minkowski space, the model of spacetime used in nongravitational physics.[21]

One can therefore consider an auxiliary theory in which "spacetime" is given by the boundary of anti-de Sitter space. This observation is the starting point for AdS/CFT corre-

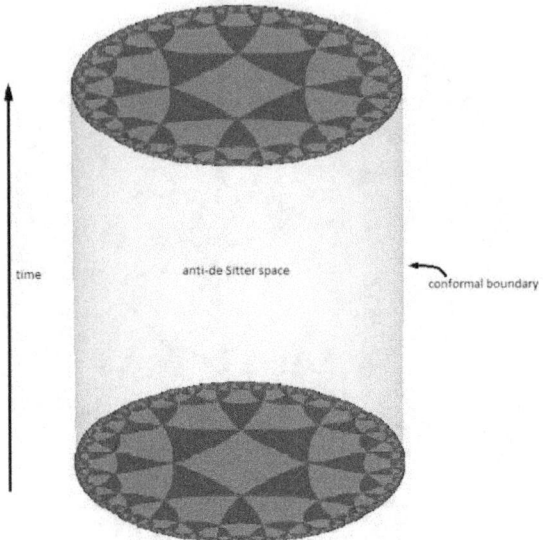

Three-dimensional anti-de Sitter space is like a stack of hyperbolic disks, each one representing the state of the universe at a given time. The resulting spacetime looks like a solid cylinder.

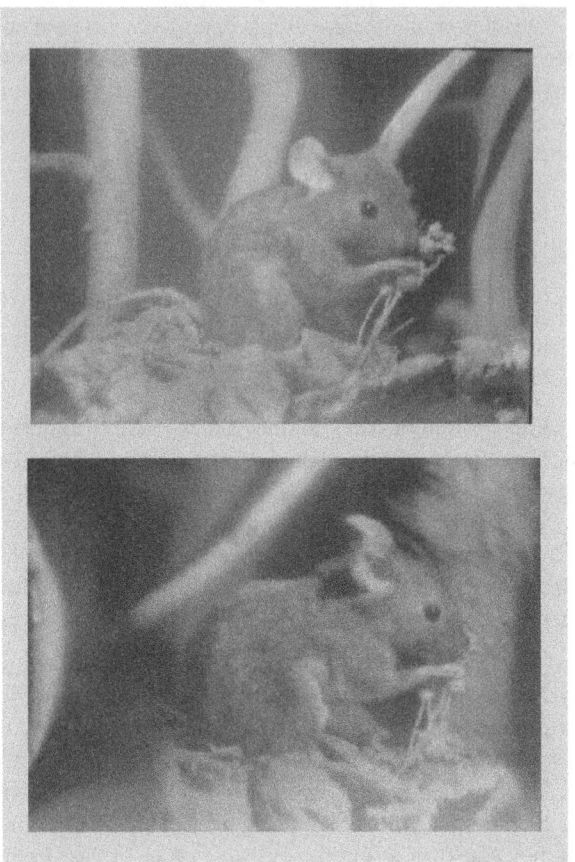

A hologram is a two-dimensional image which stores information about all three dimensions of the object it represents. The two images here are photographs of a single hologram taken from different angles.

spondence, which states that the boundary of anti-de Sitter space can be regarded as the "spacetime" for a conformal field theory. The claim is that this conformal field theory is equivalent to the gravitational theory on the bulk anti-de Sitter space in the sense that there is a "dictionary" for translating calculations in one theory into calculations in the other. Every entity in one theory has a counterpart in the other theory. For example, a single particle in the gravitational theory might correspond to some collection of particles in the boundary theory. In addition, the predictions in the two theories are quantitatively identical so that if two particles have a 40 percent chance of colliding in the gravitational theory, then the corresponding collections in the boundary theory would also have a 40 percent chance of colliding.[22]

Notice that the boundary of anti-de Sitter space has fewer dimensions than anti-de Sitter space itself. For instance, in the three-dimensional example illustrated above, the boundary is a two-dimensional surface. The AdS/CFT correspondence is often described as a "holographic duality" because this relationship between the two theories is similar to the relationship between a three-dimensional object and its image as a hologram.[23] Although a hologram is two-dimensional, it encodes information about all three dimensions of the object it represents. In the same way, theories which are related by the AdS/CFT correspondence are conjectured to be *exactly* equivalent, despite living in different numbers of dimensions. The conformal field theory is like a hologram which captures information about the higher-dimensional quantum gravity theory.[19]

34.2.3   Examples of the correspondence

Following Maldacena's insight in 1997, theorists have discovered many different realizations of the AdS/CFT correspondence. These relate various conformal field theories to compactifications of string theory and M-theory in various numbers of dimensions. The theories involved are generally not viable models of the real world, but they have certain features, such as their particle content or high degree of symmetry, which make them useful for solving problems in quantum field theory and quantum gravity.[24]

The most famous example of the AdS/CFT correspondence states that type IIB string theory on the product space $AdS_5 \times S^5$ is equivalent to N = 4 supersymmetric Yang–Mills theory on the four-dimensional boundary.[25] In this example, the spacetime on which the gravitational theory lives is effectively five-dimensional (hence the notation AdS_5), and there are five additional "compact" dimensions (encoded by the S^5 factor). In the real world, spacetime is four-dimensional, at least macroscopically, so this version of the correspondence does not provide a realistic model of gravity. Likewise, the dual theory is not a viable model of any real-world system as it assumes a large amount of supersymmetry. Nevertheless, as explained below, this boundary theory shares some features in common with quantum chromodynamics, the fundamental theory of the strong force. It describes particles similar to the gluons of quantum chromodynamics together with certain fermions.[7] As a result, it has found applications in nuclear physics, particularly in the study of the quark–gluon plasma.[26]

Another realization of the correspondence states that M-theory on $AdS_7 \times S^4$ is equivalent to the so-called (2,0)-theory in six dimensions.[27] In this example, the spacetime of the gravitational theory is effectively seven-dimensional. The existence of the (2,0)-theory that appears on one side of the duality is predicted by the classification of superconformal field theories. It is still poorly understood because it is a quantum mechanical theory without a classical limit.[28] Despite the inherent difficulty in studying this theory, it is considered to be an interesting object for a variety of reasons, both physical and mathematical.[29]

Yet another realization of the correspondence states that M-theory on $AdS_4 \times S^7$ is equivalent to the ABJM superconformal field theory in three dimensions.[30] Here the gravitational theory has four noncompact dimensions, so this version of the correspondence provides a somewhat more realistic description of gravity.[31]

34.3 Applications to quantum gravity

34.3.1 A non-perturbative formulation of string theory

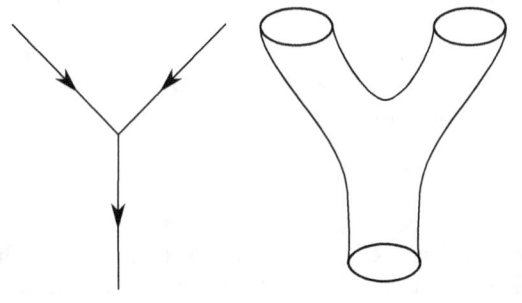

Interaction in the quantum world: world lines of point-like particles or a world sheet swept up by closed strings in string theory.

In quantum field theory, one typically computes the probabilities of various physical events using the techniques of perturbation theory. Developed by Richard Feynman and others in the first half of the twentieth century, perturbative quantum field theory uses special diagrams called Feynman diagrams to organize computations. One imagines that these diagrams depict the paths of point-like particles and their interactions.[32] Although this formalism is extremely useful for making predictions, these predictions are only possible when the strength of the interactions, the coupling constant, is small enough to reliably describe the theory as being close to a theory without interactions.[33]

The starting point for string theory is the idea that the point-like particles of quantum field theory can also be modeled as one-dimensional objects called strings. The interaction of strings is most straightforwardly defined by generalizing the perturbation theory used in ordinary quantum field theory. At the level of Feynman diagrams, this means replacing the one-dimensional diagram representing the path of a point particle by a two-dimensional surface representing the motion of a string. Unlike in quantum field theory, string theory does not yet have a full non-perturbative definition, so many of the theoretical questions that physicists would like to answer remain out of reach.[34]

The problem of developing a non-perturbative formulation of string theory was one of the original motivations for studying the AdS/CFT correspondence.[35] As explained above, the correspondence provides several examples of quantum field theories which are equivalent to string theory on anti-de Sitter space. One can alternatively view this correspondence as providing a *definition* of string theory in the special case where the gravitational field is asymptotically anti-de Sitter (that is, when the gravitational field resembles that of anti-de Sitter space at spatial infinity). Physically interesting quantities in string theory are defined in terms of quantities in the dual quantum field theory.[19]

34.3.2 Black hole information paradox

Main article: Black hole information paradox

In 1975, Stephen Hawking published a calculation which suggested that black holes are not completely black but emit a dim radiation due to quantum effects near the event horizon.[36] At first, Hawking's result posed a problem for theorists because it suggested that black holes destroy information. More precisely, Hawking's calculation seemed to conflict with one of the basic postulates of quantum mechanics, which states that physical systems evolve in time according to the Schrödinger equation. This property is usually referred to as unitarity of time evolution. The apparent contradiction between Hawking's calculation and the unitarity postulate of quantum mechanics came to be known as the black hole information paradox.[37]

The AdS/CFT correspondence resolves the black hole information paradox, at least to some extent, because it shows how a black hole can evolve in a manner consistent with quantum mechanics in some contexts. Indeed, one can consider black holes in the context of the AdS/CFT correspondence, and any such black hole corresponds to a configuration of particles on the boundary of anti-de Sitter space.[38] These particles obey the usual rules of quantum mechanics and in particular evolve in a unitary fashion, so the black hole must also evolve in a unitary fashion, respecting the principles of quantum mechanics.[39] In 2005, Hawking announced that the paradox had been settled in favor of information conservation by the AdS/CFT correspondence, and he suggested a concrete mechanism by which black holes might preserve information.[40]

34.4 Applications to quantum field theory

34.4.1 Nuclear physics

Main article: AdS/QCD

One physical system which has been studied using the AdS/CFT correspondence is the quark–gluon plasma, an exotic state of matter produced in particle accelerators. This state of matter arises for brief instants when heavy ions such as gold or lead nuclei are collided at high energies. Such collisions cause the quarks that make up atomic nuclei to deconfine at temperatures of approximately two trillion kelvins, conditions similar to those present at around 10^{-11} seconds after the Big Bang.[41]

The physics of the quark–gluon plasma is governed by quantum chromodynamics, but this theory is mathematically intractable in problems involving the quark–gluon plasma.[42] In an article appearing in 2005, Đàm Thanh Sơn and his collaborators showed that the AdS/CFT correspondence could be used to understand some aspects of the quark–gluon plasma by describing it in the language of string theory.[26] By applying the AdS/CFT correspondence, Sơn and his collaborators were able to describe the quark gluon plasma in terms of black holes in five-dimensional spacetime. The calculation showed that the ratio of two quantities associated with the quark–gluon plasma, the shear viscosity η and volume density of entropy s, should be approximately equal to a certain universal constant:

$$\frac{\eta}{s} \approx \frac{\hbar}{4\pi k}$$

where $\hbar$ denotes the reduced Planck's constant and k is Boltzmann's constant.[43] In addition, the authors conjectured that this universal constant provides a lower bound for η/s in a large class of systems. In 2008, the predicted value of this ratio for the quark–gluon plasma was confirmed at the Relativistic Heavy Ion Collider at Brookhaven National Laboratory.[44]

Another important property of the quark–gluon plasma is that very high energy quarks moving through the plasma are stopped or "quenched" after traveling only a few femtometers. This phenomenon is characterized by a number $\widehat{q}$ called the jet quenching parameter, which relates the energy loss of such a quark to the squared distance traveled through the plasma. Calculations based on the AdS/CFT correspondence have allowed theorists to estimate $\widehat{q}$, and the results agree roughly with the measured value of this parameter, suggesting that the AdS/CFT correspondence will be useful for developing a deeper understanding of this phenomenon.[45]

34.4.2 Condensed matter physics

A magnet levitating above a high-temperature superconductor. Today some physicists are working to understand high-temperature superconductivity using the AdS/CFT correspondence.[46]

Main article: AdS/CMT

Over the decades, experimental condensed matter physicists have discovered a number of exotic states of matter, including superconductors and superfluids. These states are described using the formalism of quantum field theory, but some phenomena are difficult to explain using standard field theoretic techniques. Some condensed matter theorists including Subir Sachdev hope that the AdS/CFT correspondence will make it possible to describe these systems in the language of string theory and learn more about their behavior.[47]

So far some success has been achieved in using string theory methods to describe the transition of a superfluid to an insulator. A superfluid is a system of electrically neutral atoms that flows without any friction. Such systems are often produced in the laboratory using liquid helium, but recently experimentalists have developed new ways of producing artificial superfluids by pouring trillions of cold atoms into a lattice of criss-crossing lasers. These atoms initially behave as a superfluid, but as experimentalists increase the intensity of the lasers, they become less mobile and then suddenly transition to an insulating state. During the transition, the atoms behave in an unusual way. For example, the atoms slow to a halt at a rate that depends on the temperature and on Planck's constant, the fundamental parameter of quantum mechanics, which does not enter into the description of the other phases. This behavior has recently been understood by considering a dual description where properties of the fluid are described in terms of a higher dimensional black hole.[48]

34.4.3 Criticism

With many physicists turning towards string-based methods to attack problems in nuclear and condensed matter physics, some theorists working in these areas have expressed doubts about whether the AdS/CFT correspondence can provide the tools needed to realistically model real-world systems. In a talk at the Quark Matter conference in 2006,[49] Larry McLerran pointed out that the N=4 super Yang–Mills theory that appears in the AdS/CFT correspondence differs significantly from quantum chromodynamics, making it difficult to apply these methods to nuclear physics. According to McLerran,

> $N = 4$ supersymmetric Yang–Mills is not QCD ... It has no mass scale and is conformally invariant. It has no confinement and no running coupling constant. It is supersymmetric. It has no chiral symmetry breaking or mass generation. It has six scalar and fermions in the adjoint representation ... It may be possible to correct some or all of the above problems, or, for various physical problems, some of the objections may not be relevant. As yet there is not consensus nor compelling arguments for the conjectured fixes or phenomena which would insure that the $N = 4$ supersymmetric Yang Mills results would reliably reflect QCD.[49]

In a letter to Physics Today, Nobel laureate Philip W. Anderson voiced similar concerns about applications of AdS/CFT to condensed matter physics, stating

> As a very general problem with the AdS/CFT approach in condensed-matter theory, we can point to those telltale initials "CFT"—conformal field theory. Condensed-matter problems are, in general, neither relativistic nor conformal. Near a quantum critical point, both time and space may be scaling, but even there we still have a preferred coordinate system and, usually, a lattice. There is some evidence of other linear-T phases to the left of the strange metal about which they are welcome to speculate, but again in this case the condensed-matter problem is overdetermined by experimental facts.[50]

34.5 History and development

Gerard 't Hooft obtained results related to the AdS/CFT correspondence in the 1970s by studying analogies between string theory and nuclear physics.

34.5.1   String theory and nuclear physics

Main articles: History of string theory and 1/N expansion

The discovery of the AdS/CFT correspondence in late 1997 was the culmination of a long history of efforts to relate string theory to nuclear physics.[51] In fact, string theory was originally developed during the late 1960s and early 1970s as a theory of hadrons, the subatomic particles like the proton and neutron that are held together by the strong nuclear force. The idea was that each of these particles could be viewed as a different oscillation mode of a string. In the late 1960s, experimentalists had found that hadrons fall into families called Regge trajectories with squared energy proportional to angular momentum, and theorists showed that this relationship emerges naturally from the physics of a rotating relativistic string.[52]

On the other hand, attempts to model hadrons as strings faced serious problems. One problem was that string theory includes a massless spin-2 particle whereas no such particle appears in the physics of hadrons.[51] Such a particle would mediate a force with the properties of gravity. In 1974, Joel Scherk and John Schwarz suggested that string theory was therefore not a theory of nuclear physics as many theorists had thought but instead a theory of quantum gravity.[53] At the same time, it was realized that hadrons are actually made of quarks, and the string theory approach was abandoned in favor of quantum chromodynamics.[51]

In quantum chromodynamics, quarks have a kind of charge that comes in three varieties called colors. In a paper from 1974, Gerard 't Hooft studied the relationship between string theory and nuclear physics from another point of view by considering theories similar to quantum chromodynamics, where the number of colors is some arbitrary number N, rather than three. In this article, 't Hooft considered a certain limit where N tends to infinity and argued that in this limit certain calculations in quantum field theory resemble calculations in string theory.[54]

34.5.2   Black holes and holography

Main articles: Black hole information paradox, Thorne–Hawking–Preskill bet and Holographic principle

In 1975, Stephen Hawking published a calculation which suggested that black holes are not completely black but emit a dim radiation due to quantum effects near the event horizon.[36] This work extended previous results of Jacob Bekenstein who had suggested that black holes have a well defined entropy.[55] At first, Hawking's result appeared to contradict one of the main postulates of quantum mechanics, namely the unitarity of time evolution. Intuitively, the

Stephen Hawking predicted in 1975 that black holes emit radiation due to quantum effects.

unitarity postulate says that quantum mechanical systems do not destroy information as they evolve from one state to another. For this reason, the apparent contradiction came to be known as the black hole information paradox.[56]

Later, in 1993, Gerard 't Hooft wrote a speculative paper on quantum gravity in which he revisited Hawking's work on black hole thermodynamics, concluding that the total number of degrees of freedom in a region of spacetime surrounding a black hole is proportional to the surface area of the horizon.[57] This idea was promoted by Leonard Susskind and is now known as the holographic principle.[58] The holographic principle and its realization in string theory through the AdS/CFT correspondence have helped elucidate the mysteries of black holes suggested by Hawking's work and are believed to provide a resolution of the black hole information paradox.[39] In 2004, Hawking conceded that black holes do not violate quantum mechanics,[59] and he suggested a concrete mechanism by which they might preserve information.[40]

Juan Maldacena first proposed the AdS/CFT correspondence in late 1997.

Leonard Susskind made early contributions to the idea of holography in quantum gravity.

34.5.3 Maldacena's paper

In late 1997, Juan Maldacena published a landmark paper that initiated the study of AdS/CFT.[27] According to Alexander Markovich Polyakov, "[Maldacena's] work opened the flood gates."[60] The conjecture immediately excited great interest in the string theory community[39] and was considered in articles by Steven Gubser, Igor Klebanov and Polyakov,[61] and by Edward Witten.[62] These papers made Maldacena's conjecture more precise and showed that the conformal field theory appearing in the correspondence lives on the boundary of anti-de Sitter space.[60]

One special case of Maldacena's proposal says that N=4 super Yang–Mills theory, a gauge theory similar in some ways to quantum chromodynamics, is equivalent to string theory in five-dimensional anti-de Sitter space.[30] This result helped clarify the earlier work of 't Hooft on the relationship between string theory and quantum chromodynamics, taking string theory back to its roots as a theory of nuclear physics.[52] Maldacena's results also provided a concrete realization of the holographic principle with important implications for quantum gravity and black hole physics.[1]

By the year 2015, Maldacena's paper had become the most highly cited paper in high energy physics with over 10,000 citations.[3] These subsequent articles have provided considerable evidence that the correspondence is correct, although so far it has not been rigorously proved.[63]

34.5.4 AdS/CFT finds applications

Main articles: AdS/QCD and AdS/CMT

In 1999, after taking a job at Columbia University, nuclear physicist Đàm Thanh Sơn paid a visit to Andrei Starinets, a friend from Sơn's undergraduate days who happened to be doing a Ph.D. in string theory at New York University.[64] Although the two men had no intention of collaborating, Sơn soon realized that the AdS/CFT calculations Starinets was doing could shed light on some aspects of the quark–gluon plasma, an exotic state of matter produced when heavy ions are collided at high energies. In collaboration with Starinets and Pavel Kovtun, Sơn was able to use the AdS/CFT correspondence to calculate a key parameter of the plasma.[26] As Sơn later recalled, "We turned the calculation on its head to give us a prediction for the value of the shear viscosity of a plasma ... A friend of mine in nuclear physics joked that ours was the first useful paper to come out of string theory."[47]

Today physicists continue to look for applications of the AdS/CFT correspondence in quantum field theory.[65] In addition to the applications to nuclear physics advocated by Đàm Thanh Sơn and his collaborators, condensed matter physicists such as Subir Sachdev have used string theory methods to understand some aspects of condensed matter physics. A notable result in this direction was the description, via the AdS/CFT correspondence, of the transition of a superfluid to an insulator.[48] Another emerging subject is the fluid/gravity correspondence, which uses the AdS/CFT correspondence to translate problems in fluid dynamics into problems in general relativity.[66]

34.6 Generalizations

34.6.1 Three-dimensional gravity

Main article: (2+1)-dimensional topological gravity

In order to better understand the quantum aspects of gravity in our four-dimensional universe, some physicists have considered a lower-dimensional mathematical model in which spacetime has only two spatial dimensions and one time dimension.[67] In this setting, the mathematics describing the gravitational field simplifies drastically, and one can study quantum gravity using familiar methods from quantum field theory, eliminating the need for string theory or other more radical approaches to quantum gravity in four dimensions.[68]

Beginning with the work of J. D. Brown and Marc Henneaux in 1986,[69] physicists have noticed that quantum gravity in a three-dimensional spacetime is closely related to two-dimensional conformal field theory. In 1995, Henneaux and his coworkers explored this relationship in more detail, suggesting that three-dimensional gravity in anti-de Sitter space is equivalent to the conformal field theory known as Liouville field theory.[70] Another conjecture formulated by Edward Witten states that three-dimensional gravity in anti-de Sitter space is equivalent to a conformal field theory with monster group symmetry.[71] These conjectures provide examples of the AdS/CFT correspondence that do not require the full apparatus of string or M-theory.[72]

34.6.2 dS/CFT correspondence

Main article: dS/CFT correspondence

Unlike our universe, which is now known to be expanding at

an accelerating rate, anti-de Sitter space is neither expanding nor contracting. Instead it looks the same at all times.[18] In more technical language, one says that anti-de Sitter space corresponds to a universe with negative cosmological constant, whereas the real universe has a small positive cosmological constant.[73]

Although the properties of gravity at short distances should be somewhat independent of the value of the cosmological constant,[74] it is desirable to have a version of the AdS/CFT correspondence for positive cosmological constant. In 2001, Andrew Strominger introduced a version of the duality called the dS/CFT correspondence.[75] This duality involves a model of spacetime called de Sitter space with a positive cosmological constant. Such a duality is interesting from the point of view of cosmology since many cosmologists believe that the very early universe was close to being de Sitter space.[18] Our universe may also resemble de Sitter space in the distant future.[18]

34.6.3 Kerr/CFT correspondence

Main article: Kerr/CFT correspondence

Although the AdS/CFT correspondence is often useful for studying the properties of black holes,[76] most of the black holes considered in the context of AdS/CFT are physically unrealistic. Indeed, as explained above, most versions of the AdS/CFT correspondence involve higher-dimensional models of spacetime with unphysical supersymmetry.

In 2009, Monica Guica, Thomas Hartman, Wei Song, and Andrew Strominger showed that the ideas of AdS/CFT could nevertheless be used to understand certain astrophysical black holes. More precisely, their results apply to black holes that are approximated by extremal Kerr black holes, which have the largest possible angular momentum compatible with a given mass.[77] They showed that such black holes have an equivalent description in terms of conformal field theory. The Kerr/CFT correspondence was later extended to black holes with lower angular momentum.[78]

34.6.4 Higher spin gauge theories

The AdS/CFT correspondence is closely related to another duality conjectured by Igor Klebanov and Alexander Markovich Polyakov in 2002.[79] This duality states that certain "higher spin gauge theories" on anti-de Sitter space are equivalent to conformal field theories with O(N) symmetry. Here the theory in the bulk is a type of gauge theory describing particles of arbitrarily high spin. It is similar to string theory, where the excited modes of vibrating strings correspond to particles with higher spin, and it may help to better understand the string theoretic versions of AdS/CFT and possibly even prove the correspondence.[80] In 2010, Simone Giombi and Xi Yin obtained further evidence for this duality by computing quantities called three-point functions.[81]

34.7 See also

- Algebraic holography

- Ambient construction

- Randall–Sundrum model

34.8 Notes

[1] de Haro et al. 2013, p. 2

[2] Klebanov and Maldacena 2009

[3] "Top Cited Articles of All Time (2014 edition)". INSPIRE-HEP. Retrieved 26 December 2015.

[4] A standard textbook on general relativity is Wald 1984.

[5] Maldacena 2005, p. 58

[6] Griffiths 2004

[7] Maldacena 2005, p. 62

[8] See the subsection entitled "Examples of the correspondence". For examples which do not involve string theory or M-theory, see the section entitled "Generalizations".

[9] Wald 1984, p. 4

[10] Zwiebach 2009, p. 8

[11] Zwiebach 2009, pp. 7–8

[12] This analogy is used for example in Greene 2000, p. 186.

[13] A standard text is Peskin and Schroeder 1995.

[14] For an introduction to the applications of quantum field theory to condensed matter physics, see Zee 2010.

[15] Conformal field theories are characterized by their invariance under conformal transformations.

[16] For an introduction to conformal field theory emphasizing its applications to perturbative string theory, see Volume II of Deligne et al. 1999.

[17] Klebanov and Maldacena 2009, p. 28

[18] Maldacena 2005, p. 60

[19] Maldacena 2005, p. 61

[20] The mathematical relationship between the interior and boundary of anti-de Sitter space is related to the ambient construction of Charles Fefferman and Robin Graham. For details see Fefferman and Graham 1985, Fefferman and Graham 2011.

[21] Zwiebach 2009, p. 552

[22] Maldacena 2005, pp. 61–62

[23] Maldacena 2005, p. 57

[24] The known realizations of AdS/CFT typically involve unphysical numbers of spacetime dimensions and unphysical supersymmetries.

[25] This example is the main subject of the three pioneering articles on AdS/CFT: Maldacena 1998; Gubser, Klebanov, and Polyakov 1998; and Witten 1998.

[26] Merali 2011, p. 303; Kovtun, Son, and Starinets 2001

[27] Maldacena 1998

[28] For a review of the (2,0)-theory, see Moore 2012.

[29] See Moore 2012 and Alday, Gaiotto, and Tachikawa 2010.

[30] Aharony et al. 2008

[31] Aharony et al. 2008, sec. 1

[32] A standard textbook introducing the formalism of Feynman diagrams is Peskin and Schroeder 1995.

[33] Zee 2010, p. 43

[34] Zwiebach 2009, p. 12

[35] Maldacena 1998, sec. 6

[36] Hawking 1975

[37] For an accessible introduction to the black hole information paradox, and the related scientific dispute between Hawking and Leonard Susskind, see Susskind 2008.

[38] Zwiebach 2009, p. 554

[39] Maldacena 2005, p. 63

[40] Hawking 2005

[41] Zwiebach 2009, p. 559

[42] More precisely, one cannot apply the methods of perturbative quantum field theory.

[43] Zwiebach 2009, p. 561; Kovtun, Son, and Starinets 2001

[44] Merali 2011, p. 303; Luzum and Romatschke 2008

[45] Zwiebach 2009, p. 561

[46] Merali 2011

[47] Merali 2011, p. 303

[48] Sachdev 2013, p. 51

[49] McLerran 2007

[50] Anderson, Philip. "Strange connections to strange metals". *Physics Today.* Retrieved 14 August 2013.

[51] Zwiebach 2009, p. 525

[52] Aharony et al. 2008, sec. 1.1

[53] Scherk and Schwarz 1974

[54] 't Hooft 1974

[55] Bekenstein 1973

[56] Susskind 2008

[57] 't Hooft 1993

[58] Susskind 1995

[59] Susskind 2008, p. 444

[60] Polyakov 2008, p. 6

[61] Gubser, Klebanov, and Polyakov 1998

[62] Witten 1998

[63] Maldacena 2005, p. 63; Cowen 2013

[64] Merali 2011, pp. 302–303

[65] Merali 2011; Sachdev 2013

[66] Rangamani 2009

[67] For a review, see Carlip 2003.

[68] According to the results of Witten 1988, three-dimensional quantum gravity can be understood by relating it to Chern–Simons theory.

[69] Brown and Henneaux 1986

[70] Coussaert, Henneaux, and van Driel 1995

[71] Witten 2007

[72] Guica et al. 2009, p. 1

[73] Perlmutter 2003

[74] Biquard 2005, p. 33

[75] Strominger 2001

[76] See the subsection entitled "Black hole information paradox".

[77] Guica et al. 2009

[78] Castro, Maloney, and Strominger 2010

[79] Klebanov and Polyakov 2002

[80] See the Introduction in Klebanov and Polyakov 2002.

[81] Giombi and Yin 2010

34.9   References

• Aharony, Ofer; Bergman, Oren; Jafferis, Daniel Louis; Maldacena, Juan (2008). "$N = 6$ superconformal Chern-Simons-matter theories, M2-branes and their gravity duals". *Journal of High Energy Physics* **2008** (10): 091. arXiv:0806.1218. Bibcode:2008JHEP...10..091A. doi:10.1088/1126-6708/2008/10/091.

• Aharony, Ofer; Gubser, Steven; Maldacena, Juan; Ooguri, Hirosi; Oz, Yaron (2000). "Large N Field Theories, String Theory and Gravity". *Phys. Rept.* **323** (3–4): 183–386. arXiv:hep-th/9905111. Bibcode:1999PhR...323..183A. doi:10.1016/S0370-1573(99)00083-6.

• Alday, Luis; Gaiotto, Davide; Tachikawa, Yuji (2010). "Liouville correlation functions from four-dimensional gauge theories". *Letters in Mathematical Physics* **91** (2): 167–197. arXiv:0906.3219. Bibcode:2010LMaPh..91..167A. doi:10.1007/s11005-010-0369-5.

• Bekenstein, Jacob (1973). "Black holes and entropy". *Physical Review D* **7** (8): 2333–2346. Bibcode:1973PhRvD...7.2333B. doi:10.1103/PhysRevD.7.2333.

• Biquard, Olivier (2005). *AdS/CFT Correspondence: Einstein Metrics and Their Conformal Boundaries.* European Mathematical Society. ISBN 978-3-03719-013-5.

• Brown, J. David; Henneaux, Marc (1986). "Central charges in the canonical realization of asymptotic symmetries: an example from three dimensional gravity". *Communications in Mathematical Physics* **104** (2): 207–226. Bibcode:1986CMaPh.104..207B. doi:10.1007/BF01211590.

• Carlip, Steven (2003). *Quantum Gravity in 2+1 Dimensions.* Cambridge Monographs on Mathematical Physics. ISBN 978-0-521-54588-4.

• Castro, Alejandra; Maloney, Alexander; Strominger, Andrew (2010). "Hidden conformal symmetry of the Kerr black hole". *Physical Review D* **82** (2). arXiv:1004.0996. Bibcode:2010PhRvD..82b4008C. doi:10.1103/PhysRevD.82.024008.

• Coussaert, Olivier; Henneaux, Marc; van Driel, Peter (1995). "The asymptotic dynamics of three-dimensional Einstein gravity with a negative cosmological constant". *Classical and Quantum Gravity* **12** (12): 2961–2966. arXiv:gr-qc/9506019.

Bibcode:1995CQGra..12.2961C. doi:10.1088/0264-9381/12/12/012.

- Cowen, Ron (2013). "Simulations back up theory that Universe is a hologram". *Nature News & Comment.* doi:10.1038/nature.2013.14328. Retrieved 21 December 2013.

- de Haro, Sebastian; Dieks, Dennis; 't Hooft, Gerard; Verlinde, Erik (2013). "Forty Years of String Theory Reflecting on the Foundations". *Foundations of Physics* **43** (1): 1–7. Bibcode:2013FoPh...43....1D. doi:10.1007/s10701-012-9691-3.

- Deligne, Pierre; Etingof, Pavel; Freed, Daniel; Jeffery, Lisa; Kazhdan, David; Morgan, John; Morrison, David; Witten, Edward, eds. (1999). *Quantum Fields and Strings: A Course for Mathematicians.* American Mathematical Society. ISBN 978-0-8218-2014-8.

- Fefferman, Charles; Graham, Robin (1985). "Conformal invariants". *Asterisque*: 95–116.

- Fefferman, Charles; Graham, Robin (2011). *The Ambient Metric.* Princeton University Press. ISBN 978-1-4008-4058-8.

- Giombi, Simone; Yin, Xi (2010). "Higher spin gauge theory and holography: the three-point functions". *Journal of High Energy Physics* **2010** (9): 1–80. arXiv:0912.3462. Bibcode:2010JHEP...09..115G. doi:10.1007/JHEP09(2010)115.

- Greene, Brian (2000). *The Elegant Universe: Superstrings, Hidden Dimensions, and the Quest for the Ultimate Theory.* Random House. ISBN 978-0-9650888-0-0.

- Griffiths, David (2004). *Introduction to Quantum Mechanics.* Pearson Prentice Hall. ISBN 978-0-13-111892-8.

- Gubser, Steven; Klebanov, Igor; Polyakov, Alexander (1998). "Gauge theory correlators from non-critical string theory". *Physics Letters B* **428**: 105–114. arXiv:hep-th/9802109. Bibcode:1998PhLB..428..105G. doi:10.1016/S0370-2693(98)00377-3.

- Guica, Monica; Hartman, Thomas; Song, Wei; Strominger, Andrew (2009). "The Kerr/CFT Correspondence". *Physical Review D* **80** (12). arXiv:0809.4266. Bibcode:2009PhRvD..80l4008G. doi:10.1103/PhysRevD.80.124008.

- Hawking, Stephen (1975). "Particle creation by black holes". *Communications in Mathematical Physics* **43** (3): 199–220. Bibcode:1975CMaPh..43..199H. doi:10.1007/BF02345020.

- Hawking, Stephen (2005). "Information loss in black holes". *Physical Review D* **72** (8). arXiv:hep-th/0507171. Bibcode:2005PhRvD..72h4013H. doi:10.1103/PhysRevD.72.084013.

- Klebanov, Igor; Maldacena, Juan (2009). "Solving Quantum Field Theories via Curved Spacetimes" (PDF). *Physics Today* **62**: 28–33. Bibcode:2009PhT....62a..28K. doi:10.1063/1.3074260. Retrieved May 2013.

- Klebanov, Igor; Polyakov, Alexander (2002). "The AdS dual of the critical O(N) vector model". *Physics Letters B* **550** (3–4): 213–219. arXiv:hep-th/0210114. Bibcode:2002PhLB..550..213K. doi:10.1016/S0370-2693(02)02980-5.

- Kovtun, P. K.; Son, Dam T.; Starinets, A. O. (2001). "Viscosity in strongly interacting quantum field theories from black hole physics". *Physical Review Letters* **94** (11): 111601. arXiv:hep-th/0405231. Bibcode:2005PhRvL..94k1601K. doi:10.1103/PhysRevLett.94.111601. PMID 15903845.

- Luzum, Matthew; Romatschke, Paul (2008). "Conformal relativistic viscous hydrodynamics: Applications to RHIC results at $\sqrt{s_{NN}}$ = 200 GeV". *Physical Review C* **78** (3). arXiv:0804.4015. doi:10.1103/PhysRevC.78.034915.

- Maldacena, Juan (1998). "The Large N limit of superconformal field theories and supergravity". *Advances in Theoretical and Mathematical Physics* **2**: 231–252. arXiv:hep-th/9711200. Bibcode:1998AdTMP...2..231M. doi:10.1023/A:1026654312961.

- Maldacena, Juan (2005). "The Illusion of Gravity" (PDF). *Scientific American* **293** (5): 56–63. Bibcode:2005SciAm.293e..56M. doi:10.1038/scientificamerican1105-56. PMID 16318027. Retrieved July 2013.

- McLerran, Larry (2007). "Theory Summary : Quark Matter 2006". *Journal of Physics G: Nuclear and Particle Physics* **34** (8): S583. arXiv:hep-ph/0702004. Bibcode:2007JPhG...34..583M. doi:10.1088/0954-3899/34/8/S50.

- Merali, Zeeya (2011). "Collaborative physics: string theory finds a bench mate". *Nature* **478** (7369): 302–304. Bibcode:2011Natur.478..302M. doi:10.1038/478302a. PMID 22012369.

- Moore, Gregory (2012). "Lecture Notes for Felix Klein Lectures" (PDF). Retrieved 14 August 2013.

- Perlmutter, Saul (2003). "Supernovae, dark energy, and the accelerating universe". *Physics Today* **56** (4): 53–62. Bibcode:2003PhT....56d..53P. doi:10.1063/1.1580050.

- Peskin, Michael; Schroeder, Daniel (1995). *An Introduction to Quantum Field Theory*. Westview Press. ISBN 978-0-201-50397-5.

- Polyakov, Alexander (2008). "From Quarks to Strings". arXiv:0812.0183 [hep-th].

- Rangamani, Mukund (2009). "Gravity and Hydrodynamics: Lectures on the fluid-gravity correspondence". *Classical and quantum gravity* **26** (22): 4003. arXiv:0905.4352. Bibcode:2009CQGra..26v4003R. doi:10.1088/0264-9381/26/22/224003.

- Sachdev, Subir (2013). "Strange and stringy". *Scientific American* **308** (44): 44–51. Bibcode:2012SciAm.308a..44S. doi:10.1038/scientificamerican0113-44.

- Scherk, Joel; Schwarz, John (1974). "Dual models for non-hadrons". *Nuclear Physics B* **81** (1): 118–144. Bibcode:1974NuPhB..81..118S. doi:10.1016/0550-3213(74)90010-8.

- Strominger, Andrew (2001). "The dS/CFT correspondence". *Journal of High Energy Physics* **2001** (10): 034. arXiv:hep-th/0106113. Bibcode:2001JHEP...10..034S. doi:10.1088/1126-6708/2001/10/034.

- Susskind, Leonard (1995). "The World as a Hologram". *Journal of Mathematical Physics* **36** (11): 6377–6396. arXiv:hep-th/9409089. Bibcode:1995JMP....36.6377S. doi:10.1063/1.531249.

- Susskind, Leonard (2008). *The Black Hole War: My Battle with Stephen Hawking to Make the World Safe for Quantum Mechanics*. Little, Brown and Company. ISBN 978-0-316-01641-4.

- 't Hooft, Gerard (1974). "A planar diagram theory for strong interactions". *Nuclear Physics B* **72** (3): 461–473. Bibcode:1974NuPhB..72..461T. doi:10.1016/0550-3213(74)90154-0.

- 't Hooft, Gerard (1993). "Dimensional Reduction in Quantum Gravity". arXiv:gr-qc/9310026.

- Wald, Robert (1984). *General Relativity*. University of Chicago Press. ISBN 978-0-226-87033-5.

- Witten, Edward (1988). "2+1 dimensional gravity as an exactly soluble system". *Nuclear Physics B*

- **311** (1): 46–78. Bibcode:1988NuPhB.311...46W. doi:10.1016/0550-3213(88)90143-5.

- Witten, Edward (1998). "Anti-de Sitter space and holography". *Advances in Theoretical and Mathematical Physics* **2**: 253–291. arXiv:hep-th/9802150. Bibcode:1998AdTMP...2..253W.

- Witten, Edward (2007). "Three-dimensional gravity revisited". arXiv:0706.3359 [hep-th].

- Zee, Anthony (2010). *Quantum Field Theory in a Nutshell* (2nd ed.). Princeton University Press. ISBN 978-0-691-14034-6.

- Zwiebach, Barton (2009). *A First Course in String Theory*. Cambridge University Press. ISBN 978-0-521-88032-9.

Chapter 35

Quasiparticle

In physics, **quasiparticles** and **collective excitations** (which are closely related) are emergent phenomena that occur when a microscopically complicated system such as a solid behaves *as if* it contained different weakly interacting particles in free space. For example, as an electron travels through a semiconductor, its motion is disturbed in a complex way by its interactions with all of the other electrons and nuclei; however it *approximately* behaves like an electron with a *different mass* traveling unperturbed through free space. This "electron" with a different mass is called an "electron quasiparticle".[1] In another example, the aggregate motion of electrons in the valence band of a semiconductor is the same as if the semiconductor contained instead positively charged quasiparticles called holes. Other quasiparticles or collective excitations include phonons (particles derived from the vibrations of atoms in a solid), plasmons (particles derived from plasma oscillations), and many others.

These particles are typically called "quasiparticles" if they are related to fermions (like electrons and holes), and called "collective excitations" if they are related to bosons (like phonons and plasmons),[1] although the precise distinction is not universally agreed upon.[2]

The quasiparticle concept is most important in condensed matter physics, since it is one of the few known ways of simplifying the quantum mechanical many-body problem.

35.1 Overview

35.1.1 General introduction

Solids are made of only three kinds of particles: Electrons, protons, and neutrons. Quasiparticles are none of these; instead they are an *emergent phenomenon* that occurs inside the solid. Therefore, while it is quite possible to have a single particle (electron or proton or neutron) floating in space, a *quasi*particle can instead only exist inside the solid.

Motion in a solid is extremely complicated: Each electron and proton gets pushed and pulled (by Coulomb's law) by all the other electrons and protons in the solid (which may themselves be in motion). It is these strong interactions that make it very difficult to predict and understand the behavior of solids (see many-body problem). On the other hand, the motion of a *non-interacting* particle is quite simple: In classical mechanics, it would move in a straight line, and in quantum mechanics, it would move in a superposition of plane waves. This is the motivation for the concept of quasiparticles: The complicated motion of the *actual* particles in a solid can be mathematically transformed into the much simpler motion of imagined *quasi*particles, which behave more like non-interacting particles.

In summary, quasiparticles are a mathematical tool for simplifying the description of solids. They are not "real" particles inside the solid. Instead, saying "A quasiparticle is present" or "A quasiparticle is moving" is shorthand for saying "A large number of electrons and nuclei are moving in a specific coordinated way."

35.1.2 Relation to many-body quantum mechanics

The principal motivation for quasiparticles is that it is almost impossible to *directly* describe every particle in a macroscopic system. For example, a barely-visible (0.1mm) grain of sand contains around 10^{17} atoms and 10^{18} electrons. Each of these attracts or repels every other by Coulomb's law. In quantum mechanics, a system is described by a wavefunction, which, if the particles are interacting (as they are in our case), depends on the position of every particle in the system. So, each particle adds three independent variables to the wavefunction, one for each coordinate needed to describe the position of that particle. Because of this, directly approaching the many-body problem of 10^{18} interacting electrons by straightforwardly trying to solve the appropriate Schrödinger equation is impossible in practice, since it amounts to solving a partial differential equation not just in three dimensions, but in 3×10^{18} di-

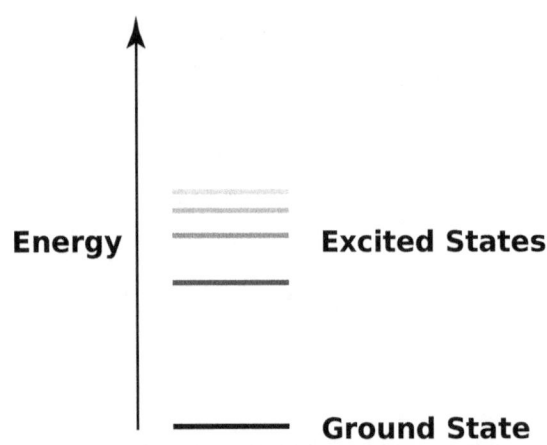

Energy   **Excited States**

Ground State

Any system, no matter how complicated, has a ground state along with an infinite series of higher-energy excited states.

mensions – one for each component of the position of each particle.

One simplifying factor is that the system as a whole, like any quantum system, has a ground state and various excited states with higher and higher energy above the ground state. In many contexts, only the "low-lying" excited states, with energy reasonably close to the ground state, are relevant. This occurs because of the Boltzmann distribution, which implies that very-high-energy thermal fluctuations are unlikely to occur at any given temperature.

Quasiparticles and collective excitations are a type of low-lying excited state. For example, a crystal at absolute zero is in the ground state, but if one phonon is added to the crystal (in other words, if the crystal is made to vibrate slightly at a particular frequency) then the crystal is now in a low-lying excited state. The single phonon is called an *elementary excitation*. More generally, low-lying excited states may contain any number of elementary excitations (for example, many phonons, along with other quasiparticles and collective excitations).[3]

When the material is characterized as having "several elementary excitations", this statement presupposes that the different excitations can be combined together. In other words, it presupposes that the excitations can coexist simultaneously and independently. This is never *exactly* true. For example, a solid with two identical phonons does not have exactly twice the excitation energy of a solid with just one phonon, because the crystal vibration is slightly anharmonic. However, in many materials, the elementary excitations are very *close* to being independent. Therefore, as a *starting point*, they are treated as free, independent entities, and then corrections are included via interactions between the elementary excitations, such as "phonon-phonon scattering".

Therefore, using quasiparticles / collective excitations, instead of analyzing 10^{18} particles, one needs to deal with only a handful of somewhat-independent elementary excitations. It is therefore a very effective approach to simplify the many-body problem in quantum mechanics. This approach is not useful for *all* systems however: In strongly correlated materials, the elementary excitations are so far from being independent that it is not even useful as a starting point to treat them as independent.

35.1.3 Distinction between quasiparticles and collective excitations

Usually, an elementary excitation is called a "quasiparticle" if it is a fermion and a "collective excitation" if it is a boson.[1] However, the precise distinction is not universally agreed upon.[2]

There is a difference in the way that quasiparticles and collective excitations are intuitively envisioned.[2] A quasiparticle is usually thought of as being like a dressed particle: It is built around a real particle at its "core", but the behavior of the particle is affected by the environment. A standard example is the "electron quasiparticle": A real electron particle, in a crystal, behaves as if it had a different mass. On the other hand, a collective excitation is usually imagined to be a reflection of the aggregate behavior of the system, with no single real particle at its "core". A standard example is the phonon, which characterizes the vibrational motion of every atom in the crystal.

However, these two visualizations leave some ambiguity. For example, a magnon in a ferromagnet can be considered in one of two perfectly equivalent ways: (a) as a mobile defect (a misdirected spin) in a perfect alignment of magnetic moments or (b) as a quantum of a collective spin wave that involves the precession of many spins. In the first case, the magnon is envisioned as a quasiparticle, in the second case, as a collective excitation. However, both (a) and (b) are equivalent and correct descriptions. As this example shows, the intuitive distinction between a quasiparticle and a collective excitation is not particularly important or fundamental.

The problems arising from the collective nature of quasiparticles have also been discussed within the philosophy of science, notably in relation to the identity conditions of quasiparticles and whether they should be considered "real" by the standards of, for example, entity realism.[4][5]

35.1.4 Effect on bulk properties

By investigating the properties of individual quasiparticles, it is possible to obtain a great deal of information about low-energy systems, including the flow properties and heat capacity.

In the heat capacity example, a crystal can store energy by forming phonons, and/or forming excitons, and/or forming plasmons, etc. Each of these is a separate contribution to the overall heat capacity.

35.1.5 History

The idea of quasiparticles originated in Lev Landau's theory of Fermi liquids, which was originally invented for studying liquid helium-3. For these systems a strong similarity exists between the notion of quasi-particle and dressed particles in quantum field theory. The dynamics of Landau's theory is defined by a kinetic equation of the mean-field type. A similar equation, the Vlasov equation, is valid for a plasma in the so-called plasma approximation. In the plasma approximation, charged particles are considered to be moving in the electromagnetic field collectively generated by all other particles, and hard collisions between the charged particles are neglected. When a kinetic equation of the mean-field type is a valid first-order description of a system, second-order corrections determine the entropy production, and generally take the form of a Boltzmann-type collision term, in which figure only "far collisions" between virtual particles. In other words, every type of mean-field kinetic equation, and in fact every mean-field theory, involves a quasi-particle concept.

35.2 Examples of quasiparticles and collective excitations

This section contains examples of quasiparticles and collective excitations. The first subsection below contains common ones that occur in a wide variety of materials under ordinary conditions; the second subsection contains examples that arise in particular, special contexts.

35.2.1 More common examples

See also: List of quasiparticles

- In solids, an **electron quasiparticle** is an electron as affected by the other forces and interactions in the solid. The electron quasiparticle has the same charge and spin as a "normal" (elementary particle) electron, and like a normal electron, it is a fermion. However, its mass can differ substantially from that of a normal electron; see the article effective mass.[1] Its

electric field is also modified, as a result of electric field screening. In many other respects, especially in metals under ordinary conditions, these so-called Landau quasiparticles closely resemble familiar electrons; as Crommie's "quantum corral" showed, an STM can clearly image their interference upon scattering.

- A **hole** is a quasiparticle consisting of the lack of an electron in a state; it is most commonly used in the context of empty states in the valence band of a semiconductor.[1] A hole has the opposite charge of an electron.

- A **phonon** is a collective excitation associated with the vibration of atoms in a rigid crystal structure. It is a quantum of a sound wave.

- A **magnon** is a collective excitation[1] associated with the electrons' spin structure in a crystal lattice. It is a quantum of a spin wave.

- A **roton** is a collective excitation associated with the rotation of a fluid (often a superfluid). It is a quantum of a vortex.

- In materials, a **photon** quasiparticle is a photon as affected by its interactions with the material. In particular, the photon quasiparticle has a modified relation between wavelength and energy (dispersion relation), as described by the material's index of refraction. It may also be termed a **polariton**, especially near a resonance of the material. For example, an **exciton-polariton** is a superposition of an exciton and a photon; a **phonon-polariton** is a superposition of a phonon and a photon.

- A **plasmon** is a collective excitation, which is the quantum of plasma oscillations (wherein all the electrons simultaneously oscillate with respect to all the ions).

- A **polaron** is a quasiparticle which comes about when an electron interacts with the polarization of its surrounding ions.

- An **exciton** is an electron and hole bound together.

- A **plasmariton** is a coupled optical phonon and dressed photon consisting of a plasmon and photon.

35.2.2 More specialized examples

- Composite fermions arise in a two-dimensional system subject to a large magnetic field, most famously those systems that exhibit the fractional quantum Hall effect.[6] These quasiparticles are quite unlike normal particles in two ways. First, their charge can be less than the electron charge *e*. In fact, they have been observed with charges of e/3, e/4, e/5, and e/7.[7] Second, they can be anyons, an exotic type of particle that is neither a fermion nor boson.[8]

- Stoner excitations in ferromagnetic metals

- Bogoliubov quasiparticles in superconductors. Superconductivity is carried by Cooper pairs—usually described as pairs of electrons—that move through the crystal lattice without resistance. A broken Cooper pair is called a Bogoliubov quasiparticle.[9] It differs from the conventional quasiparticle in metal because it combines the properties of a negatively charged electron and a positively charged hole (an electron void). Physical objects like impurity atoms, from which quasiparticles scatter in an ordinary metal, only weakly affect the energy of a Cooper pair in a conventional superconductor. In conventional superconductors, interference between Bogoliubov quasiparticles is tough for an STM to see. Because of their complex global electronic structures, however, high-Tc cuprate superconductors are another matter. Thus Davis and his colleagues were able to resolve distinctive patterns of quasiparticle interference in Bi-2212.[10]

- A Majorana fermion is a particle which equals its own antiparticle, and can emerge as a quasiparticle in certain superconductors, or in a quantum spin liquid.[11]

- Magnetic monopoles arise in condensed matter systems such as spin ice and carry an effective magnetic charge as well as being endowed with other typical quasiparticle properties such as an effective mass. They may be formed through spin flips in frustrated pyrochlore ferromagnets and interact through a Coulomb potential.

- Skyrmions

- Spinon is represented by quasiparticle produced as a result of electron spin-charge separation, and can form both quantum spin liquid and strongly correlated quantum spin liquid in some minerals like Herbertsmithite.[12]

35.3　See also

- Fractionalization

- List of quasiparticles

- Mean field theory

- Pseudoparticle

35.4　References

[1] E. Kaxiras, *Atomic and Electronic Structure of Solids*, ISBN 0-521-52339-7, pages 65–69.

[2] *A guide to Feynman diagrams in the many-body problem*, by Richard D. Mattuck, p10. "As we have seen, the quasi particle consists of the original real, individual particle, plus a cloud of disturbed neighbors. It behaves very much like an individual particle, except that it has an effective mass and a lifetime. But there also exist other kinds of fictitious particles in many-body systems, i.e. 'collective excitations'. These do not center around individual particles, but instead involve collective, wavelike motion of *all* the particles in the system simultaneously."

[3] *Principles of Nanophotonics* by Motoichi Ohtsu, p205 google books link

[4] A. Gelfert, 'Manipulative Success and the Unreal', *International Studies in the Philosophy of Science* Vol. 17, 2003, 245–263

[5] B. Falkenburg, *Particle Metaphysics* (The Frontiers Collection), Berlin: Springer 2007, esp. pp. 243–46

[6] Physics Today Article

[7] Cosmos magazine June 2008

[8] Nature article

[9] "Josephson Junctions". *Science and Technology Review*. Lawrence Livermore National Laboratory.

[10] J. E. Hoffman; McElroy, K; Lee, DH; Lang, KM; Eisaki, H; Uchida, S; Davis, JC; et al. (2002). "Imaging Quasiparticle Interference in Bi2Sr2CaCu2O8+". *Science* **297** (5584): 1148–51. arXiv:cond-mat/0209276. Bibcode:2002Sci...297.1148H. doi:10.1126/science.1072640. PMID 12142440.

[11] Banerjee, A.; Bridges, C. A.; Yan, J.-Q.; et al. (4 April 2016). "Proximate Kitaev quantum spin liquid behaviour in a honeycomb magnet". *Nature Materials*. doi:10.1038/nmat4604. (subscription required (help)).

[12] Shaginyan, V. R.; et al. (2012). "Identification of Strongly Correlated Spin Liquid in Herbertsmithite". *EPL* **97** (5): 56001. arXiv:1111.0179. Bibcode:2012EL.....9756001S. doi:10.1209/0295-5075/97/56001.

35.5　Further reading

- L. D. Landau, *Soviet Phys. JETP.* 3:920 (1957)

- L. D. Landau, *Soviet Phys. JETP.* 5:101 (1957)

- A. A. Abrikosov, L. P. Gor'kov, and I. E. Dzyaloshinski, *Methods of Quantum Field Theory in Statistical Physics* (1963, 1975). Prentice-Hall, New Jersey; Dover Publications, New York.

- D. Pines, and P. Nozières, *The Theory of Quantum Liquids* (1966). W.A. Benjamin, New York. *Volume I: Normal Fermi Liquids* (1999). Westview Press, Boulder.

- J. W. Negele, and H. Orland, *Quantum Many-Particle Systems* (1998). Westview Press, Boulder

- Amusia, M., Popov, K., Shaginyan, V., Stephanovich, V. (2014). *Theory of Heavy-Fermion Compounds - Theory of Strongly Correlated Fermi-Systems*. Springer. ISBN 978-3-319-10825-4.

35.6 External links

- PhysOrg.com – Scientists find new 'quasiparticles'

- Curious 'quasiparticles' baffle physicists by Jacqui Hayes, Cosmos 6 June 2008. Accessed June 2008

Chapter 36

Vacuum solution

A **vacuum solution** is a solution of a field equation in which the sources of the field are taken to be identically zero. That is, such field equations are written without matter interaction (i.e.- set to zero).

36.1 Examples

- In Maxwell's theory of electromagnetism, a vacuum solution would represent the electromagnetic field in a region of space where there are no electromagnetic sources (charges and electric currents), i.e. where the current 4-vector vanishes:[1]

$$J^a = 0$$

- Einstein's theory of general relativity where a vacuum solution[2] would represent the gravitational field in a region of spacetime where there are no gravitational sources (masses), i.e. where the energy-momentum tensor vanishes:[3]

$$T_{ab} = 0$$

- Kaluza–Klein vacuum (static) field equations[4]

- Kasner vacuum solution[5]

36.2 See also

- Einstein field equations

- Kasner metric

- Kerr metric

- Maxwell's equations

- Vacuum solution (general relativity)

36.3 Notes

[1] Esposito, S. (1997), "Classical vgr? c solutions of Maxwell's equations and the photon tunneling effect", *Physics Letters A* **225** (4-6): 203–209, arXiv:physics/9611018, Bibcode:1997PhLA..225..203E, doi:10.1016/S0375-9601(96)00872-9, retrieved 2009-07-04

[2] Stephani, H. (2003), *Exact solutions of Einstein's field equations* (PDF), retrieved 2009-07-04

[3] Quevedo, H. (1990), "Multipole Moments in General Relativity-Static and Stationary Vacuum Solutions", *Fortschritte der Physik* **38** (10): 733, Bibcode:1990ForPh..38..733Q, doi:10.1002/prop.2190381002, retrieved 2009-07-04

[4] Sorkin, R.D. (1983), "Kaluza-klein monopole", *Physical Review Letters* **51** (2): 87–90, Bibcode:1983PhRvL..51...87S, doi:10.1103/PhysRevLett.51.87

[5] Chodos, A.; Detweiler, S. (1980), "Where has the fifth dimension gone?", *Physical Review D* **21** (8): 2167–2170, Bibcode:1980PhRvD..21.2167C, doi:10.1103/PhysRevD.21.2167

36.4 References

- Gott, J.R.; Richard, J. (1985), "Gravitational lensing effects of vacuum strings- Exact solutions", *Astrophysical Journal* **288** (Part 1), Bibcode:1985ApJ...288..422G, doi:10.1086/162808

- Friedrich, H.; Nagy, G. (1999), "The initial boundary value problem for Einstein's vacuum field equation" (PDF), *Communications in Mathematical Physics* **201** (3): 619–655, Bibcode:1999CMaPh.201..619F, doi:10.1007/s002200050571, retrieved 2009-07-04

- Appelquist, T.; Chodos, A. (1983), "Quantum dynamics of Kaluza-Klein theories", *Physical Review D* **28** (4): 772–784, Bibcode:1983PhRvD..28..772A, doi:10.1103/PhysRevD.28.772

Chapter 37

Einstein field equations

The **Einstein field equations** (**EFE**; also known as "**Einstein's equations**") are the set of 10 equations in Albert Einstein's general theory of relativity that describes the fundamental interaction of gravitation as a result of spacetime being curved by matter and energy.[1] First published by Einstein in 1915[2] as a tensor equation, the EFE equate local spacetime curvature (expressed by the Einstein tensor) with the local energy and momentum within that spacetime (expressed by the stress–energy tensor).[3]

Similar to the way that electromagnetic fields are determined using charges and currents via Maxwell's equations, the EFE are used to determine the spacetime geometry resulting from the presence of mass–energy and linear momentum, that is, they determine the metric tensor of spacetime for a given arrangement of stress–energy in the spacetime. The relationship between the metric tensor and the Einstein tensor allows the EFE to be written as a set of nonlinear partial differential equations when used in this way. The solutions of the EFE are the components of the metric tensor. The inertial trajectories of particles and radiation (geodesics) in the resulting geometry are then calculated using the geodesic equation.

As well as obeying local energy–momentum conservation, the EFE reduce to Newton's law of gravitation where the gravitational field is weak and velocities are much less than the speed of light.[4]

Exact solutions for the EFE can only be found under simplifying assumptions such as symmetry. Special classes of exact solutions are most often studied as they model many gravitational phenomena, such as rotating black holes and the expanding universe. Further simplification is achieved in approximating the actual spacetime as flat spacetime with a small deviation, leading to the linearised EFE. These equations are used to study phenomena such as gravitational waves.

37.1   Mathematical form

The Einstein field equations (EFE) may be written in the form:[5][1]

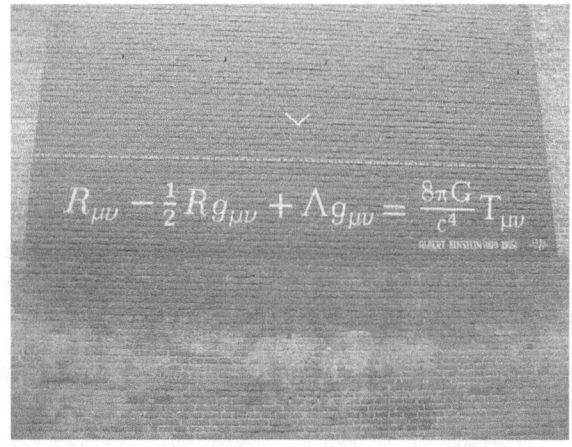

EFE on a wall in Leiden

where $R_{\mu\nu}$ is the Ricci curvature tensor, R is the scalar curvature, $g_{\mu\nu}$ is the metric tensor, Λ is the cosmological constant, G is Newton's gravitational constant, c is the speed of light in vacuum, and $T_{\mu\nu}$ is the stress–energy tensor.

The EFE is a tensor equation relating a set of symmetric 4×4 tensors. Each tensor has 10 independent components. The four Bianchi identities reduce the number of independent equations from 10 to 6, leaving the metric with four gauge fixing degrees of freedom, which correspond to the freedom to choose a coordinate system.

Although the Einstein field equations were initially formulated in the context of a four-dimensional theory, some theorists have explored their consequences in n dimensions. The equations in contexts outside of general relativity are still referred to as the Einstein field equations. The vacuum field equations (obtained when T is identically zero) define Einstein manifolds.

Despite the simple appearance of the equations they are actually quite complicated. Given a specified distribution of matter and energy in the form of a stress–energy tensor, the EFE are understood to be equations for the metric tensor $g_{\mu\nu}$, as both the Ricci tensor and scalar curvature depend on the metric in a complicated nonlinear manner. In fact, when fully written out, the EFE are a system of 10 coupled, nonlinear, hyperbolic-elliptic partial differential equations.

One can write the EFE in a more compact form by defining the Einstein tensor

$$G_{\mu\nu} = R_{\mu\nu} - \tfrac{1}{2} R\, g_{\mu\nu},$$

which is a symmetric second-rank tensor that is a function of the metric. The EFE can then be written as

$$G_{\mu\nu} + \Lambda g_{\mu\nu} = \frac{8\pi G}{c^4} T_{\mu\nu}.$$

Using geometrized units where $G = c = 1$, this can be rewritten as

$$G_{\mu\nu} + \Lambda g_{\mu\nu} = 8\pi T_{\mu\nu}.$$

The expression on the left represents the curvature of spacetime as determined by the metric; the expression on the right represents the matter/energy content of spacetime. The EFE can then be interpreted as a set of equations dictating how matter/energy determines the curvature of spacetime.

These equations, together with the geodesic equation,[6] which dictates how freely-falling matter moves through space-time, form the core of the mathematical formulation of general relativity.

37.1.1   Sign convention

The above form of the EFE is the standard established by Misner, Thorne, and Wheeler.[7] The authors analyzed all conventions that exist and classified according to the following three signs (S1, S2, S3):

$$g_{\mu\nu} = [S1] \times \operatorname{diag}(-1, +1, +1, +1)$$

$$R^{\mu}{}_{\alpha\beta\gamma} = [S2] \times (\Gamma^{\mu}_{\alpha\gamma,\beta} - \Gamma^{\mu}_{\alpha\beta,\gamma} + \Gamma^{\mu}_{\sigma\beta}\Gamma^{\sigma}_{\gamma\alpha} - \Gamma^{\mu}_{\sigma\gamma}\Gamma^{\sigma}_{\beta\alpha})$$

$$G_{\mu\nu} = [S3] \times \frac{8\pi G}{c^4} T_{\mu\nu}$$

The third sign above is related to the choice of convention for the Ricci tensor:

$$R_{\mu\nu} = [S2] \times [S3] \times R^{\alpha}{}_{\mu\alpha\nu}$$

With these definitions Misner, Thorne, and Wheeler classify themselves as (+ + +), whereas Weinberg (1972)[8] is (+ − −), Peebles (1980) and Efstathiou (1990) are (− + +), while Peacock (1994), Rindler (1977), Atwater (1974), Collins Martin & Squires (1989) are (− + −).

Authors including Einstein have used a different sign in their definition for the Ricci tensor which results in the sign of the constant on the right side being negative

$$R_{\mu\nu} - \frac{1}{2} R\, g_{\mu\nu} - \Lambda g_{\mu\nu} = -\frac{8\pi G}{c^4} T_{\mu\nu}.$$

The sign of the (very small) cosmological term would change in both these versions, if the (+ − − −) metric sign convention is used rather than the MTW (− + + +) metric sign convention adopted here.

37.1.2   Equivalent formulations

Taking the trace with respect to the metric of both sides of the EFE one gets

$$R - \frac{D}{2} R + D\Lambda = \frac{8\pi G}{c^4} T$$

where D is the spacetime dimension. This expression can be rewritten as

$$-R + \frac{D\Lambda}{\left(\frac{D}{2} - 1\right)} = \frac{8\pi G}{c^4} \frac{T}{\frac{D}{2} - 1}.$$

If one adds $-1/2 g_{\mu\nu}$ times this to the EFE, one gets the following equivalent "trace-reversed" form

$$R_{\mu\nu} - \frac{\Lambda g_{\mu\nu}}{\frac{D}{2} - 1} = \frac{8\pi G}{c^4} \left(T_{\mu\nu} - \frac{1}{D-2} T\, g_{\mu\nu} \right).$$

For example, in $D = 4$ dimensions this reduces to

$$R_{\mu\nu} - \Lambda g_{\mu\nu} = \frac{8\pi G}{c^4} \left(T_{\mu\nu} - \tfrac{1}{2} T\, g_{\mu\nu} \right).$$

Reversing the trace again would restore the original EFE. The trace-reversed form may be more convenient in some cases (for example, when one is interested in weak-field limit and can replace $g_{\mu\nu}$ in the expression on the right with the Minkowski metric without significant loss of accuracy).

37.2 The cosmological constant

Main article: Cosmological constant

Einstein modified his original field equations to include a cosmological constant term Λ proportional to the metric

$$R_{\mu\nu} - \frac{1}{2} R\, g_{\mu\nu} + \Lambda g_{\mu\nu} = \frac{8\pi G}{c^4} T_{\mu\nu}\,.$$

Since Λ is constant, the energy conservation law is unaffected.

The cosmological constant term was originally introduced by Einstein to allow for a universe that is not expanding or contracting. This effort was unsuccessful because:

- the universe described by this theory was unstable, and

- observations by Edwin Hubble confirmed that our universe is expanding.

So, Einstein abandoned Λ, calling it the "biggest blunder [he] ever made".[9]

Despite Einstein's motivation for introducing the cosmological constant term, there is nothing inconsistent with the presence of such a term in the equations. For many years the cosmological constant was almost universally considered to be 0. However, recent improved astronomical techniques have found that a positive value of Λ is needed to explain the accelerating universe.[10][11]

Einstein thought of the cosmological constant as an independent parameter, but its term in the field equation can also be moved algebraically to the other side, written as part of the stress–energy tensor:

$$T_{\mu\nu}^{(\text{vac})} = -\frac{\Lambda c^4}{8\pi G} g_{\mu\nu}\,.$$

The resulting vacuum energy is constant and given by

$$\rho_{\text{vac}} = \frac{\Lambda c^2}{8\pi G}$$

The existence of a cosmological constant is thus equivalent to the existence of a non-zero vacuum energy. Thus, the terms "cosmological constant" and "vacuum energy" are now used interchangeably in general relativity.

37.3 Features

37.3.1 Conservation of energy and momentum

General relativity is consistent with the local conservation of energy and momentum expressed as

$$\nabla_\beta T^{\alpha\beta} = T^{\alpha\beta}{}_{;\beta} = 0$$

which expresses the local conservation of stress–energy. This conservation law is a physical requirement. With his field equations Einstein ensured that general relativity is consistent with this conservation condition.

37.3.2 Nonlinearity

The nonlinearity of the EFE distinguishes general relativity from many other fundamental physical theories. For example, Maxwell's equations of electromagnetism are linear in the electric and magnetic fields, and charge and current distributions (i.e. the sum of two solutions is also a solution); another example is Schrödinger's equation of quantum mechanics which is linear in the wavefunction.

37.3.3 The correspondence principle

The EFE reduce to Newton's law of gravity by using both the weak-field approximation and the slow-motion approximation. In fact, the constant G appearing in the EFE is determined by making these two approximations.

37.4 Vacuum field equations

If the energy-momentum tensor $T\mu\nu$ is zero in the region under consideration, then the field equations are also referred to as the vacuum field equations. By setting $T\mu\nu = 0$ in the trace-reversed field equations, the vacuum equations can be written as

$$R_{\mu\nu} = 0\,.$$

In the case of nonzero cosmological constant, the equations are

$$R_{\mu\nu} = \frac{\Lambda}{\frac{D}{2} - 1} g_{\mu\nu}\,.$$

A Swiss commemorative coin from 1979, showing the vacuum field equations with zero cosmological constant (top).

The solutions to the vacuum field equations are called vacuum solutions. Flat Minkowski space is the simplest example of a vacuum solution. Nontrivial examples include the Schwarzschild solution and the Kerr solution.

Manifolds with a vanishing Ricci tensor, $R\mu\nu = 0$, are referred to as Ricci-flat manifolds and manifolds with a Ricci tensor proportional to the metric as Einstein manifolds.

37.5   Einstein–Maxwell equations

See also: Maxwell's equations in curved spacetime

If the energy-momentum tensor $T\mu\nu$ is that of an electromagnetic field in free space, i.e. if the electromagnetic stress–energy tensor

$$T^{\alpha\beta} = -\frac{1}{\mu_0}\left(F^{\alpha\psi}F_\psi{}^\beta + \frac{1}{4}g^{\alpha\beta}F_{\psi\tau}F^{\psi\tau}\right)$$

is used, then the Einstein field equations are called the *Einstein–Maxwell equations* (with cosmological constant Λ, taken to be zero in conventional relativity theory):

$$R^{\alpha\beta} - \frac{1}{2}Rg^{\alpha\beta} + \Lambda g^{\alpha\beta} =$$
$$\frac{8\pi G}{c^4\mu_0}\left(F^{\alpha\psi}F_\psi{}^\beta + \frac{1}{4}g^{\alpha\beta}F_{\psi\tau}F^{\psi\tau}\right).$$

Additionally, the covariant Maxwell Equations are also applicable in free space:

$$F^{\alpha\beta}{}_{;\beta} = 0$$
$$F_{[\alpha\beta;\gamma]} = \frac{1}{3}\left(F_{\alpha\beta;\gamma} + F_{\beta\gamma;\alpha} + F_{\gamma\alpha;\beta}\right) =$$
$$\frac{1}{3}\left(F_{\alpha\beta,\gamma} + F_{\beta\gamma,\alpha} + F_{\gamma\alpha,\beta}\right) = 0.$$

where the semicolon represents a covariant derivative, and the brackets denote anti-symmetrization. The first equation asserts that the 4-divergence of the two-form F is zero, and the second that its exterior derivative is zero. From the latter, it follows by the Poincaré lemma that in a coordinate chart it is possible to introduce an electromagnetic field potential $A\alpha$ such that

$$F_{\alpha\beta} = A_{\alpha;\beta} - A_{\beta;\alpha} = A_{\alpha,\beta} - A_{\beta,\alpha}$$

in which the comma denotes a partial derivative. This is often taken as equivalent to the covariant Maxwell equation from which it is derived.[12] However, there are global solutions of the equation which may lack a globally defined potential.[13]

37.6   Solutions

Main article: Solutions of the Einstein field equations

The solutions of the Einstein field equations are metrics of spacetime. These metrics describe the structure of the spacetime including the inertial motion of objects in the spacetime. As the field equations are non-linear, they cannot always be completely solved (i.e. without making approximations). For example, there is no known complete solution for a spacetime with two massive bodies in it (which is a theoretical model of a binary star system, for example). However, approximations are usually made in these cases. These are commonly referred to as post-Newtonian approximations. Even so, there are numerous cases where the field equations have been solved completely, and those are called exact solutions.[14]

The study of exact solutions of Einstein's field equations is one of the activities of cosmology. It leads to the prediction of black holes and to different models of evolution of the universe.

One can also discover new solutions of the Einstein field equations via the method of orthonormal frames as pioneered by Ellis and MacCallum.[15] In this approach, the Einstein field equations are reduced to a set of coupled, non-linear, ordinary differential equations. As discussed by Hsu and Wainwright,[16] self-similar solutions to the Einstein field equations are fixed points of the resulting dynamical system. New solutions have been discovered using these methods by LeBlanc [17] and Kohli and Haslam.[18]

37.7 The linearised EFE

Main article: Linearized gravity

The nonlinearity of the EFE makes finding exact solutions difficult. One way of solving the field equations is to make an approximation, namely, that far from the source(s) of gravitating matter, the gravitational field is very weak and the spacetime approximates that of Minkowski space. The metric is then written as the sum of the Minkowski metric and a term representing the deviation of the true metric from the Minkowski metric, with terms that are quadratic in or higher powers of the deviation being ignored. This linearisation procedure can be used to investigate the phenomena of gravitational radiation.

37.8 Polynomial form

One might think that EFE are non-polynomial since they contain the inverse of the metric tensor. However, the equations can be arranged so that they contain only the metric tensor and not its inverse. First, the determinant of the metric in 4 dimensions can be written:

$$\det(g) = \frac{1}{24}\varepsilon^{\alpha\beta\gamma\delta}\varepsilon^{\kappa\lambda\mu\nu}g_{\alpha\kappa}g_{\beta\lambda}g_{\gamma\mu}g_{\delta\nu}$$

using the Levi-Civita symbol; and the inverse of the metric in 4 dimensions can be written as:

$$g^{\alpha\kappa} = \frac{1}{6}\varepsilon^{\alpha\beta\gamma\delta}\varepsilon^{\kappa\lambda\mu\nu}g_{\beta\lambda}g_{\gamma\mu}g_{\delta\nu}/\det(g).$$

Substituting this definition of the inverse of the metric into the equations then multiplying both sides by $\det(g)$ until there are none left in the denominator results in polynomial equations in the metric tensor and its first and second derivatives. The action from which the equations are derived can also be written in polynomial form by suitable redefinitions of the fields.[19]

37.9 See also

- Einstein–Hilbert action

- Equivalence principle

- Exact solutions in general relativity

- General relativity resources

- History of general relativity

- Hamilton–Jacobi–Einstein equation

- Mathematics of general relativity

- Ricci calculus

37.10 Notes

[1] Einstein, Albert (1916). "The Foundation of the General Theory of Relativity". *Annalen der Physik* **354** (7): 769. Bibcode:1916AnP...354..769E. doi:10.1002/andp.19163540702. Archived from the original (PDF) on 2012-02-06.

[2] Einstein, Albert (November 25, 1915). "Die Feldgleichungen der Gravitation". *Sitzungsberichte der Preussischen Akademie der Wissenschaften zu Berlin*: 844–847. Retrieved 2006-09-12.

[3] Misner, Charles W.; Thorne, Kip S.; Wheeler, John Archibald (1973). *Gravitation*. San Francisco: W. H. Freeman. ISBN 978-0-7167-0344-0. Chapter 34, p. 916.

[4] Carroll, Sean (2004). *Spacetime and Geometry – An Introduction to General Relativity*. pp. 151–159. ISBN 0-8053-8732-3.

[5] Grøn, Øyvind; Hervik, Sigbjorn (2007). *Einstein's General Theory of Relativity: With Modern Applications in Cosmology* (illustrated ed.). Springer Science & Business Media. p. 180. ISBN 978-0-387-69200-5. Extract of page 180

[6] Weinberg, Steven (1993). *Dreams of a Final Theory: the search for the fundamental laws of nature*. Vintage Press. pp. 107, 233. ISBN 0-09-922391-0.

[7] Misner, Thorne & Wheeler 1973

[8] Weinberg 1972

[9] Gamow, George (April 28, 1970). *My World Line : An Informal Autobiography*. Viking Adult. ISBN 0-670-50376-2. Retrieved 2007-03-14.

[10] Wahl, Nicolle (2005-11-22). "Was Einstein's 'biggest blunder' a stellar success?". Archived from the original on 2007-03-07. Retrieved 2007-03-14.

[11] Turner, Michael S. (May 2001). "Making Sense of the New Cosmology". *Int.J.Mod.Phys.* *A17S1* **17**: 180–196. arXiv:astro-ph/0202008. Bibcode:2002IJMPA..17S.180T. doi:10.1142/S0217751X02013113.

[12] Brown, Harvey (2005). *Physical Relativity*. Oxford University Press. p. 164. ISBN 978-0-19-927583-0.

[13] Trautman, Andrzej (1977). "Solutions of the Maxwell and Yang-Mills equations associated with hopf fibrings". *International Journal of Theoretical Physics* **16** (9): 561–565. Bibcode:1977IJTP...16..561T. doi:10.1007/BF01811088..

[14] Stephani, Hans; D. Kramer; M. MacCallum; C. Hoense-
laers; E. Herlt (2003). *Exact Solutions of Einstein's Field
Equations*. Cambridge University Press. ISBN 0-521-
46136-7.

[15] Ellis, GFR and MacCallum, M, "A class of homogeneous
cosmological models", Comm. Math. Phys. Volume 12,
Number 2 (1969), 108-141.

[16] Hsu, L and Wainwright, J, "Self-similar spatially homogeneous cosmologies: orthogonal perfect fluid and vacuum solutions", Class. Quantum Grav. 3 (1986) 1105-1124"

[17] LeBlanc, V.G, "Asymptotic states of magnetic Bianchi I
cosmologies", 1997 Class. Quantum Grav. 14 2281

[18] Kohli, Ikjyot Singh and Haslam, Michael C, "Dynamical systems approach to a Bianchi type I viscous magnetohydrody-
namic model", Phys. Rev. D 88, 063518 (2013)

[19] Einstein's Field Equations in Polynomial Form

37.11   References

See General relativity resources.

- Charles Misner; Kip S Thorne; John Archibald
  Wheeler (1973), *Gravitation*, San Francisco: W. H.
  Freeman, p. 501ff, ISBN 0-7167-0344-0

- Weinberg, Steven (1972), *Gravitation and Cosmology*,
  John Wiley & Sons, Inc, ISBN 0-471-92567-5

37.12   External links

- Hazewinkel, Michiel, ed.   (2001), "Einstein equations", *Encyclopedia of Mathematics*, Springer, ISBN
  978-1-55608-010-4

- Caltech Tutorial on Relativity — A simple introduction to Einstein's Field Equations.

- The Meaning of Einstein's Equation — An explanation of Einstein's field equation, its derivation, and
  some of its consequences

- Video Lecture on Einstein's Field Equations by MIT
  Physics Professor Edmund Bertschinger.

- Arch and scaffold: How Einstein found his field equations Physics Today November 2015, History of the
  Development of the Field Equations

Chapter 38

Anti-de Sitter space

In mathematics and physics, n-dimensional **anti-de Sitter space** (AdSn) is a maximally symmetric Lorentzian manifold with constant negative scalar curvature. It is the Lorentzian analogue of hyperbolic space, just as Minkowski space is the analogue of Euclidean space and de Sitter space is the analogue of elliptical space.

It is best known for its role in the AdS/CFT correspondence. The anti-de Sitter space and de Sitter space are named after Willem de Sitter (1872–1934), professor of astronomy at Leiden University and director of the Leiden Observatory. Willem de Sitter and Albert Einstein worked together closely in the 1920s in Leiden on the spacetime structure of the universe.

In the language of general relativity, anti-de Sitter space is a maximally symmetric vacuum solution of Einstein's field equation with a negative (attractive) cosmological constant Λ, corresponding to a negative energy density and positive pressure of the vacuum.

In mathematics, anti-de Sitter space is sometimes defined more generally as a space of arbitrary metric signature (p, q). In physics, often only the case of one timelike dimension is considered. This corresponds to the equivalent metric signatures $(n-1, 1)$ and $(1, n-1)$, where the choice is by a sign convention.

38.1 Non-technical explanation

This non-technical explanation first defines the terms used in the introductory material of this entry. Then, it briefly sets forth the underlying idea of a general relativity-like spacetime. Then it discusses how de Sitter space describes a distinct variant of the ordinary spacetime of general relativity (called Minkowski space) related to the cosmological constant, and how anti-de Sitter space differs from de Sitter space. It also explains that Minkowski space, de Sitter space and anti-de Sitter space, as applied to general relativity, can all be thought of as being embedded in a flat five-dimensional spacetime. Finally, it offers some caveats that describe in general terms how this non-technical explanation fails to capture the full detail of the mathematical concept.

38.1.1 Technical terms translated

A maximally symmetric Lorentzian manifold corresponds to a general relativity-like spacetime in which time and space in all directions are mathematically equivalent.

A constant scalar curvature means a general relativity gravity-like bending of spacetime that has a curvature described by a single number that is the same everywhere in spacetime in the absence of matter or energy.

Negative curvature means curved hyperbolically, like a saddle surface or the Gabriel's Horn surface, similar to that of a trumpet bell. It might be described as being the "opposite" of the surface of a sphere which has a positive curvature. A negative curvature corresponds to an attractive force, while a positive curvature such as a sphere corresponds to a repulsive force.

The AdS/CFT (anti-de Sitter space/conformal field theory) correspondence is an idea originally proposed by Juan Maldacena in late 1997. The AdS/CFT correspondence is the idea that it is possible in general to describe a force in quantum mechanics (like electromagnetism, the weak force or the strong force) in a certain number of dimensions (for example four) with a string theory where the strings exist in an anti-de Sitter space, with one additional dimension.

A quantum field theory is a set of equations and rules for using them of the kind used in quantum mechanics to describe forces (such as electromagnetism, the weak force and the strong force) in a way that is mathematically unstable (see stability theory).

A conformal field theory is basically a quantum field theory that is scale invariant. Thus, the equations work the same way provided inputs have consistent units, even if the unit in question is unknown. In contrast, a scale variant quantum field theory would describe a force that behaves in a qualitatively different way at short distances compared to long distances.

The AdS/CFT correspondence is notable because it is not obvious that quantum field theories can be represented geometrically. Quantum field theories involve quantities that when explained to non-experts are commonly described as representing intangible ideas like probabilities and possible paths that a quantum could take to get from one place to another. The connection of quantum field theories to a physical geometric description is less obvious than the connection between the classical equations (i.e. non-quantum mechanical descriptions of gravity and electromagnetism) and geometry. There are no non-quantum mechanical equations for the weak nuclear force and the strong nuclear force, the other two fundamental forces.

38.1.2   Spacetime in general relativity

General relativity is a theory of the nature of time, space and gravity in which gravity is a curvature of space and time that results from the presence of matter or energy. Energy and matter are equivalent (as expressed in the equation $E = mc^2$), and space and time can be translated into equivalent units based on the speed of light (c in the $E = mc^2$ equation).

A common analogy involves the way that a dip in a flat sheet of rubber, caused by a heavy object sitting on it, influences the path taken by small objects rolling nearby, causing them to deviate inward from the path they would have followed had the heavy object been absent. Of course, in general relativity, both the small and large objects mutually influence the curvature of spacetime.

The attractive force of gravity created by matter is due to a negative curvature of spacetime, represented in the rubber sheet analogy by the negatively curved (trumpet-bell-like) dip in the sheet.

A key feature of general relativity is that it describes gravity not as a conventional force like electromagnetism, but as a change in the geometry of spacetime that results from the presence of matter or energy.

The analogy used above describes the curvature of a two-dimensional space caused by gravity in general relativity in a three-dimensional superspace in which the third dimension corresponds to the effect of gravity. A geometrical way of thinking about general relativity describes the effects of the gravity in the real world four-dimensional space geometrically by projecting that space into a five-dimensional superspace with the fifth dimension corresponding to the curva-

ture in spacetime that is produced by gravity and gravity-like effects in general relativity.

As a result, in general relativity, the familiar Newtonian equation of gravity $F = G\frac{m_1 m_2}{r^2}$ (i.e. gravitation pull between two objects equals the gravitational constant times the product of their masses divided by the square of the distance between them) is merely an approximation of the gravity-like effects seen in general relativity. However this approximation becomes inaccurate in extreme physical situations. For example, in general relativity, objects in motion have a slightly different gravitation effect than objects at rest.

Some of the differences between the familiar Newtonian equation of gravity and the predictions of general relativity flow from the fact that gravity in general relativity bends both time and space, not just space. In normal circumstances, gravity bends time so slightly that the differences between Newtonian gravity and general relativity are impossible to detect without precise instruments.

38.1.3   de Sitter space distinguished from spacetime in general relativity

Fundamentally, the key concept behind the idea of de Sitter space is that it involves a variation on the spacetime of general relativity in which spacetime is itself slightly curved even in the absence of matter or energy.

The relationship of the normal idea of the spacetime in which general relativity operates to the de Sitter space is analogous to the relationship between Euclidean geometry (i.e. in two dimensions, the geometry of flat surfaces) and non-Euclidean geometry (i.e. in two dimensions, the geometries of surfaces that are not flat).

An inherent curvature of spacetime even in the absence of matter or energy is another way of thinking about the idea of the cosmological constant in general relativity. An inherent curvature of spacetime and the cosmological constant are also equivalent to the idea that a vacuum (i.e. empty space without any matter or energy in it) has a fundamental energy of its own.

In the common analogy of an object causing a dip in a flat cloth, normal de Sitter space has a curvature analogous to a flat cloth sitting atop a sphere with a very slight curvature because it is so large. Empty de Sitter space is slightly repulsive; it has a slight natural curvature in the opposite direction of the curvature in spacetime created by a massive object. It is a way of saying that gravity plays out against the background of a slightly anti-gravitational empty space.

Normal de Sitter space corresponds to the positive cosmological constant that is observed in reality, with the size of the cosmological constant being equivalent to the curvature of the de Sitter space.

de Sitter space can also be thought of as a general relativity-like spacetime in which empty space itself has some energy, which causes this spacetime (i.e. the universe) to expand at an ever greater rate.

38.1.4 anti-de Sitter space distinguished from de Sitter space

An anti-de Sitter space, in contrast, is a general relativity-like spacetime, where in the absence of matter or energy, the curvature of spacetime is naturally hyperbolic.

In the common analogy of an object causing a dip in a flat cloth, anti-de Sitter space has a curvature analogous to a flat cloth sitting on a saddle, with a very slight curvature because it is so large. This would correspond to a negative cosmological constant (something not observed in the real life cosmos). Anti-de Sitter space can also be thought of as a general relativity like spacetime in which empty space itself has negative energy, which causes this spacetime (i.e. the universe) to collapse in on itself at an ever greater rate.

In an anti-de Sitter space, as in a de Sitter space, the extent of inherent spacetime curvature corresponds to the magnitude of the negative cosmological constant to which it is equivalent.

38.1.5 de Sitter space and anti-de Sitter space as five-dimensional geometries

As noted above, the analogy used above describes curvature of a two-dimensional space caused by gravity in general relativity in a three-dimensional superspace in which the third dimension corresponds to the effect of gravity. More generally, a geometrical approach to general relativity describes the effect of gravity as a curvature of the four dimensions of spacetime in a fifth dimension that corresponds to gravity and gravity-like effects in general relativity. When this five-dimensional superspace describes a version of general relativity without a cosmological constant, it is called Minkowski space.

The concepts of de Sitter space and anti-de Sitter space describe the effects of the cosmological constant in the real world four-dimensional space geometrically by projecting that space into a five-dimensional superspace with the fifth dimension corresponding to the curvature in time and space that is produced by gravity and gravity-like effects in general relativity such as the cosmological constant.

While anti-de Sitter space does not correspond to gravity in general relativity with the observed cosmological constant,

an anti-de Sitter space is believed to correspond to other forces in quantum mechanics (like electromagnetism, the weak nuclear force and the strong nuclear force) described via string theory. This is called the AdS/CFT correspondence.

Note also that while an anti-de Sitter space would describe general relativity with a negative cosmological constant in five dimensions (four for spacetime and one for the effect of the cosmological constant), the idea is actually more general. One can have an anti-de Sitter space (or a de Sitter space) in an arbitrary number of dimensions. The generality of the concepts of de Sitter space and anti-de Sitter space make them useful in theoretical physics, particularly in string theory, that often assume a world with more than four dimensions.

38.1.6 Caveats

Naturally, as the remainder of this article explains in technical detail, the general concepts described in this non-technical explanation of anti-de Sitter space have a much more rigorous and precise mathematical and physical description. People are ill suited to visualizing things in five or more dimensions, but mathematical equations are not similarly challenged and can represent five-dimensional concepts in a way just as appropriate as the methods that mathematical equations use to describe easier to visualize three and four-dimensional concepts.

There is a particularly important implication of the more precise mathematical description that differs from the analogy-based heuristic description of de Sitter space and anti-de Sitter space above. The mathematical description of anti-de Sitter space generalizes the idea of curvature. In the mathematical description, curvature is a property of a particular point and can be divorced from some invisible surface to which curved points in spacetime meld themselves. So, for example, concepts like singularities (the most widely known of which in general relativity is the black hole) which cannot be expressed completely in a real world geometry, can correspond to particular states of a mathematical equation.

The full mathematical description also captures some subtle distinctions made in general relativity between space-like dimensions and time-like dimensions.

38.2 Definition and properties

Much as elliptical and hyperbolic spaces can be visualized by an isometric embedding in a flat space of one higher dimension (as the sphere and pseudosphere respectively), anti-de Sitter space can be visualized as the Lorentzian analogue of a sphere in a space of one additional dimension. To a physicist the extra dimension is timelike, while to a mathematician it is negative; in this article we adopt the convention that timelike dimensions are negative so that these notions coincide.

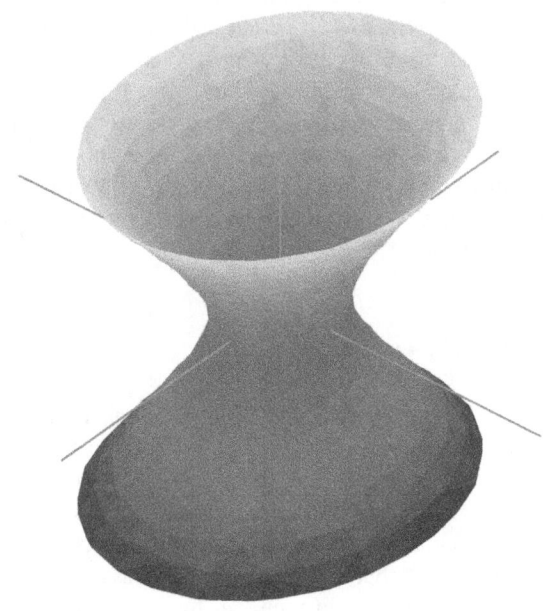

Image of (1 + 1)-dimensional anti-de Sitter space embedded in flat (1 + 2)-dimensional space. The t_1 and t_2 axes lie in the plane of rotational symmetry, and the x_1 axis is normal to that plane. The embedded surface contains closed timelike curves circling the x_1 axis, but these can be eliminated by "unrolling" the embedding (more precisely, by taking the universal cover).

The anti-de Sitter space of signature (p,q) can then be isometrically embedded in the space $\mathbb{R}^{p,q+1}$ with coordinates $(x_1, ..., x_p, t_1, ..., t_{q+1})$ and the pseudometric

$$ds^2 = \sum_{i=1}^{p} dx_i^2 - \sum_{j=1}^{q+1} dt_j^2$$

as the sphere

$$\sum_{i=1}^{p} x_i^2 - \sum_{j=1}^{q+1} t_j^2 = -\alpha^2$$

where α is a nonzero constant with dimensions of length (the radius of curvature). Note that this is a sphere in the sense that it is a collection of points at constant metric distance from the origin, but visually it is a hyperboloid, as in the image shown.

The metric on anti-de Sitter space is the metric induced from the ambient metric. One can check that the induced metric is nondegenerate and has Lorentzian signature.

When $q = 0$, this construction gives ordinary hyperbolic space. The remainder of the discussion applies when $q \geq 1$.

38.2.1 Closed timelike curves and the universal cover

When $q \geq 1$, the embedding above has closed timelike curves; for example, the path parameterized by $t_1 = \alpha \sin(\tau), t_2 = \alpha \cos(\tau)$, and all other coordinates zero, is such a curve. When $q \geq 2$ these curves are inherent to the geometry (unsurprisingly, as any space with more than one temporal dimension will contain closed timelike curves), but when $q = 1$, they can be eliminated by passing to the universal covering space, effectively "unrolling" the embedding. A similar situation occurs with the pseudosphere, which curls around on itself although the hyperbolic plane does not; as a result it contains self-intersecting straight lines (geodesics) while the hyperbolic plane does not. Some authors define anti-de Sitter space as equivalent to the embedded sphere itself, while others define it as equivalent to the universal cover of the embedding. Generally the latter definition is the one of interest in physics.

38.2.2 Symmetries

If the universal cover is not taken, (p,q) anti-de Sitter space has O(p,q+1) as its isometry group. If the universal cover is taken the isometry group is a cover of O(p,q+1). This is most easily understood by defining anti-de Sitter space as a symmetric space, using the quotient space construction, given below.

38.3 Coordinate patches

A coordinate patch covering part of the space gives the half-space coordinatization of anti-de Sitter space. The metric for this patch is

$$ds^2 = \frac{1}{y^2} \left(-dt^2 + dy^2 + \sum_i dx_i^2 \right),$$

with $y > 0$ giving the half-space. We easily see that this metric is conformally equivalent to a flat half-space Minkowski spacetime.

The constant time slices of this coordinate patch are hyperbolic spaces in the Poincaré half-plane metric. In the limit as $y \to 0$, this half-space metric is conformally equivalent to the Minkowski metric $ds^2 = -dt^2 + \sum_i dx_i^2$. Thus, the anti-de Sitter space contains a conformal Minkowski space at infinity ("infinity" having y-coordinate zero in this patch).

In AdS space time is periodic, and the universal cover has non-periodic time. The coordinate patch above covers half of a single period of the spacetime.

Because the conformal infinity of AdS is timelike, specifying the initial data on a spacelike hypersurface would not determine the future evolution uniquely (*i.e.* deterministically) unless there are boundary conditions associated with the conformal infinity.

The "half-space" region of anti-deSitter space and its boundary.

Another commonly used coordinate system which covers the entire space is given by the coordinates t, $r \geqslant 0$ and the hyper-polar coordinates α, θ and φ.

$$ds^2 = -\left(k^2 r^2 + 1\right) dt^2 + \frac{1}{k^2 r^2 + 1} dr^2 + r^2 d\Omega^2$$

The image on the right represents the "half-space" region of anti-deSitter space and its boundary. The interior of the cylinder corresponds to anti-de Sitter spacetime, while its cylindrical boundary corresponds to its conformal boundary. The green shaded region in the interior corresponds to the region of AdS covered by the half-space coordinates and it is bounded by two null, *aka* lightlike, geodesic hyperplanes; the green shaded area on the surface corresponds to the region of conformal space covered by Minkowski space.

The green shaded region covers half of the AdS space and half of the conformal spacetime; the left ends of the green discs will touch in the same fashion as the right ends.

38.4 As a homogeneous, symmetric space

In the same way that the 2-sphere

$$S^2 = \frac{O(3)}{O(2)}$$

is a quotient of two orthogonal groups, anti-de Sitter with parity (reflectional symmetry) and time reversal symmetry can be seen as a quotient of two generalized orthogonal groups

$$\mathrm{AdS}_n = \frac{O(2, n-1)}{O(1, n-1)}$$

whereas AdS without P or C can be seen as the quotient

$$\frac{\mathrm{Spin}^+(2, n-1)}{\mathrm{Spin}^+(1, n-1)}$$

of spin groups.

This quotient formulation gives AdS_n the structure of a homogeneous space. The Lie algebra of the generalized orthogonal group $o(1, n)$ is given by matrices

$$\mathcal{H} = \begin{pmatrix} 0 & 0 & \begin{pmatrix} \cdots 0 \cdots \\ \leftarrow v^t \rightarrow \end{pmatrix} \\ \begin{pmatrix} \vdots & \uparrow \\ 0 & v \\ \vdots & \downarrow \end{pmatrix} & & B \end{pmatrix}$$

where B is a skew-symmetric matrix. A complementary generator in the Lie algebra of $\mathcal{G} = O(2, n)$ is

$$\mathcal{Q} = \begin{pmatrix} 0 & a & \begin{pmatrix} \leftarrow w^t \rightarrow \\ \cdots 0 \cdots \end{pmatrix} \\ -a & 0 & \\ \begin{pmatrix} \uparrow & \vdots \\ w & 0 \\ \downarrow & \vdots \end{pmatrix} & & 0 \end{pmatrix}.$$

These two fulfill $\mathcal{G} = \mathcal{H} \oplus \mathcal{Q}$. Explicit matrix computation shows that $[\mathcal{H}, \mathcal{Q}] \subseteq \mathcal{Q}$ and $[\mathcal{Q}, \mathcal{Q}] \subseteq \mathcal{H}$. Thus, anti-de Sitter is a reductive homogeneous space, and a non-Riemannian symmetric space.

38.5   A simple definition for anti-de Sitter space and its properties

AdS$_n$ is an n-dimensional solution for the theory of gravitation with Einstein–Hilbert action with negative cosmological constant Λ, ($\Lambda < 0$), i.e. the theory described by the following Lagrangian density:

$$\mathcal{L} = \frac{1}{16\pi G_{(n)}}(R - 2\Lambda)$$

Therefore, it is a solution of the Einstein field equations:

$$G_{\mu\nu} + \Lambda g_{\mu\nu} = 0$$

where $G_{\mu\nu}$ is Einstein tensor and $g_{\mu\nu}$ is the metric of the space-time. Introducing the radius α as $\Lambda = \frac{-(n-1)(n-2)}{2\alpha^2}$ this solution can be immersed in a $n+1$ dimensional space-time with signature $(-,-,+,\cdots,+)$ by the following constraint:

$$-X_1^2 - X_2^2 + \sum_{i=3}^{n+1} X_i^2 = -\alpha^2$$

38.5.1   Global coordinates

AdS$_n$ is parametrized in global coordinates by the parameters $(\tau, \rho, \theta, \varphi_1, \cdots, \varphi_{n-3})$ as:

$$\begin{cases} X_1 = \alpha \cosh \rho \cos \tau \\ X_2 = \alpha \cosh \rho \sin \tau \\ X_i = \alpha \sinh \rho \, \hat{x}_i \qquad \sum_i \hat{x}_i^2 = 1 \end{cases}$$

where $\hat{x}_i$ parametrize a S^{n-2} sphere. i.e. we have $\hat{x}_1 = \sin \theta \sin \varphi_1 \ldots \sin \varphi_{n-3}$, $\quad \hat{x}_2 = \sin \theta \sin \varphi_1 \ldots \cos \varphi_{n-3}$ etc. The AdS$_n$ metric in these coordinates is:

$$ds^2 = \alpha^2(-\cosh^2 \rho \, d\tau^2 + d\rho^2 + \sinh^2 \rho \, d\Omega_{n-2}^2)$$

where $\tau \in [0, 2\pi]$ and $\rho \in \mathbb{R}^+$. Considering the periodicity of time τ and in order to avoid closed timelike curves (CTC), one should take the universal cover $\tau \in \mathbb{R}$. In the limit $\rho \to \infty$ one can approach to the boundary of this space-time usually called AdS$_n$ conformal boundary.

With the transformations $r \equiv \alpha \sinh \rho$ and $t \equiv \alpha\tau$ we can have the usual AdS$_n$ metric in global coordinates:

$$ds^2 = -f(r) \, dt^2 + \frac{1}{f(r)} \, dr^2 + r^2 \, d\Omega_{n-2}^2$$

where $f(r) = 1 + \frac{r^2}{\alpha^2}$

38.5.2   Poincaré coordinates

By the following parametrization:

$$\begin{cases} X_1 = \frac{\alpha^2}{2r}(1 + \frac{r^2}{\alpha^4}(\alpha^2 + \vec{x}^2 - t^2)) \\ X_2 = \frac{r}{\alpha}t \\ X_i = \frac{r}{\alpha}x_i \qquad i \in \{3, \cdots, n\} \\ X_{n+1} = \frac{\alpha^2}{2r}(1 - \frac{r^2}{\alpha^4}(\alpha^2 - \vec{x}^2 + t^2)) \end{cases}$$

the AdS$_n$ metric in the Poincaré coordinates is:

$$ds^2 = -\frac{r^2}{\alpha^2} \, dt^2 + \frac{\alpha^2}{r^2} \, dr^2 + \frac{r^2}{\alpha^2} \, d\vec{x}^2$$

in which $0 \leq r$. The codimension 2 surface $r = 0$ is Poincaré Killing horizon and $r \to \infty$ approaches to the boundary of AdS$_n$ space-time, so unlike the global coordinates, the Poincaré coordinates do not cover all AdS$_n$ manifold. Using $u \equiv \frac{r}{\alpha^2}$ this metric can be written in the following way:

$$ds^2 = \alpha^2 \left(\frac{du^2}{u^2} + u^2(dx_\mu \, dx^\mu) \right)$$

where $x^\mu = (t, \vec{x})$. By the transformation $z \equiv \frac{1}{u}$ also it can be written as:

$$ds^2 = \frac{\alpha^2}{z^2}(dz^2 + dx_\mu \, dx^\mu)$$

38.5.3   Geometric properties

AdS$_n$ metric with radius α is one of the maximal symmetric n-dimensional spacetimes. It has the following geometric properties:

Riemann curvature tensor:

$$R_{\mu\nu\alpha\beta} = \frac{-1}{\alpha^2}(g_{\mu\alpha}g_{\nu\beta} - g_{\mu\beta}g_{\nu\alpha})$$

Ricci curvature:

$$R_{\mu\nu} = \frac{-(n-1)}{\alpha^2} g_{\mu\nu}$$

Scalar curvature:

$$R = \frac{-n(n-1)}{\alpha^2}$$

38.6 References

- Bengtsson, Ingemar: Anti-de Sitter space. Lecture notes.

- Qingming Cheng (2001), "Anti-de Sitter space", in Hazewinkel, Michiel, *Encyclopedia of Mathematics*, Springer, ISBN 978-1-55608-010-4

- Ellis, G. F. R.; Hawking, S. W. *The large scale structure of space-time.* Cambridge university press (1973). (see pages 131–134).

- Frances, C: The conformal boundary of anti-de Sitter space-times. AdS/CFT correspondence: Einstein metrics and their conformal boundaries, 205–216, IRMA Lect. Math. Theor. Phys., 8, Eur. Math. Soc., Zürich, 2005.

- Matsuda, H. *A note on an isometric imbedding of upper half-space into the anti-de Sitter space.* Hokkaido Mathematical Journal Vol.13 (1984) p. 123–132.

- Wolf, Joseph A. *Spaces of constant curvature.* (1967) p. 334.

Chapter 39

Minkowski space

In mathematical physics, **Minkowski space** or **Minkowski spacetime** is a combination of Euclidean space and time into a four-dimensional manifold where the spacetime interval between any two events is independent of the inertial frame of reference in which they are recorded. Although initially developed by mathematician Hermann Minkowski for Maxwell's equations of electromagnetism, the mathematical structure of Minkowski spacetime was shown to be an immediate consequence of the postulates of special relativity.[1]

Minkowski space is closely associated with Einstein's theory of special relativity, and is the most common mathematical structure on which special relativity is formulated. While the individual components in Euclidean space and time will often differ due to length contraction and time dilation, in Minkowski spacetime, all frames of reference will agree on the total distance in spacetime between events.[nb 1] Because it treats time differently than the three spatial dimensions, Minkowski space differs from four-dimensional Euclidean space.[nb 2]

In Euclidean space, the isometry group (the maps preserving the regular inner product) is the Euclidean group. The analogous isometry group for Minkowski space, preserving intervals of spacetime equipped with the associated non-positive definite bilinear form (here called the **Minkowski inner product**,[nb 3]) is the Poincaré group. The Minkowski inner product is defined as to yield the spacetime interval between two events when given their coordinate difference vector as argument.

Hermann Minkowski (1864 – 1909) found that the theory of special relativity, introduced by his former student Albert Einstein, could best be understood in a four-dimensional space, since known as the Minkowski spacetime.

39.1 History

39.1.1 Four-dimensional Euclidean spacetime

See also: Four-dimensional space

In 1905, and later published in 1906, Henri Poincaré showed that by taking time to be an imaginary fourth spacetime coordinate ($\sqrt{-1}\, c\, t$), a Lorentz transformation can be regarded as a rotation of coordinates in a four-dimensional Euclidean space with three real coordinates representing space, and one imaginary coordinate, representing time, as the fourth dimension. Since the space is then a pseudo-Euclidean space, the rotation is a representation of a hyperbolic rotation, although Poincaré did not give this interpretation, his purpose being only to explain the Lorentz transformation in terms of the familiar Euclidean rotation.[2]

This idea was elaborated by Hermann Minkowski,[3] who used it to restate the Maxwell equations in four dimensions, showing directly their invariance under the Lorentz transformation. He further reformulated in four dimensions the then-recent theory of special relativity of Einstein. From this he concluded that time and space should be treated equally, and so arose his concept of events taking place in a unified four-dimensional spacetime continuum.

39.1.2 Minkowski space

In a further development,[4] he gave an alternative formulation of this idea that used a real time coordinate instead of an imaginary one, representing the four variables (x, y, z, t) of space and time in coordinate form in a four dimensional affine space. Points in this space correspond to events in spacetime. In this space, there is a defined light-cone associated with each point, and events not on the light-cone are classified by their relation to the apex as *spacelike* or *timelike*. It is principally this view of spacetime that is current nowadays, although the older view involving imaginary time has also influenced special relativity. Minkowski, aware of the fundamental restatement of the theory which he had made, said

> The views of space and time which I wish to lay before you have sprung from the soil of experimental physics, and therein lies their strength. They are radical. Henceforth space by itself, and time by itself, are doomed to fade away into mere shadows, and only a kind of union of the two will preserve an independent reality.
> — Hermann Minkowski, 1907[4]

For further historical information see references Galison (1979), Corry (1997) and Walter (1999).

39.2 Mathematical structure

For an overview, Minkowski space is a 4-dimensional real vector space equipped with a nondegenerate, symmetric bi-

linear form on the tangent space at each point in spacetime, here simply called the Minkowski inner product, with signature either $(+,-,-,-)$ or $(-,+,+,+)$. In practice, one need not be concerned with the tangent spaces. The vector space nature of Minkowski space allows for the canonical identification of vectors in tangent spaces at points (events) with vectors (points, events) in Minkowski space itself.[5] For some purposes it is desirable to identify tangent vectors at a point p with *displacement vectors* at p, which is, of course, admissible by essentially the same canonical identification.[6]

The signature refers to which sign the Minkowski inner product yields when given space and time basis vectors as arguments. In general, mathematicians and general relativists prefer the former while particle physicists tend to use the latter. Arguments for the former (pure space vectors yield positive "norm-squared") include "continuity" from the Euclidean case corresponding to the non-relativistic limit $c \to \infty$. Arguments for the latter (pure space vectors yield negative "norm-squared") include that otherwise ubiquitous minus signs in particle physics go away.

Mathematically associated to this bilinear form is a tensor of type (0,2) at each point in spacetime, called the Minkowski metric. The Minkowski metric, the bilinear form, and the Minkowski inner product are actually all the very same object. In coordinates, this is the 4×4 matrix representing the bilinear form. Keeping this in mind may facilitate reading what follows.

For comparison, in general relativity, a Lorentzian manifold L is likewise equipped with a metric tensor g, which is a nondegenerate symmetric bilinear form on the tangent space T_pL at each point p of L. In coordinates, it may be represented by a 4×4 matrix *depending on spacetime position*. Minkowski space is thus a comparatively simple special case of a Lorentzian manifold. Its metric tensor, called the Minkowski metric, is in coordinates the same symmetric matrix at every point of M, and its arguments can, per above, be taken as vectors in spacetime itself.

Introducing more terminology (but not more structure), Minkowski space is thus a pseudo-Euclidean space with total dimension $n = 4$ and signature (3, 1) or (1, 3). Elements of Minkowski space are called events. Minkowski space is often denoted $\mathbf{R}^{3,1}$ or $\mathbf{R}^{1,3}$ to emphasize the chosen signature, or just M. It is perhaps the simplest example of a pseudo-Riemannian manifold.

39.2.1 Pseudo-Euclidean metric generalities

Main article: Pseudo-Euclidean space

The Minkowski metric[nb 4] η is the metric tensor of

Minkowski space. It is a Pseudo-Euclidean metric. As such it is a nondegenerate symmetric bilinear form, a type (0,2) tensor. It accepts two arguments up, vp, vectors in $TpM, p \in M$, the tangent space at p in M. Due to the above-mentioned canonical identification of TpM with M itself, it accepts arguments u, v with both u and v in M.

As a notational convention, vectors v in M, called 4-vectors, are denoted in sans-serif italics, and not, as is common in the Euclidean setting, with boldface **v**. The latter is generally reserved for the 3-vector part (to be introduced below) of a 4-vector.

The definition

$$u \cdot v = \eta(u, v)$$

yields an inner product-like structure on M, previously and also henceforth, called the Minkowski inner product, similar to the Euclidean inner product, but it describes a different geometry. It has the following properties.

- $\eta(au + v, w) = a\eta(u, w) + \eta(v, w), \quad \forall u, v \in M, \forall a \in \mathbb{R}$ slot) first in (linearity

- $\eta(u, v) = \eta(v, u)$ (symmetry)

- $\eta(u, v) = 0 \quad \forall v \in M \Rightarrow u = 0$ (non-degeneracy)

The first two conditions imply bilinearity. The defining *difference* between a pseudo-inner product and an inner product proper is that the former is *not* required to be positive definite, that is, $\eta(u, u) < 0$ is allowed.

Two vectors v and w are said to be orthogonal if $\eta(v, w) = 0$.

A vector e is called a unit vector if $\eta(e, e) = \pm 1$. A basis for M consisting of mutually orthogonal unit vectors is called an orthonormal basis.

For a given inertial frame, an orthonormal basis in space, combined by the unit time vector, forms an orthonormal basis in Minkowski space. The number of positive and negative unit vectors in any such basis is a fixed pair of numbers, equal to the signature of the bilinear form associated with the inner product. This is Sylvester's law of inertia.

More terminology (but not more structure): The Minkowski metric is a pseudo-Riemannian metric, more specifically, a Lorentzian metric, even more specifically, *the* Lorentz metric, reserved for 4-dimensional flat spacetime with the remaining ambiguity only being the signature convention.

39.2.2   Minkowski metric

From the second postulate of special relativity, together with homogeneity of spacetime and isotropy of space, follows that the spacetime interval between two events 1, 2,

$$\pm \left[c^2(t_1 - t_2)^2 - (x_1 - x_2)^2 - (y_1 - y_2)^2 - (z_1 - z_2)^2 \right],$$

is independent of the inertial frame chosen, as is shown here. The factor ± simply means that the choice of signature is left open. The numerical values of η, viewed as a matrix representing the Minkowski inner product, follow from the theory of bilinear forms.

Just as the signature of the metric is differently defined in the literature, this quantity is not consistently named. The interval (as defined here) is sometimes referred to as the interval squared.[7] Even the square root of the present interval occurs.[8] When signature and interval are fixed, ambiguity still remains as which coordinate is the time coordinate. It may be the fourth, or it may be the zeroth. This is not an exhaustive list of notational inconsistencies. It is a fact of life that one has to check out the definitions first thing when one consults the relativity literature.

The invariance of the interval under coordinate transformations between inertial frames follows from the invariance of

$$\pm \left[c^2 t^2 - x^2 - y^2 - z^2 \right]$$

(with either sign ± preserved), provided the transformations are linear. This quadratic form can be used to define a bilinear form

$$u \cdot v = \pm \left[c^2 t_1 t_2 - x_1 x_2 - y_1 y_2 - z_1 z_2 \right].$$

via the polarization identity. This bilinear form can in turn be written as

$$u \cdot v = u^{\mathsf{T}}[\eta]v,$$

where $[\eta]$ is a 4×4 matrix associated with η. Possibly confusingly, denote $[\eta]$ with just η as is common practice. The matrix is read off from the explicit bilinear form as

$$\eta = \pm \begin{pmatrix} -1 & 0 & 0 & 0 \\ 0 & 1 & 0 & 0 \\ 0 & 0 & 1 & 0 \\ 0 & 0 & 0 & 1 \end{pmatrix},$$

and the bilinear form

$$u \cdot v = \eta(u, v),$$

with which this section started by assuming its existence, is now identified.

For definiteness and shorter presentation, the signature $(-,+,+,+)$ is adopted below. The choice has no (known) physical implications. The symmetry group preserving the bilinear form with one choice of signature is isomorphic (under the map given here) with the symmetry group preserving the other choice of signature. This means that both choices are in accord with the two postulates of relativity.

39.2.3 Standard basis

A standard basis for Minkowski space is a set of four mutually orthogonal vectors { e_0, e_1, e_2, e_3 } such that

$$-\eta(e_0, e_0) = \eta(e_1, e_1) = \eta(e_2, e_2) = \eta(e_3, e_3) = 1.$$

These conditions can be written compactly in the form

$$\eta(e_\mu, e_\nu) = \eta_{\mu\nu}.$$

Relative to a standard basis, the components of a vector v are written (v^0, v^1, v^2, v^3) where the Einstein notation is used to write $v = v^\mu e_\mu$. The component v^0 is called the **timelike component** of v while the other three components are called the **spatial components**. The spatial components of a 4-vector v may be identified with a 3-vector $\mathbf{v} = (v_1, v_2, v_3)$.

In terms of components, the Minkowski inner product between two vectors v and w is given by

$$\eta(v, w) = \eta_{\mu\nu} v^\mu w^\nu = v^0 w_0 + v^1 w_1 + v^2 w_2 + v^3 w_3 = v^\mu w_\mu =$$

and

$$\eta(v, v) = \eta_{\mu\nu} v^\mu v^\nu = v^0 v_0 + v^1 v_1 + v^2 v_2 + v^3 v_3 = v^\mu v_\mu.$$

Here **lowering of an index** with the metric was used. Technically, a non-degenerate bilinear form provides a map between a vector space and its dual, in this context, the map is between the tangent spaces of M and the cotangent spaces of M. At a point in M, the tangent and cotangent spaces are dual. Just as an authentic inner product on a vector space with one argument fixed, by Riesz representation theorem, may be expressed as the action of a linear functional on the

vector space, the same holds for the Minkowski inner product of Minkowski space.

Thus if v^μ are the components of a vector in a tangent space, then $\eta_{\mu\nu} v v^\mu = v v$ are the components of a vector in the cotangent space (a linear functional). Due to the identification of vectors in tangent spaces with vectors in M itself, this is mostly ignored, and vectors with lower indices are referred to as **covariant vectors**. In this latter interpretation, the covariant vectors are (almost always implicitly) identified with vectors (linear functionals) in the dual of Minkowski space. The ones with upper indices are **contravariant vectors**. In the same fashion, the inverse of the map from tangent to cotangent spaces, explicitly given by the inverse of η in matrix representation, can be used to define **raising of an index**. The components of this inverse are denoted $\eta^{\mu\nu}$. It happens that $\eta^{\mu\nu} = \eta_{\mu\nu}$. These maps between a vector space and its dual can be denoted $\eta^\flat$ (eta-flat) and $\eta^\sharp$ (eta-sharp) by the musical analogy.[9]

The time-proven robustness of the formalism itself, sometimes referred to as index gymnastics, ensures that moving vectors around and changing from contravariant to covariant vectors and vice versa is mathematically sound. Incorrect expressions tend to reveal themselves quickly.

39.2.4 Geometry

39.3 Lorentz transformations and symmetry

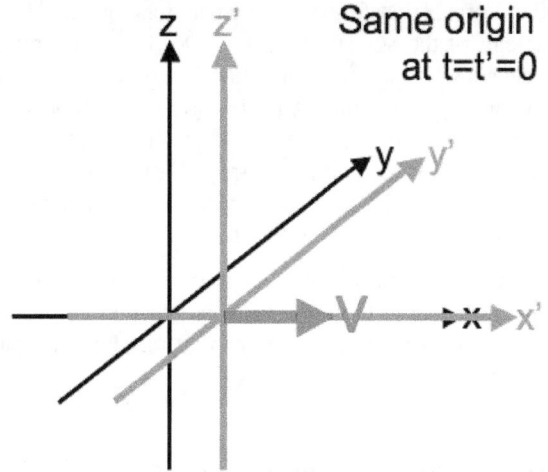

Standard configuration of coordinate systems for Lorentz transformations.

The Poincaré group is the group of all transformations preserving the interval. The interval is quite easily seen to be preserved by the translation group in 4 dimensions. The other transformations are those that preserve the interval and leave the origin fixed. Given the bilinear form associated with the Minkowski metric, the appropriate group follows directly from the theory (in particular the definition) of classical groups. In the linked article, one should identify η (in its a matrix representation) with the matrix Φ.

The appropriate group is O(3,1), in this context called the Lorentz group. Its elements are called (homogeneous) Lorentz transformations. For other methods of derivation, with a more physical twist, see derivations of the Lorentz transformations.

Among the simplest Lorentz transformations is a Lorentz boost. For reference, a boost in the *x*-direction is given by

$$
\begin{bmatrix} U_0' \\ U_1' \\ U_2' \\ U_3' \end{bmatrix} = \begin{bmatrix} \gamma & -\beta\gamma & 0 & 0 \\ -\beta\gamma & \gamma & 0 & 0 \\ 0 & 0 & 1 & 0 \\ 0 & 0 & 0 & 1 \end{bmatrix} \begin{bmatrix} U_0 \\ U_1 \\ U_2 \\ U_3 \end{bmatrix},
$$

where

$$
\gamma = \frac{1}{\sqrt{1 - \frac{v^2}{c^2}}}
$$

is the Lorentz factor, and

$$
\beta = \frac{v}{c}.
$$

Other Lorentz transformations are pure rotations, and hence elements of the SO(3) subgroup of O(3,1). A general homogeneous Lorentz transformation is a product of a pure boost and a pure rotation. An *inhomogeneous* Lorentz transformation is a homogeneous transformation followed by a translation in space and time. Special transformations are those that invert the space coordinates (P) and time coordinate (T) respectively, or both (PT).

All four-vectors in Minkowski space transform, by definition, according to the same formula under Lorentz transformations. Minkowski diagrams illustrate Lorentz transformations.

39.4   Causal structure

Main article: Causal structure

Vectors $v = (ct, x, y, z) = (ct, \mathbf{r})$ are classified according to the sign of $c^2t^2 - r^2$. A vector is **timelike** if $c^2t^2 > r^2$, **spacelike**

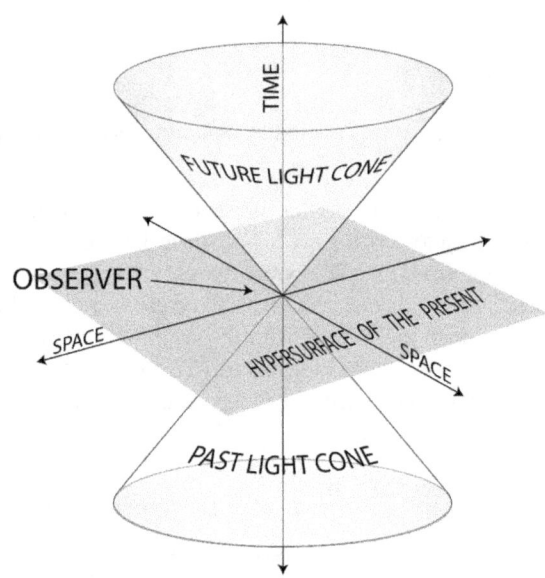

*Subdivision of Minkowski spacetime with respect to an event in four disjoint sets. The light cone, the **absolute future**, the **absolute past**, and **elsewhere**. The terminology is from Sard (1970).*

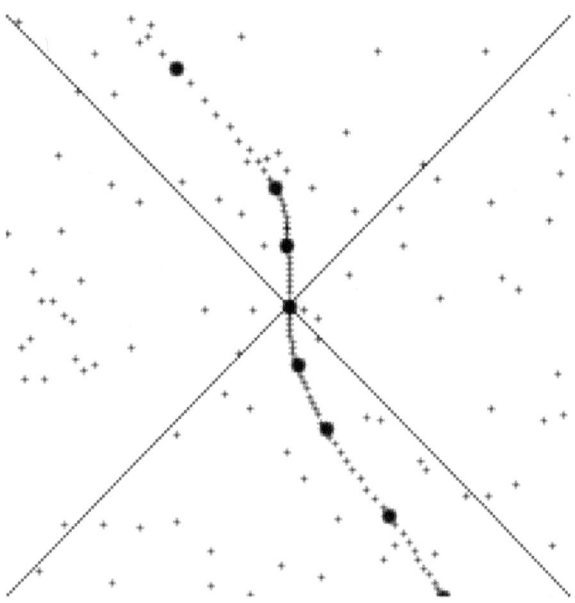

The momentarily co-moving inertial frames along the trajectory ("world line") of a rapidly accelerating observer (center). The vertical direction indicates time, while the horizontal indicates distance, the dashed line is the spacetime of the observer. The small dots are specific events in spacetime. Note how the momentarily co-moving inertial frame changes when the observer accelerates.

if $c^2t^2 < r^2$, and **null** or **lightlike** if $c^2t^2 = r^2$. This can be expressed in terms of the sign of $\eta(v,v)$ as well, but depends on the signature. The classification of any vector will be the same in all frames of reference, because of the invariance of the interval.

The set of all null vectors at an event[nb 5] of Minkowski space constitutes the light cone of that event. Given a timelike vector v, there is a worldline of constant velocity associated with it, represented by a straight line in a Minkowski diagram.

Once a direction of time is chosen,[nb 6] timelike and null vectors can be further decomposed into various classes. For timelike vectors one has

1. future-directed timelike vectors whose first component is positive, (tip of vector located in absolute future in figure) and

2. past-directed timelike vectors whose first component is negative (absolute past).

Null vectors fall into three classes:

1. the zero vector, whose components in any basis are (0,0,0,0) (origin),

2. future-directed null vectors whose first component is positive (upper light cone), and

3. past-directed null vectors whose first component is negative (lower light cone).

Spacelike vectors are in elsewhere. The terminology stems from the fact that spacelike separated events are connected by vectors requiring faster-than-light travel, and so cannot possibly influence each other. Together with spacelike and lightlike vectors there are 7 classes in all.

An orthonormal basis for Minkowski space necessarily consists of one timelike and three spacelike unit vectors. If one wishes to work with non-orthonormal bases it is possible to have other combinations of vectors. For example, one can easily construct a (non-orthonormal) basis consisting entirely of null vectors, called a **null basis**. Over the reals, if two null vectors are orthogonal (zero Minkowski tensor value), then they must be proportional. However, allowing complex numbers, one can obtain a null tetrad, which is a basis consisting of null vectors, some of which are orthogonal to each other.

Vector fields are called timelike, spacelike or null if the associated vectors are timelike, spacelike or null at each point where the field is defined.

39.4.1   Chronological and causality relations

Let $x, y \in M$. We say that

1. x *chronologically precedes* y if $y - x$ is future-directed timelike. This relation has the transitive property and so can be written x < y.

2. x *causally precedes* y if $y - x$ is future-directed null or future-directed timelike. It gives a partial ordering of space-time and so can be written x ≤ y.

39.4.2   Reversed triangle inequality

If v and w are both future-directed timelike four-vectors, then in the (+ - - -) sign convention for norm,

$$\|v + w\| \geq \|v\| + \|w\|.$$

39.5   Relationships to other formulations

39.5.1   Different number of dimensions

Strictly speaking, Minkowski space refers to a mathematical formulation in four dimensions. However, the mathematics can easily be extended or simplified to create an analogous "Minkowski space" in any number of dimensions. If $n \geq 2$, n-dimensional Minkowski space is a vector space of real dimension n on which there is a constant Lorentz metric of signature $(n - 1, 1)$ or $(1, n - 1)$. These generalizations are used in theories where spacetime is assumed to have more or less than 4 dimensions. String theory and M-theory are two examples where $n > 4$. In string theory, there appears conformal field theories with $1 + 1$ spacetime dimensions.

39.5.2   Flat versus curved space

As a *flat spacetime*, the three spatial components of Minkowski spacetime always obey the Pythagorean Theorem. Minkowski space is a suitable basis for special relativity, a good description of physical systems over finite distances in systems without significant gravitation. However, in order to take gravity into account, physicists use the theory of general relativity, which is formulated in the mathematics of a non-Euclidean geometry. When this geometry is used as a model of physical space, it is known as curved space.

Even in curved space, Minkowski space is still a good description in an infinitesimal region surrounding any point (barring gravitational singularities).[nb 7] More abstractly, we say that in the presence of gravity spacetime is described by a curved 4-dimensional manifold for which the tangent space to any point is a 4-dimensional Minkowski space. Thus, the structure of Minkowski space is still essential in the description of general relativity.

39.6   See also

- Causal structure
- Euclidean space
- Four vector
- Hyperboloid model
- Introduction to mathematics of general relativity
- Lorentzian manifold
- Metric tensor
- Minkowski diagram
- Minkowski plane
- Speed of light
- Super Minkowski space
- World line

39.7   Remarks

[1] This makes spacetime distance an invariant.

[2] Minkowski space can be formulated as an equivalent 4-D Euclidean space if you assume time is always an imaginary number. This is how the spacetime was first formulated, but since Minkowski reworked the structure, time is almost always required to be a real number.

[3] Consistent use of the term "Minkowski inner product" is intended for the bilinear form here, since it is in widespread use. It is by no means "standard" in the literature, but no such standard seems to exist.

[4] The Minkowski inner product is not an inner product, since it is not positive-definite, i.e. the quadratic form $\eta(v, v)$ need not be positive for nonzero v. The positive-definite condition has been replaced by the weaker condition of non-degeneracy. The bilinear form is said to be *indefinite*.

[5] Translate the coordinate system so that the event is the new origin.

[6] This corresponds to the time coordinate either increasing or decreasing when proper time for any particle increases. An application of T flips this direction.

[7] This similarity between flat and curved space at infinitesimally small distance scales is foundational to the definition of a manifold in general.

39.8   Notes

[1] Landau & Lifshitz 2002, p. 5

[2] Poincaré 1905–1906, pp. 129–176 Wikisource translation: On the Dynamics of the Electron

[3] Minkowski 1907–1908, pp. 53–111 *Wikisource translation: The Fundamental Equations for Electromagnetic Processes in Moving Bodies.

[4] Minkowski 1907–1909, pp. 75–88 Various English translations on Wikisource: Space and Time.

[5] Lee 2003, Proposition 3.8. The identification is routinely done in mathematics.

[6] Lee 2003, See Lee's discussion on geometric tangent vectors early in chapter 3.

[7] Sard 1970, p. 71

[8] Landau & Lifshitz 2002, p. 4

[9] Lee 2003, The tangent-cotangent isomorphism p. 282.

39.9   References

- Corry, L. (1997). "Hermann Minkowski and the postulate of relativity". *Arch. Hist. Exact Sci.* (Springer-Verlag) **51** (4): 273–314. doi:10.1007/BF00518231. ISSN 0003-9519. (subscription required (help)).

- Catoni, F.; et al. (2008). *Mathematics of Minkowski Space*. Frontiers in Mathematics. Basel: Birkhäuser Verlag. doi:10.1007/978-3-7643-8614-6. ISBN 978-3-7643-8613-9. ISSN 1660-8046.

- Galison, P. L. (1979). R McCormach; et al., eds. *Minkowski's Space-Time: from visual thinking to the absolute world*. Historical Studies in the Physical Sciences **10**. Johns Hopkins University Press. pp. 85–121. doi:10.2307/27757388. (subscription required (help)).

- Landau, L.D.; Lifshitz, E.M. (2002) [1939]. *The Classical Theory of Fields*. Course of Theoretical Physics **2** (4th ed.). Butterworth–Heinemann. ISBN 0 7506 2768 9.

- Lee, J. M. (2003). *Introduction to Smooth manifolds*. Springer Graduate Texts in Mathematics **218**. ISBN 0-387-95448-1.

- Minkowski, Hermann (1907–1908), "Die Grundgleichungen für die elektromagnetischen Vorgänge in bewegten Körpern" [The Fundamental Equations for Electromagnetic Processes in Moving Bodies],

Nachrichten von der Gesellschaft der Wissenschaften zu Göttingen, Mathematisch-Physikalische Klasse: 53–111 *Wikisource translation: The Fundamental Equations for Electromagnetic Processes in Moving Bodies

- Minkowski, Hermann (1907–1909), "Raum und Zeit" [Space and Time], *Physikalische Zeitschrift* **10**: 75–88 Various English translations on Wikisource: Space and Time

- Naber, G. L. (1992). *The Geometry of Minkowski Spacetime*. New York: Springer-Verlag. ISBN 0-387-97848-8.

- Penrose, Roger (2005). "18 Minkowskian geometry". *Road to Reality : A Complete Guide to the Laws of the Universe*. Alfred A. Knopf. ISBN 9780679454434.

- Poincaré, Henri (1905–1906), "Sur la dynamique de l'électron" [On the Dynamics of the Electron], *Rendiconti del Circolo matematico di Palermo* **21**: 129–176, doi:10.1007/BF03013466 Wikisource translation: On the Dynamics of the Electron

- Sard, R. D. (1970). *Relativistic Mechanics - Special Relativity and Classical Particle Dynamics*. New York: W. A. Benjamin. ISBN 978-0805384918.

- Shaw, R. (1982). "§ 6.6 Minkowski space, § 6.7,8 Canonical forms pp 221–242". *Linear Algebra and Group Representations*. Academic Press. ISBN 0-12-639201-3.

- Walter, Scott (1999). "Minkowski, Mathematicians, and the Mathematical Theory of Relativity". In Goenner, Hubert (ed.); et al. *The Expanding Worlds of General Relativity*. Boston: Birkhäuser. pp. 45–86. ISBN 0-8176-4060-6.

39.10 External links

Media related to Minkowski diagrams at Wikimedia Commons

- Animation clip on YouTube visualizing Minkowski space in the context of special relativity.

- The Geometry of Special Relativity: The Minkowski Space - Time Light Cone

Chapter 40

6D (2,0) superconformal field theory

In theoretical physics, the **six-dimensional (2,0)-superconformal field theory** is a quantum field theory whose existence is predicted by arguments in string theory. It is still poorly understood because there is no known description of the theory in terms of an action functional. Despite the inherent difficulty in studying this theory, it is considered to be an interesting object for a variety of reasons, both physical and mathematical.[1]

40.1 Applications

The (2,0)-theory has proven to be important for studying the general properties of quantum field theories. Indeed, this theory subsumes a large number of mathematically interesting effective quantum field theories and points to new dualities relating these theories. For example, Luis Alday, Davide Gaiotto, and Yuji Tachikawa showed that by compactifying this theory on a surface, one obtains a four-dimensional quantum field theory, and there is a duality known as the AGT correspondence which relates the physics of this theory to certain physical concepts associated with the surface itself.[2] More recently, theorists have extended these ideas to study the theories obtained by compactifying down to three dimensions.[3]

In addition to its applications in quantum field theory, the (2,0)-theory has spawned a number of important results in pure mathematics. For example, the existence of the (2,0)-theory was used by Witten to give a "physical" explanation for a conjectural relationship in mathematics called the geometric Langlands correspondence.[4] In subsequent work, Witten showed that the (2,0)-theory could be used to understand a concept in mathematics called Khovanov homology.[5] Developed by Mikhail Khovanov around 2000, Khovanov homology provides a tool in knot theory, the branch of mathematics that studies and classifies the different shapes of knots.[6] Another application of the (2,0)-theory in mathematics is the work of Davide Gaiotto, Greg Moore, and Andrew Neitzke, which used physical ideas to derive new results in hyperkähler geometry.[7]

40.2 See also

- ABJM superconformal field theory
- N = 4 supersymmetric Yang–Mills theory

40.3 Notes

[1] Moore 2012

[2] Alday, Gaiotto, and Tachikawa 2010

[3] Dimofte, Gaiotto, Gukov 2010

[4] Witten 2009

[5] Witten 2012

[6] Khovanov 2000

[7] Gaiotto, Moore, Neitzke 2013

40.4 References

- Alday, Luis; Gaiotto, Davide; Tachikawa, Yuji (2010). "Liouville correlation functions from four-dimensional gauge theories". *Letters in Mathematical Physics* **91** (2): 167–197. arXiv:0906.3219. Bibcode:2010LMaPh..91..167A. doi:10.1007/s11005-010-0369-5.

- Dimofte, Tudor; Gaiotto, Davide; Gukov, Sergei (2010). "Gauge theories labelled by three-manifolds". *Communications in Mathematical Physics* **325** (2): 367–419. Bibcode:2014CMaPh.325..367D. doi:10.1007/s00220-013-1863-2.

- Gaiotto, Davide; Moore, Gregory; Neitzke, Andrew (2013). "Wall-crossing, Hitchin systems, and the WKB approximation". *Advances in Mathematics* **2341**: 239–403. arXiv:0907.3987. doi:10.1016/j.aim.2012.09.027.

- Khovanov, Mikhail (2000). "A categorification of the Jones polynomial". *Duke Mathematical Journal* **101** (3): 359–426. doi:10.1215/s0012-7094-00-10131-7.

- Moore, Gregory (2012). "Lecture Notes for Felix Klein Lectures" (PDF). Retrieved 14 August 2013.

- Witten, Edward (2009). "Geometric Langlands from six dimensions". arXiv:0905.2720 [hep-th].

- Witten, Edward (2012). "Fivebranes and knots". *Quantum Topology* **3** (1): 1–137. doi:10.4171/qt/26.

Chapter 41

ABJM superconformal field theory

In theoretical physics, **ABJM theory** is a quantum field theory studied by Ofer Aharony, Oren Bergman, Daniel Jafferis, and Juan Maldacena. It provides a holographic dual to M-theory on $AdS_4 \times S^7$. The ABJM theory is also closely related to Chern-Simons theory, and it serves as a useful toy model for solving problems that arise in condensed matter physics.[1]

41.1   See also

- 6D (2,0) superconformal field theory

41.2   Notes

[1] Aharony et al. 2008

41.3   References

- Aharony, Ofer; Bergman, Oren; Jafferis, Daniel Louis; Maldacena, Juan (2008). "N=6 superconformal Chern-Simons-matter theories, M2-branes and their gravity duals". *Journal of High Energy Physics* **2008** (10): 091. arXiv:0806.1218. Bibcode:2008JHEP...10..091A. doi:10.1088/1126-6708/2008/10/091.

Chapter 42

Chern–Simons theory

The **Chern–Simons theory**, named after Shiing-Shen Chern and James Harris Simons, is a 3-dimensional topological quantum field theory of Schwarz type, developed by Edward Witten. It is so named because its action is proportional to the integral of the Chern–Simons 3-form.

In condensed matter physics, Chern–Simons theory describes the topological order in fractional quantum Hall effect states. In mathematics, it has been used to calculate knot invariants and three-manifold invariants such as the Jones polynomial.

Particularly, Chern–Simons theory is specified by a choice of simple Lie group G known as the gauge group of the theory and also a number referred to as the *level* of the theory, which is a constant that multiplies the action. The action is gauge dependent, however the partition function of the quantum theory is well-defined when the level is an integer and the gauge field strength vanishes on all boundaries of the 3-dimensional spacetime.

42.1 The classical theory

42.1.1 Mathematical origin

In the 1940s S. S. Chern and A. Weil studied the global curvature properties of smooth manifolds M as de Rham cohomology (Chern–Weil theory), which is an important step in the theory of characteristic classes in differential geometry. Given a flat G-principal bundle P on M there exists a unique homomorphism, called Chern–Weil homomorphism, from the algebra of G-adjoint invariant polynomial on g (Lie algebra of G) to the cohomology $H^*(M,\mathbb{R})$. If the invariant polynomial is homogeneous one can write down concretely any k-form of the closed connection ω as some $2k$-form of the associated curvature form Ω of ω.

In 1974 S. S. Chern and J. H. Simons had concretely constructed a $(2k-1)$-form $df(\omega)$ such that

$$dTf(\omega) = f(\Omega^k)$$

where T is the Chern–Weil homomorphism. This form is called Chern–Simons form. If $df(\omega)$ is closed one can integrate the above formula

$$Tf(\omega) = \int_C f(\Omega^k)$$

where C is a $(2k-1)$-dimensional cycle on M. This invariant is called **Chern–Simons invariant**. As pointed out in the introduction of the Chern–Simons paper, the Chern–Simons invariant $CS(M)$ is the boundary term that cannot be determined by any pure combinatorial formulation. It also can be defined as

$$CS(M) = \int_{s(M)} \tfrac{1}{2}Tp_1 \in \mathbb{R}/\mathbb{Z}$$

where p_1 is the first Pontryagin number and $s(M)$ is the section of the normal orthogonal bundle P. Moreover, the Chern–Simons term is described as the eta invariant defined by Atiyah, Patodi and Singer.

The gauge invariance and the metric invariance can be viewed as the invariance under the adjoint Lie group action in the Chern–Weil theory. The action integral (path integral) of the field theory in physics is viewed as the Lagrangian integral of the Chern–Simons form and Wilson loop, holonomy of vector bundle on M. These explain why the Chern–Simons theory is closely related to topological field theory.

42.1.2 Configurations

Chern–Simons theories can be defined on any topological 3-manifold M, with or without boundary. As these theories are Schwarz-type topological theories, no metric needs to be introduced on *M*.

Chern–Simons theory is a gauge theory, which means that a classical configuration in the Chern–Simons theory on *M* with gauge group *G* is described by a principal *G*-bundle on *M*. The connection of this bundle is characterized by a connection one-form *A* which is valued in the Lie algebra **g** of the Lie group *G*. In general the connection *A* is only defined on individual coordinate patches, and the values of *A* on different patches are related by maps known as gauge transformations. These are characterized by the assertion that the covariant derivative, which is the sum of the exterior derivative operator *d* and the connection *A*, transforms in the adjoint representation of the gauge group *G*. The square of the covariant derivative with itself can be interpreted as a **g**-valued 2-form *F* called the curvature form or field strength. It also transforms in the adjoint representation.

42.1.3   Dynamics

The action *S* of Chern–Simons theory is proportional to the integral of the Chern–Simons 3-form

$$S = \frac{k}{4\pi} \int_M \mathrm{tr}\left(A \wedge dA + \tfrac{2}{3} A \wedge A \wedge A\right).$$

The constant *k* is called the *level* of the theory. The classical physics of Chern–Simons theory is independent of the choice of level *k*.

Classically the system is characterized by its equations of motion which are the extrema of the action with respect to variations of the field *A*. In terms of the field curvature

$$F = dA + A \wedge A$$

the field equation is explicitly

$$0 = \frac{\delta S}{\delta A} = \frac{k}{2\pi} F.$$

The classical equations of motion are therefore satisfied if and only if the curvature vanishes everywhere, in which case the connection is said to be *flat*. Thus the classical solutions to *G* Chern–Simons theory are the flat connections of principal *G*-bundles on *M*. Flat connections are determined entirely by holonomies around noncontractible cycles on the base *M*. More precisely, they are in one to one correspondence with equivalence classes of homomorphisms from the fundamental group of *M* to the gauge group *G* up to conjugation.

If *M* has a boundary *N* then there is additional data which describes a choice of trivialization of the principal *G*-bundle on *N*. Such a choice characterizes a map from *N* to *G*. The dynamics of this map is described by the Wess–Zumino–Witten (WZW) model on *N* at level *k*.

42.2   Quantization

To canonically quantize Chern–Simons theory one defines a state on each 2-dimensional surface Σ in M. As in any quantum field theory, the states correspond to rays in a Hilbert space. There is no preferred notion of time in a Schwarz-type topological field theory and so one can impose that Σ be Cauchy surfaces, in fact a state can be defined on any surface.

Σ is codimension one, and so one may cut M along Σ. After such a cutting M will be a manifold with boundary and in particular classically the dynamics of Σ will be described by a WZW model. Witten has shown that this correspondence holds even quantum mechanically. More precisely, he demonstrated that the Hilbert space of states is always finite-dimensional and can be canonically identified with the space of conformal blocks of the G WZW model at level k. Conformal blocks are locally holomorphic and anti-holomorphic factors whose products sum to the correlation functions of a 2-dimensional conformal field theory.

For example, when Σ is a 2-sphere, this Hilbert space is one-dimensional and so there is only one state. When Σ is a 2-torus the states correspond to the integrable representations of the affine Lie algebra corresponding to g at level k. Characterizations of the conformal blocks at higher genera are not necessary for Witten's solution of Chern–Simons theory.

42.3   Observables

42.3.1   Wilson loops

The observables of Chern–Simons theory are the *n*-point correlation functions of gauge-invariant operators. The most often studied class of gauge invariant operators are Wilson loops. A Wilson loop is the holonomy around a loop in M, traced in a given representation R of G. As we will be interested in products of Wilson loops, without loss of generality we may restrict our attention to irreducible representations R.

More concretely, given an irreducible representation R and a loop K in M, one may define the Wilson loop $W_R(K)$ by

$$W_R(K) = \mathrm{Tr}_R \, \mathcal{P} \, \exp i \oint_K A$$

where A is the connection 1-form and we take the Cauchy principal value of the contour integral and $\mathcal{P}$ exp is the path-ordered exponential.

42.3.2 HOMFLY and Jones polynomials

Consider a link L in M, which is a collection of l disjoint loops. A particularly interesting observable is the l-point correlation function formed from the product of the Wilson loops around each disjoint loop, each traced in the fundamental representation of G. One may form a normalized correlation function by dividing this observable by the partition function $Z(M)$, which is just the 0-point correlation function.

In the special case in which M is the 3-sphere, Witten has shown that these normalized correlation functions are proportional to known knot polynomials. For example, in G=U(N) Chern–Simons theory at level k the normalized correlation function is, up to a phase, equal to

$$\frac{\sin(\pi/(k+N))}{\sin(\pi N/(k+N))}$$

times the HOMFLY polynomial. In particular when $N = 2$ the HOMFLY polynomial reduces to the Jones polynomial. In the SO(N) case one finds a similar expression with the Kauffman polynomial.

The phase ambiguity reflects the fact that, as Witten has shown, the quantum correlation functions are not fully defined by the classical data. The linking number of a loop with itself enters into the calculation of the partition function, but this number is not invariant under small deformations and in particular is not a topological invariant. This number can be rendered well defined if one chooses a framing for each loop, which is a choice of preferred nonzero normal vector at each point along which one deforms the loop to calculate its self-linking number. This procedure is an example of the point-splitting regularization procedure introduced by Paul Dirac and Rudolf Peierls to define apparently divergent quantities in quantum field theory in 1934.

Sir Michael Atiyah has shown that there exists a canonical choice of framing, which is generally used in the literature today and leads to a well-defined linking number. With the canonical framing the above phase is the exponential of $2\pi i/(k + N)$ times the linking number of L with itself.

42.4 Relationships with other theories

42.4.1 Topological string theories

In the context of string theory, a $U(N)$ Chern–Simons theory on an oriented Lagrangian 3-submanifold M of a 6-manifold X arises as the string field theory of open strings ending on a D-brane wrapping X in the A-model topological string theory on X. The B-model topological open string field theory on the spacefilling worldvolume of a stack of D5-branes is a 6-dimensional variant of Chern–Simons theory known as holomorphic Chern–Simons theory.

42.4.2 WZW and matrix models

Chern–Simons theories are related to many other field theories. For example, if one considers a Chern–Simons theory with gauge group G on a manifold with boundary then all of the 3-dimensional propagating degrees of freedom may be gauged away, leaving a 2-dimensional conformal field theory known as a G Wess–Zumino–Witten model on the boundary. In addition the $U(N)$ and SO(N) Chern–Simons theories at large N are well approximated by matrix models.

42.4.3 Chern–Simons, the Kodama wavefunction and loop quantum gravity

Main article: Kodama state

Edward Witten argued that the Kodama state in loop quantum gravity is unphysical due to an analogy to Chern–Simons state resulting in negative helicity and energy. Witten (2003)

42.4.4 Chern–Simons gravity theory

In 1982, S. Deser, R. Jackiw and S. Templeton proposed the Chern–Simons gravity theory in three dimensions, in which the Einstein–Hilbert action in gravity theory is modified by adding the Chern–Simons term.Deser, Jackiw & Templeton (1982)

In 2003, R. Jackiw and S. Y. Pi extended this theory to four dimensions Jackiw & Pi (2003) and Chern–Simons gravity theory has some considerable affects not only to fundamental physics but also condensed matter theory and astronomy.

The four-dimensional case is very analogous to the three-dimensional case. In three dimensions, the gravitational Chern–Simons term is

$$CS(\Gamma) = \frac{1}{2\pi^2} \int d^3x \epsilon^{ijk} \left(\Gamma^p_{iq}\partial_j\Gamma^q_{kp} + \frac{2}{3}\Gamma^p_{iq}\Gamma^q_{jr}\Gamma^r_{kp} \right).$$

This variation gives the Cotton tensor

$$= -\frac{1}{2\sqrt{g}} \left(\epsilon^{mij} D_i R^n_j + \epsilon^{nij} D_i R^m_j \right).$$

Then, Chern–Simons modification of three-dimensional gravity is made by adding the above Cotton tensor to the field equation, which can be obtained as the vacuum solution by varying the Einstein–Hilbert action.

See also (2+1)–dimensional topological gravity.

42.4.5   Chern–Simons matter theories

In 2013 Kenneth A. Intriligator and Nathan Seiberg solved these 3d Chern–Simons gauge theories and their phases using monopoles carrying extra degrees of freedom. The Witten index of the many vacua discovered was computed by compactifying the space by turning on mass parameters and then computing the index. In some vacua, supersymmetry was computed to be broken. These monopoles were related to condensed matter vortices. (Intriligator & Seiberg (2013))

The N=6 Chern-Simons-Matter theory is the holographic dual of M-theory on $AdS_4 \times S_7$.

42.5   Chern–Simons terms in other theories

The Chern–Simons term can also be added to models which aren't topological quantum field theories. In 3D, this gives rise to a massive photon if this term is added to the action of Maxwell's theory of electrodynamics. This term can be induced by integrating over a massive charged Dirac field. It also appears for example in the quantum Hall effect. Ten- and eleven-dimensional generalizations of Chern–Simons terms appear in the actions of all ten- and eleven-dimensional supergravity theories.

42.5.1   One-loop renormalization of the level

If one adds matter to a Chern–Simons gauge theory then in general it is no longer topological. However if one adds n Majorana fermions then, due to the parity anomaly, when integrated out they lead to a pure Chern–Simons theory with a one-loop renormalization of the Chern–Simons level by

$-n/2$, in other words the level k theory with n fermions is equivalent to the level $k - n/2$ theory without fermions.

42.6   See also

- Chern–Simons form
- Topological quantum field theory
- Alexander polynomial
- Jones polynomial
- 2+1D topological gravity

42.7   References

- Chern, S.-S. & Simons, J. (1974). "Characteristic forms and geometric invariants". *Annals of Mathematics* **99** (1): 48–69. doi:10.2307/1971013.

- Witten, Edward (1988). "Topological Quantum Field Theory". *Commun. Math. Phys.* **117**: 353. Bibcode:1988CMaPh.117..353W. doi:10.1007/BF01223371.http=http://projecteuclid. org/DPubS/Repository/1.0/Disseminate?view= body&id=pdf_1&handle=euclid.cmp/1104161738

- Witten, Edward (1989). "Quantum Field Theory and the Jones Polynomial". *Commun. Math. Phys.* **121** (3): 351–399. Bibcode:1989CMaPh.121..351W. doi:10.1007/BF01217730. MR 0990772.

- Witten, Edward (1995). "Chern–Simons Theory as a String Theory". *Prog. Math.* **133**: 637–678. arXiv:hep-th/9207094. Bibcode:1992hep.th....7094W.

- Witten, Edward (2003). "A Note On The Chern-Simons And Kodama Wavefunctions". arXiv:gr-qc/0306083.

- Marino, Marcos (2005). "Chern–Simons Theory and Topological Strings". *Rev. Mod. Phys.* **77** (2): 675–720. arXiv:hep-th/0406005. Bibcode:2005RvMP...77..675M. doi:10.1103/RevModPhys.77.675.

- Marino, Marcos (2005). *Chern–Simons Theory, Matrix Models, And Topological Strings*. International Series of Monographs on Physics. OUP.

- Deser, Stanley; Jackiw, Roman; Templeton, S. (1982). "Three-Dimensional Massive Gauge Theories". Phys. Rev. Lett. 48, 975–978. American Physical Society.

- Jackiw, Roman; Pi, S.-Y (2003). "Chern–Simons modification of general relativity". Phys.Rev. D68. American Physical Society.

- Intriligator, Kenneth; Seiberg, Nathan (2013). "Aspects of 3d $N = 2$ Chern–Simons-Matter Theories". *JHEP*.

42.8 External links

- Hazewinkel, Michiel, ed. (2001), "Chern-Simons functional", *Encyclopedia of Mathematics*, Springer, ISBN 978-1-55608-010-4

Chapter 43

String cosmology

String cosmology is a relatively new field that tries to apply equations of string theory to solve the questions of early cosmology. A related area of study is brane cosmology.

This approach can be dated back to a paper by Gabriele Veneziano[1] that shows how an inflationary cosmological model can be obtained from string theory, thus opening the door to a description of pre-Big Bang scenarios.

The idea is related to a property of the bosonic string in a curve background, better known as nonlinear sigma model. First calculations from this model[2] showed as the beta function, representing the running of the metric of the model as a function of an energy scale, is proportional to the Ricci tensor giving rise to a Ricci flow. As this model has conformal invariance and this must be kept to have a sensible quantum field theory, the beta function must be zero producing immediately the Einstein field equations. While Einstein equations seem to appear somewhat out of place, nevertheless this result is surely striking showing as a background two-dimensional model could produce higher-dimensional physics. An interesting point here is that such a string theory can be formulated without a requirement of criticality at 26 dimensions for consistency as happens on a flat background. This is a serious hint that the underlying physics of Einstein equations could be described by an effective two-dimensional conformal field theory. Indeed, the fact that we have evidence for an inflationary universe is an important support to string cosmology.

In the evolution of the universe, after the inflationary phase, the expansion observed today sets in that is well described by Friedmann equations. A smooth transition is expected between these two different phases. String cosmology appears to have difficulties in explaining this transition. This is known in literature as the **graceful exit problem**.

An inflationary cosmology implies the presence of a scalar field that drives inflation. In string cosmology, this arises from the so-called dilaton field. This is a scalar term entering into the description of the bosonic string that produces a scalar field term into the effective theory at low energies. The corresponding equations resemble those of a Brans–Dicke theory.

Analysis has been worked out from a critical number of dimension (26) down to four. In general one gets Friedmann equations in an arbitrary number of dimensions. The other way round is to assume that a certain number of dimensions is compactified producing an effective four-dimensional theory to work with. Such a theory is a typical Kaluza–Klein theory with a set of scalar fields arising from compactified dimensions. Such fields are called **moduli**.

43.1 Technical details

This section presents some of the relevant equations entering into string cosmology. The starting point is the Polyakov action, which can be written as:

$$S_2 = \frac{1}{4\pi\alpha'} \int d^2 z \sqrt{\gamma} \left[\gamma^{ab} G_{\mu\nu}(X) \partial_a X^\mu \partial_b X^\nu \right.$$
$$\left. + \alpha'\, {}^{(2)}R\Phi(X) \right],$$

where ${}^{(2)}R$ is the Ricci scalar in two dimensions, Φ the dilaton field, and α' the string constant. The indices a, b range over 1,2, and μ, ν over $1, \ldots, D$, where D the dimension of the target space. A further antisymmetric field could be added. This is generally considered when one wants this action generating a potential for inflation.[3] Otherwise, a generic potential is inserted by hand, as well as a cosmological constant.

The above string action has a conformal invariance. This is a property of a two dimensional Riemannian manifold. At the quantum level, this property is lost due to anomalies and the theory itself is not consistent, having no unitarity. So it is necessary to require that conformal invariance is kept at any order of perturbation theory. Perturbation theory is the only known approach to manage the quantum field theory. Indeed, the beta functions at two loops are

$$\beta^G_{\mu\nu} = R_{\mu\nu} + 2\alpha' \nabla_\mu \Phi \nabla_\nu \Phi + O(\alpha'^2),$$

and

$$\beta^{\Phi} = \frac{D-26}{6} - \frac{\alpha'}{2}\nabla^2\Phi + \alpha'\nabla_{\kappa}\Phi\nabla^{\kappa}\Phi + O(\alpha'^2).$$

The assumption that conformal invariance holds implies that

$$\beta^G_{\mu\nu} = \beta^{\Phi} = 0,$$

producing the corresponding equations of motion of low-energy physics. These conditions can only be satisfied perturbatively, but this has to hold at any order of perturbation theory. The first term in β^{Φ} is just the anomaly of the bosonic string theory in a flat spacetime. But here there are further terms that can grant a compensation of the anomaly also when $D \neq 26$, and from this cosmological models of a pre-big bang scenario can be constructed. Indeed, this low energy equations can be obtained from the following action:

$$S = \frac{1}{2\kappa_0^2}\int d^D x \sqrt{-G}e^{-2\Phi}\left[-\frac{2(D-26)}{3\alpha'}\right.$$
$$\left. + R + 4\partial_{\mu}\Phi\partial^{\mu}\Phi + O(\alpha')\right],$$

where κ_0^2 is a constant that can always be changed by redefining the dilaton field. One can also rewrite this action in a more familiar form by redefining the fields (Einstein frame) as

$$g_{\mu\nu} = e^{2\omega}G_{\mu\nu},$$
$$\omega = \frac{2(\Phi_0 - \Phi)}{D-2},$$

and using $\tilde{\Phi} = \Phi - \Phi_0$ one can write

$$S = \frac{1}{2\kappa^2}\int d^D x\sqrt{-g}\left[-\frac{2(D-26)}{3\alpha'}e^{\frac{4\tilde{\Phi}}{D-2}} + \tilde{R} - \right.$$
$$\left. \frac{4}{D-2}\partial_{\mu}\tilde{\Phi}\partial^{\mu}\tilde{\Phi} + O(\alpha')\right],$$

where

$$\tilde{R} = e^{-2\omega}[R - (D-1)\nabla^2\omega - (D-2)(D-1)\partial_{\mu}\omega\partial^{\mu}\omega].$$

This is the formula for the Einstein action describing a scalar field interacting with a gravitational field in D dimensions. Indeed, the following identity holds:

$$\kappa = \kappa_0 e^{2\Phi_0} = (8\pi G_D)^{\frac{1}{2}} = \frac{\sqrt{8\pi}}{M_p},$$

where G_D is the Newton constant in D dimensions and M_p the corresponding Planck mass. When setting $D = 4$ in this action, the conditions for inflation are not fulfilled unless a potential or antisymmetric term is added to the string action,[3] in which case power-law inflation is possible.

43.2 Notes

[1] Veneziano, G. (1991). "Scale factor duality for classical and quantum strings". *Physics Letters B* **265** (3–4): 287. Bibcode:1991PhLB..265..287V. doi:10.1016/0370-2693(91)90055-U.

[2] Friedan, D. (1980). "Nonlinear Models in 2+ε Dimensions" (PDF). *Physical Review Letters* **45** (13): 1057. Bibcode:1980PhRvL..45.1057F. doi:10.1103/PhysRevLett.45.1057.

[3] Easther, R.; Maeda, Kei-ichi; Wands, D. (1996). "Tree-level string cosmology". *Physical Review D* **53** (8): 4247. arXiv:hep-th/9509074. Bibcode:1996PhRvD..53.4247E. doi:10.1103/PhysRevD.53.4247.

43.3 References

- Polchinski, Joseph (1998a). *String Theory Vol. I: An Introduction to the Bosonic String*. Cambridge University Press. ISBN 0-521-63303-6.

- Polchinski, Joseph (1998b). *String Theory Vol. II: Superstring Theory and Beyond*. Cambridge University Press. ISBN 0-521-63304-4.

- Lidsey, James D.; Wands, David; Copeland, E. J. (2000). "Superstring Cosmology". *Physics Report* **337** (4–5): 343. arXiv:hep-th/9909061. Bibcode:2000PhR...337..343L. doi:10.1016/S0370-1573(00)00064-8.

43.4 External links

- String cosmology on arxiv.org

- Maurizio Gasperini's homepage

Chapter 44

String theory landscape

The **string theory landscape** refers to the huge number of possible false vacua in string theory.[1] The large number of theoretically allowed configurations has prompted suggestions that certain physical mysteries, particularly relating to the fine-tuning of constants like the cosmological constant or the Higgs boson mass, may be explained not by a physical mechanism but by assuming that many different vacua are physically realized.[2] The *anthropic landscape* thus refers to the collection of those portions of the landscape that are suitable for supporting intelligent life, an application of the anthropic principle that selects a subset of the otherwise possible configurations.

In string theory the number of false vacua is thought to be somewhere between 10^{10} to 10^{100}.[1] The large number of possibilities arises from different choices of Calabi–Yau manifolds and different values of generalized magnetic fluxes over different homology cycles. If one assumes that there is no structure in the space of vacua, the problem of finding one with a sufficiently small cosmological constant is NP complete,[3] being a version of the subset sum problem.

44.1 Anthropic principle

Main article: Anthropic principle

The idea of the string theory landscape has been used to propose a concrete implementation of the anthropic principle, the idea that fundamental constants may have the values they have not for fundamental physical reasons, but rather because such values are necessary for life (and hence intelligent observers to measure the constants). In 1987, Steven Weinberg proposed that the observed value of the cosmological constant was so small because it is impossible for life to occur in a universe with a much larger cosmological constant.[4] In order to implement this idea in a concrete physical theory, it is necessary to postulate a multiverse in which fundamental physical parameters can take different values. This has been realized in the context of eternal inflation.

44.2 Bayesian probability

Main article: Bayesian probability

Some physicists, starting with Weinberg, have proposed that Bayesian probability can be used to compute probability distributions for fundamental physical parameters, where the probability $P(x)$ of observing some fundamental parameters x is given by,

$$P(x) = P_{\text{prior}}(x) \times P_{\text{selection}}(x),$$

where P_{prior} is the prior probability, from fundamental theory, of the parameters x and $P_{\text{selection}}$ is the anthropic selection function, determined by the number of "observers" that would occur in the universe with parameters x. These probabilistic arguments are the most controversial aspect of the landscape. Technical criticisms of these proposals have pointed out that:

- The function P_{prior} is completely unknown in string theory and may be impossible to define or interpret in any sensible probabilistic way.

- The function $P_{\text{selection}}$ is completely unknown, since so little is known about the origin of life. Simplified criteria (such as the number of galaxies) must be used as a proxy for the number of observers. Moreover, it may never be possible to compute it for parameters radically different from those of the observable universe.

(Interpreting probability in a context where it is only possible to draw one sample from a distribution is problematic in frequentist probability but not in Bayesian probability, which is not defined in terms of the frequency of repeated events.)

Various physicists have tried to address these objections, and the ideas remain extremely controversial both within and outside the string theory community. These ideas have been reviewed by Carroll.[5]

44.3 Simplified approaches

Tegmark *et al.* have recently considered these objections and proposed a simplified anthropic scenario for axion dark matter in which they argue that the first two of these problems do not apply.[6]

Vilenkin and collaborators have proposed a consistent way to define the probabilities for a given vacuum.[7]

A problem with many of the simplified approaches people have tried is that they "predict" a cosmological constant that is too large by a factor of 10–1000 (depending on one's assumptions) and hence suggest that the cosmic acceleration should be much more rapid than is observed.[8][9][10]

44.4 Criticism

Although few dispute the idea that string theory appears to have an unimaginably large number of metastable vacua, the existence - meaning and scientific relevance of the anthropic landscape - remain highly controversial. Prominent proponents of the idea include Andrei Linde, Sir Martin Rees and especially Leonard Susskind, who advocate it as a solution to the cosmological-constant problem. Opponents, such as David Gross, suggest that the idea is inherently unscientific, unfalsifiable or premature. A famous debate on the anthropic landscape of string theory is the Smolin–Susskind debate on the merits of the landscape.

The term "landscape" comes from evolutionary biology (see *Fitness landscape*) and was first applied to cosmology by Lee Smolin in his book.[11] It was first used in the context of string theory by Susskind.

There are several popular books about the anthropic principle in cosmology.[12] The authors of two physics blogs are opposed to this use of the anthropic principle.[13]

44.5 See also

- Extra dimensions

- Compactification

44.6 References

[1] The most commonly quoted number is of the order 10^{500}. See M. Douglas, "The statistics of string / M theory vacua", *JHEP* **0305**, 46 (2003). arXiv:hep-th/0303194; S. Ashok and M. Douglas, "Counting flux vacua", *JHEP* **0401**, 060 (2004).

[2] L. Susskind, "The anthropic landscape of string theory", arXiv:hep-th/0302219.

[3] Frederik Denef; Douglas, Michael R. (2006). "Computational complexity of the landscape". *Annals of Physics* **322** (5): 1096–1142. arXiv:hep-th/0602072. Bibcode:2007AnPhy.322.1096D. doi:10.1016/j.aop.2006.07.013.

[4] S. Weinberg, "Anthropic bound on the cosmological constant", *Phys. Rev. Lett.* **59**, 2607 (1987).

[5] S. M. Carroll, "Is our universe natural?", arXiv:hep-th/0512148.

[6] M. Tegmark, A. Aguirre, M. Rees and F. Wilczek, "Dimensionless constants, cosmology and other dark matters", arXiv:astro-ph/0511774. F. Wilczek, "Enlightenment, knowledge, ignorance, temptation", arXiv:hep-ph/0512187. See also the discussion at .

[7] See, *e.g.* Alexander Vilenkin (2006). "A measure of the multiverse". *Journal of Physics A: Mathematical and Theoretical* **40** (25): 6777–6785. arXiv:hep-th/0609193. Bibcode:2007JPhA...40.6777V. doi:10.1088/1751-8113/40/25/S22.

[8] Abraham Loeb (2006). "An observational test for the anthropic origin of the cosmological constant". *JCAP* **0605**: 009. (subscription required (help)).

[9] Jaume Garriga & Alexander Vilenkin (2006). "Anthropic prediction for Lambda and the Q catastrophe". *Prog. Theor.Phys. Suppl.* **163**: 245–57. arXiv:hep-th/0508005. Bibcode:2006PThPS.163..245G. doi:10.1143/PTPS.163.245. (subscription required (help)).

[10] Delia Schwartz-Perlov & Alexander Vilenkin (2006). "Probabilities in the Bousso-Polchinski multiverse". *JCAP* **0606**: 010. (subscription required (help)).

[11] L. Smolin, "Did the universe evolve?", *Classical and Quantum Gravity* **9**, 173–191 (1992). L. Smolin, *The Life of the Cosmos* (Oxford, 1997)

[12] L. Susskind, *The cosmic landscape: string theory and the illusion of intelligent design* (Little, Brown, 2005). M. J. Rees, *Just six numbers: the deep forces that shape the universe* (Basic Books, 2001). R. Bousso and J. Polchinski, "The string theory landscape", *Sci. Am.* **291**, 60–69 (2004).

[13] Lubos Motl's blog criticized the anthropic principle and Peter Woit's blog frequently attacks the anthropic string landscape.

44.7   External links

- String landscape; moduli stabilization; flux vacua; flux compactification on arxiv.org

- Cvetič, Mirjam; García-Etxebarria, Iñaki; Halverson, James (March 2011). "On the computation of non-perturbative effective potentials in the string theory landscape". *Fortschritte der Physik* **59** (3-4): 243–283. doi:10.1002/prop.201000093.

Chapter 45

Calabi–Yau manifold

"Calabi-Yau" redirects here. For the play by Susanna Speier, see Calabi-Yau (play).

A **Calabi–Yau manifold**, also known as a **Calabi–**

A 2D slice of the 6D Calabi-Yau quintic manifold.

Yau space, is a special type of manifold that is described in certain branches of mathematics such as algebraic geometry. The Calabi–Yau manifold's properties, such as Ricci flatness, also yield applications in theoretical physics. Particularly in superstring theory, the extra dimensions of spacetime are sometimes conjectured to take the form of a 6-dimensional Calabi–Yau manifold, which led to the idea of mirror symmetry.

Calabi–Yau manifolds are complex manifolds that are generalizations of K3 surfaces in any number of complex dimensions (i.e. any even number of real dimensions). They were originally defined as compact Kähler manifolds with a vanishing first Chern class and a Ricci-flat metric, though many other similar but inequivalent definitions are sometimes used. They were named "Calabi–Yau spaces" by Candelas et al. (1985) after E. Calabi (1954, 1957) who

first conjectured that such surfaces might exist, and S. T. Yau (1978) who proved the Calabi conjecture.

45.1 Definitions

The motivational definition given by Yau is of a compact Kähler manifold with a vanishing first Chern class, that is also Ricci flat.[1] Calabi conjectured their existence and Yau proved the conjecture.

There are many other definitions of a Calabi–Yau manifold used by different authors, some inequivalent. This section summarizes some of the more common definitions and the relations between them.

A Calabi–Yau n-fold or Calabi–Yau manifold of (complex) dimension n is sometimes defined as a compact n-dimensional Kähler manifold M satisfying one of the following equivalent conditions:

- The canonical bundle of M is trivial.

- M has a holomorphic n-form that vanishes nowhere.

- The structure group of M can be reduced from U(n) to SU(n).

- M has a Kähler metric with global holonomy contained in SU(n).

These conditions imply that the first integral Chern class $c_1(M)$ of M vanishes, but the converse is not true. The simplest examples where this happens are hyperelliptic surfaces, finite quotients of a complex torus of complex dimension 2, which have vanishing first integral Chern class but non-trivial canonical bundle.

For a compact n-dimensional Kähler manifold M the following conditions are equivalent to each other, but are weaker than the conditions above, and are sometimes used as the definition of a Calabi–Yau manifold:

- M has vanishing first real Chern class.

- M has a Kähler metric with vanishing Ricci curvature.

- M has a Kähler metric with local holonomy contained in SU(n).

- A positive power of the canonical bundle of M is trivial.

- M has a finite cover that has trivial canonical bundle.

- M has a finite cover that is a product of a torus and a simply connected manifold with trivial canonical bundle.

In particular if a compact Kähler manifold is simply connected then the weak definition above is equivalent to the stronger definition. Enriques surfaces give examples of complex manifolds that have Ricci-flat metrics, but their canonical bundles are not trivial so they are Calabi–Yau manifolds according to the second but not the first definition above. Their double covers are Calabi–Yau manifolds for both definitions (in fact K3 surfaces).

By far the hardest part of proving the equivalences between the various properties above is proving the existence of Ricci-flat metrics. This follows from Yau's proof of the Calabi conjecture, which implies that a compact Kähler manifold with a vanishing first real Chern class has a Kähler metric in the same class with vanishing Ricci curvature. (The class of a Kähler metric is the cohomology class of its associated 2-form.) Calabi showed such a metric is unique.

There are many other inequivalent definitions of Calabi–Yau manifolds that are sometimes used, which differ in the following ways (among others):

- The first Chern class may vanish as an integral class or as a real class.

- Most definitions assert that Calabi–Yau manifolds are compact, but some allow them to be non-compact. In the generalization to non-compact manifolds, the difference $(\Omega \wedge \bar{\Omega} - \omega^n/n!)$ must vanish asymptotically. Here, ω is the Kähler form associated with the Kähler metric, g (Gang Tian;Shing-Tung Yau 1990, 1991).

- Some definitions put restrictions on the fundamental group of a Calabi–Yau manifold, such as demanding that it be finite or trivial. Any Calabi–Yau manifold has a finite cover that is the product of a torus and a simply-connected Calabi–Yau manifold.

- Some definitions require that the holonomy be exactly equal to SU(n) rather than a subgroup of it, which implies that the Hodge numbers $h^{i,0}$ vanish for $0 < i < \dim(M)$. Abelian surfaces have a Ricci flat metric with holonomy strictly smaller than SU(2) (in fact trivial) so

are not Calabi–Yau manifolds according to such definitions.

- Most definitions assume that a Calabi–Yau manifold has a Riemannian metric, but some treat them as complex manifolds without a metric.

- Most definitions assume the manifold is non-singular, but some allow mild singularities. While the Chern class fails to be well-defined for singular Calabi–Yau's, the canonical bundle and canonical class may still be defined if all the singularities are Gorenstein, and so may be used to extend the definition of a smooth Calabi–Yau manifold to a possibly singular Calabi–Yau variety.

45.2 Examples

The most important fundamental fact is that any smooth algebraic variety embedded in a projective space is a Kähler manifold, because there is a natural Fubini–Study metric on a projective space which one can restrict to the algebraic variety. By definition, if ω is the Kähler metric on the algebraic variety X and the canonical bundle KX is trivial, then X is Calabi–Yau. Moreover, there is unique Kähler metric ω on X such that $[\omega_0]=[\omega]\in H^2(X,\mathbf{R})$, a fact which was conjectured by Eugenio Calabi and proved by S. T. Yau (see Calabi conjecture).

In one complex dimension, the only compact examples are tori, which form a one-parameter family. The Ricci-flat metric on a torus is actually a flat metric, so that the holonomy is the trivial group SU(1). A one-dimensional Calabi–Yau manifold is a complex elliptic curve, and in particular, algebraic.

In two complex dimensions, the K3 surfaces furnish the only compact simply connected Calabi–Yau manifolds. Non simply-connected examples are given by abelian surfaces. Enriques surfaces and hyperelliptic surfaces have first Chern class that vanishes as an element of the real cohomology group, but not as an element of the integral cohomology group, so Yau's theorem about the existence of a Ricci-flat metric still applies to them but they are sometimes not considered to be Calabi–Yau manifolds. Abelian surfaces are sometimes excluded from the classification of being Calabi–Yau, as their holonomy (again the trivial group) is a proper subgroup of SU(2), instead of being isomorphic to SU(2).

In three complex dimensions, classification of the possible Calabi–Yau manifolds is an open problem, although Yau suspects that there is a finite number of families (albeit a much bigger number than his estimate from 20 years ago). In turn, it has also been conjectured by Miles Reid that the number of topological types of Calabi-Yau 3-folds is infinite, and that they can all be transformed continuously (through certain mild singularizations such as conifolds) one into another—much as Riemann surfaces can.[2] One example of a three-dimensional Calabi–Yau manifold is a nonsingular quintic threefold in $\mathbf{CP}^4$, which is the algebraic variety consisting of all of the zeros of a homogeneous quintic polynomial in the homogeneous coordinates of the $\mathbf{CP}^4$. Another example is a smooth model of the Barth–Nieto quintic. Some discrete quotients of the quintic by various $\mathbf{Z}_5$ actions are also Calabi–Yau and have received a lot of attention in the literature. One of these is related to the original quintic by mirror symmetry.

For every positive integer n, the zero set of a non-singular homogeneous degree $n+2$ polynomial in the homogeneous coordinates of the complex projective space $\mathbf{CP}^{n+1}$ is a compact Calabi–Yau n-fold. The case $n=1$ describes an elliptic curve, while for $n=2$ one obtains a K3 surface.

All hyper-Kähler manifolds are Calabi–Yau.

45.3 Applications in superstring theory

Calabi–Yau manifolds are important in superstring theory. Essentially, Calabi–Yau manifolds are shapes that satisfy the requirement of space for the six "unseen" spatial dimensions of string theory, which may be smaller than our currently observable lengths as they have not yet been detected. A popular alternative known as large extra dimensions, which often occurs in braneworld models, is that the Calabi–Yau is large but we are confined to a small subset on which it intersects a D-brane.

In the most conventional superstring models, ten conjectural dimensions in string theory are supposed to come as four of which we are aware, carrying some kind of fibration with fiber dimension six. Compactification on Calabi–Yau n-folds are important because they leave some of the original supersymmetry unbroken. More precisely, in the absence of fluxes, compactification on a Calabi–Yau 3-fold (real dimension 6) leaves one quarter of the original supersymmetry unbroken if the holonomy is the full SU(3).

More generally, a flux-free compactification on an n-manifold with holonomy SU(n) leaves 2^{1-n} of the original supersymmetry unbroken, corresponding to 2^{6-n} supercharges in a compactification of type II supergravity or 2^{5-n} supercharges in a compactification of type I. When fluxes are included the supersymmetry condition instead implies that the compactification manifold be a generalized Calabi–Yau, a notion introduced by Hitchin (2003). These models are known as flux compactifications.

F-theory compactifications on various Calabi–Yau four-folds provide physicists with a method to find a large number of classical solution in the so-called string theory landscape.

Connected with each hole in the Calabi-Yau space is a group of low-energy string vibrational patterns. Since string theory states that our familiar elementary particles correspond to low-energy string vibrations, the presence of multiple holes causes the string patterns to fall into multiple groups, or families. Although the following statement has been simplified, it conveys the logic of the argument: if the Calabi-Yau has three holes, then three families of vibrational patterns and thus three families of particles will be observed experimentally.

Logically, since strings vibrate through all the dimensions, the shape of the curled-up ones will affect their vibrations and thus the properties of the elementary particles observed. For example, Andrew Strominger and Edward Witten have shown that the masses of particles depend on the manner of the intersection of the various holes in a Calabi-Yau. In other words, the positions of the holes relative to one another and to the substance of the Calabi-Yau space was found by Strominger and Witten to affect the masses of particles in a certain way. This, of course, is true of all particle properties.[3]

45.4 See also

- G2 manifold

- Calabi–Yau algebra

45.5 References

45.5.1 Citations

[1] Yau and Nadis (2010)

[2] Reid, Miles (1987), "The Moduli Space of 3-Folds with K = 0 May Nevertheless be Irreducible", *Math. Ann.*, **278**, 329

[3] "The Shape of Curled-Up Dimensions". Archived from the original on Sep 13, 2006. Retrieved 2006.

45.5.2 Bibliography

- Besse, Arthur L. (1987), *Einstein manifolds*, Ergebnisse der Mathematik und ihrer Grenzgebiete (3) **10**, Berlin, New York: Springer-Verlag, ISBN 978-3-540-15279-8, OCLC 13793300

- Chan,Yat-Ming (2004)"Desingularization Of Calabi–Yau 3-Folds With A Conical Singularity"

- Calabi, Eugenio (1954), "The space of Kähler metrics", *Proc. Internat. Congress Math. Amsterdam* **2**, pp. 206–207

- Calabi, Eugenio (1957), "On Kähler manifolds with vanishing canonical class", in Fox, Ralph H.; Spencer, D. C.; Tucker, A. W., *Algebraic geometry and topology. A symposium in honor of S. Lefschetz*, Princeton Mathematical Series **12**, Princeton University Press, pp. 78–89, MR 0085583

- Greene, Brian "String Theory On Calabi–Yau Manifolds"

- Candelas, Philip; Horowitz, Gary; Strominger, Andrew; Witten, Edward (1985), "Vacuum configurations for superstrings", *Nuclear Physics B* **258**: 46–74, Bibcode:1985NuPhB.258...46C, doi:10.1016/0550-3213(85)90602-9

- Gross, M.; Huybrechts, D.; Joyce, Dominic (2003), *Calabi–Yau manifolds and related geometries*, Universitext, Berlin, New York: Springer-Verlag, ISBN 978-3-540-44059-8, MR 1963559, OCLC 50695398

- Hitchin, Nigel (2003), "Generalized Calabi–Yau manifolds", *The Quarterly Journal of Mathematics* **54** (3): 281–308, arXiv:math.DG/0209099, doi:10.1093/qmath/hag025, MR 2013140

- Hübsch, Tristan (1994), *Calabi–Yau Manifolds: a Bestiary for Physicists*, Singapore, New York: World Scientific, ISBN 981-02-1927-X, OCLC 34989218

- Im, Mee Seong (2008) "Singularities-in-Calabi-Yau-varieties.pdf Singularities in Calabi–Yau varieties"

- Joyce, Dominic (2000), *Compact Manifolds with Special Holonomy*, Oxford University Press, ISBN 978-0-19-850601-0, OCLC 43864470

- Tian, Gang; Yau, Shing-Tung (1990), "Complete Kähler manifolds with zero Ricci curvature, I", *Amer. Math. Soc.* **3** (3): 579–609, doi:10.2307/1990928, JSTOR 1990928

- Tian, Gang; Yau, Shing-Tung (1991), "Complete Kähler manifolds with zero Ricci curvature, II", *Invent. Math.* **106** (1): 27–60, Bibcode:1991InMat.106...27T, doi:10.1007/BF01243902

- Yau, Shing Tung (1978), "On the Ricci curvature of a compact Kähler manifold and the complex Monge-Ampère equation. I", *Communications on Pure and Applied Mathematics* **31** (3): 339–411, doi:10.1002/cpa.3160310304, MR 480350

- Yau, Shing-Tung (2009), "Surveys in differential geometry. Vol. XIII. Geometry, analysis, and algebraic geometry: forty years of the Journal of Differential Geometry", *Scholarpedia*, Surv. Differ. Geom. (Somerville, Massachusetts: Int. Press) **4** (8): 277–318, Bibcode:2009SchpJ...4.6524Y, doi:10.4249/scholarpedia.6524, MR 2537089 |chapter= ignored (help)

- Yau, Shing-Tung and Nadis, Steve; *The Shape of Inner Space*, Basic Books, 2010.

45.6   External links

- Calabi–Yau Homepage is an interactive reference which describes many examples and classes of Calabi–Yau manifolds and also the physical theories in which they appear.

- Spinning Calabi–Yau Space video.

- *Calabi–Yau Space* by Andrew J. Hanson with additional contributions by Jeff Bryant, Wolfram Demonstrations Project.

- Weisstein, Eric W., "Calabi–Yau Space", *MathWorld*.

- Yau, S. T., *Calabi–Yau manifold*, Scholarpedia (similar to (Yau 2009))

Chapter 46

Brane cosmology

Brane cosmology refers to several theories in particle physics and cosmology related to string theory, superstring theory and M-theory.

46.1 Brane and bulk

Main article: Brane

The central idea is that the visible, four-dimensional universe is restricted to a brane inside a higher-dimensional space, called the "bulk" (also known as "hyperspace"). If the additional dimensions are compact, then the observed universe contains the extra dimensions, and then no reference to the bulk is appropriate. In the bulk model, at least some of the extra dimensions are extensive (possibly infinite), and other branes may be moving through this bulk. Interactions with the bulk, and possibly with other branes, can influence our brane and thus introduce effects not seen in more standard cosmological models.

46.2 Why gravity is weak and the cosmological constant is small

Some versions of brane cosmology, based on the large extra dimension idea, can explain the weakness of gravity relative to the other fundamental forces of nature, thus solving the so-called hierarchy problem. In the brane picture, the other three forces (electromagnetism and the weak and strong nuclear forces) are localized on the brane, but gravity has no such constraint and propagates on the full spacetime, called bulk. Much of the gravitational attractive power "leaks" into the bulk. As a consequence, the force of gravity should appear significantly stronger on small (subatomic or at least sub-millimetre) scales, where less gravitational force has "leaked". Various experiments are currently under way to test this.[1] Extensions of the large extra dimension idea with supersymmetry in the bulk appears to be promising in addressing the so-called cosmological constant problem.[2][3][4]

46.3 Models of brane cosmology

One of the earliest documented attempts to apply brane cosmology as part of a conceptual theory is dated to 1983.[5]

The authors discussed the possibility that the Universe has $(3+N)+1$ dimensions, but ordinary particles are confined in a potential well which is narrow along N spatial directions and flat along three others, and proposed a particular five-dimensional model.

In 1998/99 Merab Gogberashvili published on arXiv a number of articles where he showed that if the Universe is considered as a thin shell (a mathematical synonym for "brane") expanding in 5-dimensional space then there is a possibility to obtain one scale for particle theory corresponding to the 5-dimensional cosmological constant and Universe thickness, and thus to solve the hierarchy problem.[6][7][8] It was also shown that the four-dimensionality of the Universe is the result of the stability requirement found in mathematics since the extra component of the Einstein field equations giving the confined solution for matter fields coincides with one of the conditions of stability.

In 1999 there were proposed the closely related Randall–Sundrum (RS1 and RS2; see *5 dimensional warped geometry theory* for a nontechnical explanation of RS1) scenarios. These particular models of brane cosmology have attracted a considerable amount of attention.

Later, the pre-big bang, ekpyrotic and cyclic proposals appeared. The ekpyrotic theory hypothesizes that the origin of the observable universe occurred when two parallel branes collided.[9]

46.4 Empirical tests

See also: Large extra dimension, Empirical tests

As of now, no experimental or observational evidence of large extra dimensions, as required by the Randall–Sundrum models, has been reported. An analysis of results from the Large Hadron Collider in December 2010 severely constrains theories with large extra dimensions.[10]

46.5 See also

- Kaluza–Klein theory

- Loop quantum cosmology

- M-theory

- String theory

46.6 References

[1] Session D9 - Experimental Tests of Short Range Gravitation.

[2] Aghababaie, Burgess, Parameswaran, Quevedo (2003-04-29). "Towards a naturally small cosmological constant from branes in 6-D supergravity". *Nucl.Phys. B680 (2004) 389-414.* arXiv:hep-th/0304256. Bibcode:2004NuPhB.680..389A. doi:10.1016/j.nuclphysb.2003.12.015.

[3] Burgess, van Nierop (2011-08-01). "Technically Natural Cosmological Constant From Supersymmetric 6D Brane Backreaction". *Phys.Dark Univ. 2 (2013) 1-16.* arXiv:1108.0345. Bibcode:2013PDU.....2....1B. doi:10.1016/j.dark.2012.10.001.

[4] Burgess, van Nierop, Parameswaran, Salvio, Williams (2012-10-19). "Accidental SUSY: Enhanced Bulk Supersymmetry from Brane Back-reaction". *JHEP 1302 (2013) 120.* arXiv:1210.5405. Bibcode:2013JHEP...02..120B. doi:10.1007/JHEP02(2013)120.

[5] V. A. Rubakov and M. E. Shaposhnikov, *Do we live inside a domain wall?*, Physics Letters B 125 (1983) 136–138.

[6] M. Gogberashvili, *Hierarchy problem in the shell universe model*, Arxiv:hep-ph/9812296.

[7] M. Gogberashvili, *Our world as an expanding shell*, Arxiv: hep-ph/9812365.

[8] M. Gogberashvili, *Four dimensionality in noncompact Kaluza–Klein model*, Arxiv:hep-ph/9904383.

[9] Musser, George; Minkel, JR (2002-02-11). "A Recycled Universe: Crashing branes and cosmic acceleration may power an infinite cycle in which our universe is but a phase". Scientific American Inc. Retrieved 2008-05-03.

[10] CMS Collaboration, "Search for Microscopic Black Hole Signatures at the Large Hadron Collider", http://arxiv.org/abs/1012.3375

46.7 External links

- Brax, Philippe; van de Bruck, Carsten (2003). "Cosmology and Brane Worlds: A Review". arXiv:hep-th/0303095. – Cosmological consequences of the brane world scenario are reviewed in a pedagogical manner.

- Langlois, David (2002). "Brane cosmology: an introduction". arXiv:hep-th/0209261. – These notes (32 pages) give an introductory review on brane cosmology.

- Papantonopoulos, Eleftherios (2002). "Brane Cosmology". arXiv:hep-th/0202044. – Lectures (24 pages) presented at the First Aegean Summer School on Cosmology, Samos, September 2001.

- Brane cosmology on arxiv.org

- Dimensional Shortcuts - evidence for sterile neutrino; (August 2007; Scientific American)

Chapter 47

Inflation (cosmology)

"Inflation model" and "Inflation theory" redirect here. For a general rise in the price level, see Inflation. For other uses, see Inflation (disambiguation).

In physical cosmology, **cosmic inflation**, **cosmological inflation**, or just **inflation** is a theory of exponential expansion of space in the early universe. The inflationary epoch lasted from 10^{-36} seconds after the Big Bang to sometime between 10^{-33} and 10^{-32} seconds. Following the inflationary period, the Universe continues to expand, but at a less rapid rate.[1]

Inflation theory was developed in the early 1980s. It explains the origin of the large-scale structure of the cosmos. Quantum fluctuations in the microscopic inflationary region, magnified to cosmic size, become the seeds for the growth of structure in the Universe (see galaxy formation and evolution and structure formation).[2] Many physicists also believe that inflation explains why the Universe appears to be the same in all directions (isotropic), why the cosmic microwave background radiation is distributed evenly, why the Universe is flat, and why no magnetic monopoles have been observed.

The detailed particle physics mechanism responsible for inflation is not known. The basic inflationary paradigm is accepted by most scientists, who believe a number of predictions have been confirmed by observation;[3] however, a substantial minority of scientists dissent from this position.[4][5][6] The hypothetical field thought to be responsible for inflation is called the inflaton.[7]

In 2002, three of the original architects of the theory were recognized for their major contributions; physicists Alan Guth of M.I.T., Andrei Linde of Stanford and Paul Steinhardt of Princeton shared the prestigious Dirac Prize "for development of the concept of inflation in cosmology".[8]

47.1 Overview

Main article: Metric expansion of space

An expanding universe generally has a cosmological horizon, which, by analogy with the more familiar horizon caused by the curvature of the Earth's surface, marks the boundary of the part of the Universe that an observer can see. Light (or other radiation) emitted by objects beyond the cosmological horizon never reaches the observer, because the space in between the observer and the object is expanding too rapidly.

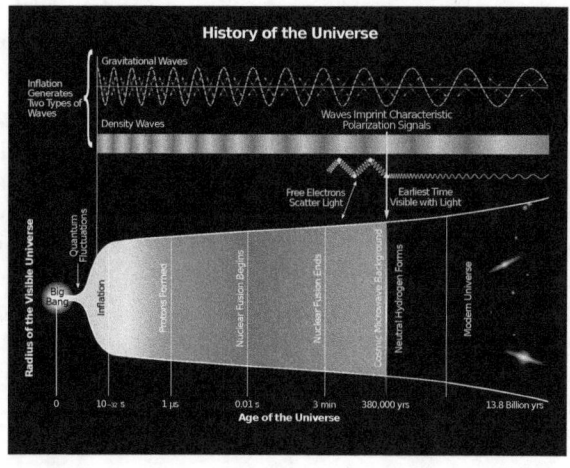

History of the Universe - gravitational waves are hypothesized to arise from cosmic inflation, a faster-than-light expansion just after the Big Bang (17 March 2014).[9][10][11]

The observable universe is one *causal patch* of a much larger unobservable universe; other parts of the Universe cannot communicate with Earth yet. These parts of the Universe are outside our current cosmological horizon. In the standard hot big bang model, without inflation, the cosmological horizon moves out, bringing new regions into view. Yet as a local observer sees such a region for the first time, it looks no different from any other region of space the local observer has already seen: its background radiation is at nearly the same temperature as the background radiation of other regions, and its space-time curvature is evolving lock-step with the others. This presents a mystery: how did these new regions know what temperature and curvature they were supposed to have? They couldn't have learned it by getting signals, because they were not previously in communication with our past light cone.[12][13]

Inflation answers this question by postulating that all the regions come from an earlier era with a big vacuum energy, or cosmological constant. A space with a cosmological constant is qualitatively different: instead of moving outward, the cosmological horizon stays put. For any one observer, the distance to the cosmological horizon is constant. With exponentially expanding space, two nearby observers are separated very quickly; so much so, that the distance between them quickly exceeds the limits of communications. The spatial slices are expanding very fast to cover huge volumes. Things are constantly moving beyond the cosmological horizon, which is a fixed distance away, and everything becomes homogeneous.

As the inflationary field slowly relaxes to the vacuum, the cosmological constant goes to zero and space begins to expand normally. The new regions that come into view during the normal expansion phase are exactly the same regions that were pushed out of the horizon during inflation, and so they are at nearly the same temperature and curvature, because they come from the same originally small patch of space.

The theory of inflation thus explains why the temperatures and curvatures of different regions are so nearly equal. It also predicts that the total curvature of a space-slice at constant global time is zero. This prediction implies that the total ordinary matter, dark matter and residual vacuum energy in the Universe have to add up to the critical density, and the evidence supports this. More strikingly, inflation allows physicists to calculate the minute differences in temperature of different regions from quantum fluctuations during the inflationary era, and many of these quantitative predictions have been confirmed.[14][15]

47.1.1 Space expands

To say that space expands exponentially means that two inertial observers are moving farther apart with accelerating velocity. In stationary coordinates for one observer, a patch of an inflating universe has the following polar metric:[16][17]

$$ds^2 = -(1 - \Lambda r^2)\, dt^2 + \frac{1}{1 - \Lambda r^2}\, dr^2 + r^2\, d\Omega^2.$$

This is just like an inside-out black hole metric—it has a zero in the dt component on a fixed radius sphere called the cosmological horizon. Objects are drawn away from the observer at $r = 0$ towards the cosmological horizon, which they cross in a finite proper time. This means that any inhomogeneities are smoothed out, just as any bumps or matter on the surface of a black hole horizon are swallowed and disappear.

Since the space–time metric has no explicit time dependence, once an observer has crossed the cosmological horizon, observers closer in take its place. This process of falling outward and replacement points closer in are always steadily replacing points further out—an exponential expansion of space–time.

This steady-state exponentially expanding spacetime is called a de Sitter space, and to sustain it there must be a cosmological constant, a vacuum energy proportional to Λ everywhere. In this case, the equation of state is $p = -\varrho$. The physical conditions from one moment to the next are stable: the rate of expansion, called the Hubble parameter, is nearly constant, and the scale factor of the Universe is proportional to e^{Ht}. Inflation is often called a period of *accelerated expansion* because the distance between two fixed observers is increasing exponentially (i.e. at an accelerating rate as they move apart), while Λ can stay approximately constant (see deceleration parameter).

47.1.2 Few inhomogeneities remain

Cosmological inflation has the important effect of smoothing out inhomogeneities, anisotropies and the curvature of space. This pushes the Universe into a very simple state, in which it is completely dominated by the inflaton field, the source of the cosmological constant, and the only significant inhomogeneities are the tiny quantum fluctuations in the inflaton. Inflation also dilutes exotic heavy particles, such as the magnetic monopoles predicted by many extensions to the Standard Model of particle physics. If the Universe was only hot enough to form such particles *before* a period of inflation, they would not be observed in nature, as they would be so rare that it is quite likely that there are none in the observable universe. Together, these effects are called the inflationary "no-hair theorem"[18] by analogy with the no hair theorem for black holes.

The "no-hair" theorem works essentially because the cosmological horizon is no different from a black-hole horizon, except for philosophical disagreements about what is on the other side. The interpretation of the no-hair theorem is that the Universe (observable and unobservable) expands by an enormous factor during inflation. In an expanding universe, energy densities generally fall, or get diluted, as the volume of the Universe increases. For example, the density of ordinary "cold" matter (dust) goes down as the inverse of the volume: when linear dimensions double, the energy density goes down by a factor of eight; the radiation energy density goes down even more rapidly as the Universe expands since the wavelength of each photon is stretched (redshifted), in addition to the photons being dispersed by the expansion. When linear dimensions are doubled, the energy density in radiation falls by a factor of sixteen (see the solution of the energy density continuity equation for an ultra-relativistic fluid). During inflation, the energy density in the inflaton field is roughly constant. However, the energy density in everything else, including inhomogeneities, curvature, anisotropies, exotic particles, and standard-model particles is falling, and through sufficient inflation these all become negligible. This leaves the Universe flat and symmetric, and (apart from the homogeneous inflaton field) mostly empty, at the moment inflation ends and reheating begins.[19]

47.1.3 Duration

A key requirement is that inflation must continue long enough to produce the present observable universe from a single, small inflationary Hubble volume. This is necessary to ensure that the Universe appears flat, homogeneous and isotropic at the largest observable scales. This requirement is generally thought to be satisfied if the Universe expanded by a factor of at least 10^{26} during inflation.[20]

47.1.4 Reheating

Inflation is a period of supercooled expansion, when the temperature drops by a factor of 100,000 or so. (The exact drop is model dependent, but in the first models it was typically from 10^{27} K down to 10^{22} K.[21]) This relatively low temperature is maintained during the inflationary phase. When inflation ends the temperature returns to the pre-inflationary temperature; this is called *reheating* or thermalization because the large potential energy of the inflaton field decays into particles and fills the Universe with Standard Model particles, including electromagnetic radiation, starting the radiation dominated phase of the Universe. Because the nature of the inflation is not known, this process is still poorly understood, although it is believed to take place through a parametric resonance.[22][23]

47.2 Motivations

Inflation resolves several problems in Big Bang cosmology that were discovered in the 1970s.[24] Inflation was first proposed by Guth while investigating the problem of why

no magnetic monopoles are seen today; he found that a positive-energy false vacuum would, according to general relativity, generate an exponential expansion of space. It was very quickly realised that such an expansion would resolve many other long-standing problems. These problems arise from the observation that to look like it does *today*, the Universe would have to have started from very finely tuned, or "special" initial conditions at the Big Bang. Inflation attempts to resolve these problems by providing a dynamical mechanism that drives the Universe to this special state, thus making a universe like ours much more likely in the context of the Big Bang theory.

47.2.1 Horizon problem

Main article: Horizon problem

The horizon problem is the problem of determining why the Universe appears statistically homogeneous and isotropic in accordance with the cosmological principle.[25][26][27] For example, molecules in a canister of gas are distributed homogeneously and isotropically because they are in thermal equilibrium: gas throughout the canister has had enough time to interact to dissipate inhomogeneities and anisotropies. The situation is quite different in the big bang model without inflation, because gravitational expansion does not give the early universe enough time to equilibrate. In a big bang with only the matter and radiation known in the Standard Model, two widely separated regions of the observable universe cannot have equilibrated because they move apart from each other faster than the speed of light and thus have never come into causal contact. In the early Universe, it was not possible to send a light signal between the two regions. Because they have had no interaction, it is difficult to explain why they have the same temperature (are thermally equilibrated). Historically, proposed solutions included the *Phoenix universe* of Georges Lemaître,[28] the related oscillatory universe of Richard Chase Tolman,[29] and the Mixmaster universe of Charles Misner. Lemaître and Tolman proposed that a universe undergoing a number of cycles of contraction and expansion could come into thermal equilibrium. Their models failed, however, because of the buildup of entropy over several cycles. Misner made the (ultimately incorrect) conjecture that the Mixmaster mechanism, which made the Universe *more* chaotic, could lead to statistical homogeneity and isotropy.[26][30]

47.2.2 Flatness problem

Main article: Flatness problem

The flatness problem is sometimes called one of the Dicke

coincidences (along with the cosmological constant problem).[31][32] It became known in the 1960s that the density of matter in the Universe was comparable to the critical density necessary for a flat universe (that is, a universe whose large scale geometry is the usual Euclidean geometry, rather than a non-Euclidean hyperbolic or spherical geometry).[33]:61

Therefore, regardless of the shape of the universe the contribution of spatial curvature to the expansion of the Universe could not be much greater than the contribution of matter. But as the Universe expands, the curvature redshifts away more slowly than matter and radiation. Extrapolated into the past, this presents a fine-tuning problem because the contribution of curvature to the Universe must be exponentially small (sixteen orders of magnitude less than the density of radiation at big bang nucleosynthesis, for example). This problem is exacerbated by recent observations of the cosmic microwave background that have demonstrated that the Universe is flat to within a few percent.[34]

47.2.3 Magnetic-monopole problem

The magnetic monopole problem, sometimes called the exotic-relics problem, says that if the early universe were very hot, a large number of very heavy, stable magnetic monopoles would have been produced. This is a problem with Grand Unified Theories, which propose that at high temperatures (such as in the early universe) the electromagnetic force, strong, and weak nuclear forces are not actually fundamental forces but arise due to spontaneous symmetry breaking from a single gauge theory.[35] These theories predict a number of heavy, stable particles that have not been observed in nature. The most notorious is the magnetic monopole, a kind of stable, heavy "charge" of magnetic field.[36][37] Monopoles are predicted to be copiously produced following Grand Unified Theories at high temperature,[38][39] and they should have persisted to the present day, to such an extent that they would become the primary constituent of the Universe.[40][41] Not only is that not the case, but all searches for them have failed, placing stringent limits on the density of relic magnetic monopoles in the Universe.[42] A period of inflation that occurs below the temperature where magnetic monopoles can be produced would offer a possible resolution of this problem: monopoles would be separated from each other as the Universe around them expands, potentially lowering their observed density by many orders of magnitude. Though, as cosmologist Martin Rees has written, "Skeptics about exotic physics might not be hugely impressed by a theoretical argument to explain the absence of particles that are themselves only hypothetical. Preventive medicine can readily seem 100 percent effective against a disease that doesn't exist!"[43]

47.3 History

47.3.1 Precursors

In the early days of General Relativity, Albert Einstein introduced the cosmological constant to allow a static solution, which was a three-dimensional sphere with a uniform density of matter. Later, Willem de Sitter found a highly symmetric inflating universe, which described a universe with a cosmological constant that is otherwise empty.[44] It was discovered that Einstein's universe is unstable, and that small fluctuations cause it to collapse or turn into a de Sitter universe.

In the early 1970s Zeldovich noticed the flatness and horizon problems of Big Bang cosmology; before his work, cosmology was presumed to be symmetrical on purely philosophical grounds. In the Soviet Union, this and other considerations led Belinski and Khalatnikov to analyze the chaotic BKL singularity in General Relativity. Misner's Mixmaster universe attempted to use this chaotic behavior to solve the cosmological problems, with limited success.

In the late 1970s, Sidney Coleman applied the instanton techniques developed by Alexander Polyakov and collaborators to study the fate of the false vacuum in quantum field theory. Like a metastable phase in statistical mechanics—water below the freezing temperature or above the boiling point—a quantum field would need to nucleate a large enough bubble of the new vacuum, the new phase, in order to make a transition. Coleman found the most likely decay pathway for vacuum decay and calculated the inverse lifetime per unit volume. He eventually noted that gravitational effects would be significant, but he did not calculate these effects and did not apply the results to cosmology.

In the Soviet Union, Alexei Starobinsky noted that quantum corrections to general relativity should be important for the early universe. These generically lead to curvature-squared corrections to the Einstein–Hilbert action and a form of $f(R)$ modified gravity. The solution to Einstein's equations in the presence of curvature squared terms, when the curvatures are large, leads to an effective cosmological constant. Therefore, he proposed that the early universe went through an inflationary de Sitter era.[45] This resolved the cosmology problems and led to specific predictions for the corrections to the microwave background radiation, corrections that were then calculated in detail.

In 1978, Zeldovich noted the monopole problem, which was an unambiguous quantitative version of the horizon problem, this time in a subfield of particle physics, which led to several speculative attempts to resolve it. In 1980 Alan Guth realized that false vacuum decay in the early universe would solve the problem, leading him to propose a scalar-driven inflation. Starobinsky's and Guth's scenarios both predicted an initial deSitter phase, differing only in mechanistic details.

47.3.2 Early inflationary models

Guth proposed inflation in January 1980 to explain the nonexistence of magnetic monopoles;[46][47] it was Guth who coined the term "inflation".[48] At the same time, Starobinsky argued that quantum corrections to gravity would replace the initial singularity of the Universe with an exponentially expanding deSitter phase.[49] In October 1980, Demosthenes Kazanas suggested that exponential expansion could eliminate the particle horizon and perhaps solve the horizon problem,[50] while Sato suggested that an exponential expansion could eliminate domain walls (another kind of exotic relic).[51] In 1981 Einhorn and Sato[52] published a model similar to Guth's and showed that it would resolve the puzzle of the magnetic monopole abundance in Grand Unified Theories. Like Guth, they concluded that such a model not only required fine tuning of the cosmological constant, but also would likely lead to a much too granular universe, i.e., to large density variations resulting from bubble wall collisions.

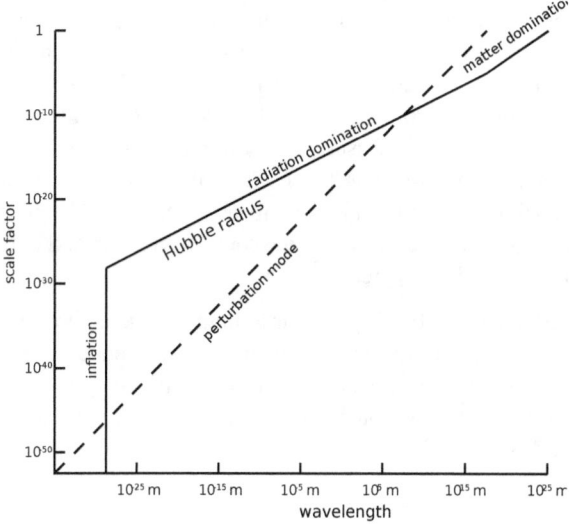

The physical size of the Hubble radius (solid line) as a function of the linear expansion (scale factor) of the universe. During cosmological inflation, the Hubble radius is constant. The physical wavelength of a perturbation mode (dashed line) is also shown. The plot illustrates how the perturbation mode grows larger than the horizon during cosmological inflation before coming back inside the horizon, which grows rapidly during radiation domination. If cosmological inflation had never happened, and radiation domination continued back until a gravitational singularity, then the mode would never have been inside the horizon in the very early universe, and no causal mechanism could have ensured that the universe was homogeneous on the scale of the perturbation mode.

Guth proposed that as the early universe cooled, it was trapped in a false vacuum with a high energy density, which is much like a cosmological constant. As the very early universe cooled it was trapped in a metastable state (it was supercooled), which it could only decay out of through the process of bubble nucleation via quantum tunneling. Bubbles of true vacuum spontaneously form in the sea of false vacuum and rapidly begin expanding at the speed of light. Guth recognized that this model was problematic because the model did not reheat properly: when the bubbles nucleated, they did not generate any radiation. Radiation could only be generated in collisions between bubble walls. But if inflation lasted long enough to solve the initial conditions problems, collisions between bubbles became exceedingly rare. In any one causal patch it is likely that only one bubble would nucleate.

47.3.3 Slow-roll inflation

The bubble collision problem was solved by Linde[53] and independently by Andreas Albrecht and Paul Steinhardt[54] in a model named *new inflation* or *slow-roll inflation* (Guth's model then became known as *old inflation*). In this model, instead of tunneling out of a false vacuum state, inflation occurred by a scalar field rolling down a potential energy hill. When the field rolls very slowly compared to the expansion of the Universe, inflation occurs. However, when the hill becomes steeper, inflation ends and reheating can occur.

47.3.4 Effects of asymmetries

Eventually, it was shown that new inflation does not produce a perfectly symmetric universe, but that quantum fluctuations in the inflaton are created. These fluctuations form the primordial seeds for all structure created in the later universe.[55] These fluctuations were first calculated by Viatcheslav Mukhanov and G. V. Chibisov in analyzing Starobinsky's similar model.[56][57][58] In the context of inflation, they were worked out independently of the work of Mukhanov and Chibisov at the three-week 1982 Nuffield Workshop on the Very Early Universe at Cambridge University.[59] The fluctuations were calculated by four groups working separately over the course of the workshop: Stephen Hawking;[60] Starobinsky;[61] Guth and So-Young Pi;[62] and Bardeen, Steinhardt and Turner.[63]

47.4 Observational status

Inflation is a mechanism for realizing the cosmological principle, which is the basis of the standard model of physical cosmology: it accounts for the homogeneity and isotropy of the observable universe. In addition, it accounts for the observed flatness and absence of magnetic monopoles. Since Guth's early work, each of these observations has received further confirmation, most impressively by the detailed observations of the cosmic microwave background made by the Wilkinson Microwave Anisotropy Probe (WMAP) spacecraft.[14] This analysis shows that the Universe is flat to within at least a few percent, and that it is homogeneous and isotropic to one part in 100,000.

In addition, inflation predicts that the structures visible in the Universe today formed through the gravitational collapse of perturbations that were formed as quantum mechanical fluctuations in the inflationary epoch. The detailed form of the spectrum of perturbations called a nearly-scale-invariant Gaussian random field (or Harrison–Zel'dovich spectrum) is very specific and has only two free parameters, the amplitude of the spectrum and the *spectral index*, which measures the slight deviation from scale invariance predicted by inflation (perfect scale invariance corresponds to the idealized de Sitter universe).[64] Inflation predicts that the observed perturbations should be in thermal equilibrium with each other (these are called *adiabatic* or *isentropic* perturbations). This structure for the perturbations has been confirmed by the WMAP spacecraft and other cosmic microwave background (CMB) experiments,[14] and galaxy surveys, especially the ongoing Sloan Digital Sky Survey.[65] These experiments have shown that the one part in 100,000 inhomogeneities observed have exactly the form predicted by theory. Moreover, there is evidence for a slight deviation from scale invariance. The *spectral index*, n_s is equal to one for a scale-invariant spectrum. The simplest inflation models predict that this quantity is between 0.92 and 0.98.[66][67][68][69] From WMAP data it can be inferred that $n_s = 0.963 \pm 0.012$,[70] implying that it differs from one at the level of two standard deviations (2σ). This is considered an important confirmation of the theory of inflation.[14]

Various inflation theories have been proposed that make radically different predictions, but they generally have much more fine tuning than should be necessary.[66][67] As a physical model, however, inflation is most valuable in that it robustly predicts the initial conditions of the Universe based on only two adjustable parameters: the spectral index (that can only change in a small range) and the amplitude of the perturbations. Except in contrived models, this is true regardless of how inflation is realized in particle physics.

Occasionally, effects are observed that appear to contradict the simplest models of inflation. The first-year WMAP data suggested that the spectrum might not be nearly scale-invariant, but might instead have a slight curvature.[71] However, the third-year data revealed that the effect was a statistical anomaly.[14] Another effect remarked upon since the first cosmic microwave background satellite, the

Cosmic Background Explorer is that the amplitude of the quadrupole moment of the CMB is unexpectedly low and the other low multipoles appear to be preferentially aligned with the ecliptic plane. Some have claimed that this is a signature of non-Gaussianity and thus contradicts the simplest models of inflation. Others have suggested that the effect may be due to other new physics, foreground contamination, or even publication bias.[72]

An experimental program is underway to further test inflation with more precise CMB measurements. In particular, high precision measurements of the so-called "B-modes" of the polarization of the background radiation could provide evidence of the gravitational radiation produced by inflation, and could also show whether the energy scale of inflation predicted by the simplest models (10^{15}–10^{16} GeV) is correct.[67][68] In March 2014, it was announced that B-mode CMB polarization consistent with that predicted from inflation had been demonstrated by a South Pole experiment.[9][10][11][73][74][75] However, on 19 June 2014, lowered confidence in confirming the findings was reported;[74][76][77] on 19 September 2014, a further reduction in confidence was reported[78][79] and, on 30 January 2015, even less confidence yet was reported.[80][81]

Other potentially corroborating measurements are expected from the Planck spacecraft, although it is unclear if the signal will be visible, or if contamination from foreground sources will interfere.[82] Other forthcoming measurements, such as those of 21 centimeter radiation (radiation emitted and absorbed from neutral hydrogen before the first stars turned on), may measure the power spectrum with even greater resolution than the CMB and galaxy surveys, although it is not known if these measurements will be possible or if interference with radio sources on Earth and in the galaxy will be too great.[83]

Dark energy is broadly similar to inflation and is thought to be causing the expansion of the present-day universe to accelerate. However, the energy scale of dark energy is much lower, 10^{-12} GeV, roughly 27 orders of magnitude less than the scale of inflation.

47.5 Theoretical status

In Guth's early proposal, it was thought that the inflaton was the Higgs field, the field that explains the mass of the elementary particles.[47] It is now believed by some that the inflaton cannot be the Higgs field[84] although the recent discovery of the Higgs boson has increased the number of works considering the Higgs field as inflaton.[85] One problem of this identification is the current tension with experimental data at the electroweak scale,[86] which is currently under study at the Large Hadron Collider (LHC). Other models of inflation relied on the properties of Grand Unified Theories.[54] Since the simplest models of grand unification have failed, it is now thought by many physicists that inflation will be included in a supersymmetric theory such as string theory or a supersymmetric grand unified theory. At present, while inflation is understood principally by its detailed predictions of the initial conditions for the hot early universe, the particle physics is largely *ad hoc* modelling. As such, although predictions of inflation have been consistent with the results of observational tests, many open questions remain.

47.5.1 Fine-tuning problem

One of the most severe challenges for inflation arises from the need for fine tuning. In new inflation, the *slow-roll conditions* must be satisfied for inflation to occur. The slow-roll conditions say that the inflaton potential must be flat (compared to the large vacuum energy) and that the inflaton particles must have a small mass.[87] New inflation requires the Universe to have a scalar field with an especially flat potential and special initial conditions. However, explanations for these fine-tunings have been proposed. For example, classically scale invariant field theories, where scale invariance is broken by quantum effects, provide an explanation of the flatness of inflationary potentials, as long as the theory can be studied through perturbation theory.[88]

Andrei Linde

Linde proposed a theory known as *chaotic inflation* in which he suggested that the conditions for inflation were actually satisfied quite generically. Inflation will occur in virtually any universe that begins in a chaotic, high energy state that has a scalar field with unbounded potential energy.[89] However, in his model the inflaton field necessarily takes values larger than one Planck unit: for this reason, these are often called *large field* models and the competing new inflation models are called *small field* models. In this situation, the predictions of effective field theory are thought to be invalid, as renormalization should cause large corrections that could prevent inflation.[90] This problem has not yet been resolved and some cosmologists argue that the small field models, in which inflation can occur at a much lower energy scale, are better models.[91] While inflation depends on quantum field theory (and the semiclassical approximation to quantum gravity) in an important way, it has not been completely reconciled with these theories.

Brandenberger commented on fine-tuning in another situation.[92] The amplitude of the primordial inhomogeneities produced in inflation is directly tied to the energy scale of inflation. This scale is suggested to be around 10^{16} GeV or 10^{-3} times the Planck energy. The natural scale is naïvely the Planck scale so this small value could be seen as another form of fine-tuning (called a hierarchy problem): the energy density given by the scalar potential is down by 10^{-12} compared to the Planck density. This is not usually considered to be a critical problem, however, because the scale of inflation corresponds naturally to the scale of gauge unification.

47.5.2 Eternal inflation

Main article: Eternal inflation

In many models, the inflationary phase of the Universe's expansion lasts forever in at least some regions of the Universe. This occurs because inflating regions expand very rapidly, reproducing themselves. Unless the rate of decay to the non-inflating phase is sufficiently fast, new inflating regions are produced more rapidly than non-inflating regions. In such models most of the volume of the Universe at any given time is inflating. All models of eternal inflation produce an infinite multiverse, typically a fractal.

Although new inflation is classically rolling down the potential, quantum fluctuations can sometimes lift it to previous levels. These regions in which the inflaton fluctuates upwards expand much faster than regions in which the inflaton has a lower potential energy, and tend to dominate in terms of physical volume. This steady state, which first developed by Vilenkin,[93] is called "eternal inflation". It has been shown that any inflationary theory with an unbounded potential is eternal.[94] It is a popular conclusion among physicists that this steady state cannot continue forever into the past.[95][96][97] Inflationary spacetime, which is similar to de Sitter space, is incomplete without a contracting region. However, unlike de Sitter space, fluctuations in a contracting inflationary space collapse to form a gravitational singularity, a point where densities become infinite. Therefore, it is necessary to have a theory for the Universe's initial conditions. Linde, however, believes inflation may be past eternal.[98]

In eternal inflation, regions with inflation have an exponentially growing volume, while regions that are not inflating don't. This suggests that the volume of the inflating part of the Universe in the global picture is always unimaginably larger than the part that has stopped inflating, even though inflation eventually ends as seen by any single preinflationary observer. Scientists disagree about how to assign a probability distribution to this hypothetical anthropic landscape. If the probability of different regions is counted by volume, one should expect that inflation will never end or applying boundary conditions that a local observer exists to observe it, that inflation will end as late as possible. Some physicists believe this paradox can be resolved by weighting observers by their pre-inflationary volume.

47.5.3   Initial conditions

Some physicists have tried to avoid the initial conditions problem by proposing models for an eternally inflating universe with no origin.[99][100][101][102] These models propose that while the Universe, on the largest scales, expands exponentially it was, is and always will be, spatially infinite and has existed, and will exist, forever.

Other proposals attempt to describe the ex nihilo creation of the Universe based on quantum cosmology and the following inflation. Vilenkin put forth one such scenario.[93] Hartle and Hawking offered the no-boundary proposal for the initial creation of the Universe in which inflation comes about naturally.[103]

Guth described the inflationary universe as the "ultimate free lunch":[104][105] new universes, similar to our own, are continually produced in a vast inflating background. Gravitational interactions, in this case, circumvent (but do not violate) the first law of thermodynamics (energy conservation) and the second law of thermodynamics (entropy and the arrow of time problem). However, while there is consensus that this solves the initial conditions problem, some have disputed this, as it is much more likely that the Universe came about by a quantum fluctuation. Don Page was an outspoken critic of inflation because of this anomaly.[106] He stressed that the thermodynamic arrow of time necessitates low entropy initial conditions, which would be highly unlikely. According to them, rather than solving this problem, the inflation theory aggravates it – the reheating at the end of the inflation era increases entropy, making it necessary for the initial state of the Universe to be even more orderly than in other Big Bang theories with no inflation phase.

Hawking and Page later found ambiguous results when they attempted to compute the probability of inflation in the Hartle-Hawking initial state.[107] Other authors have argued that, since inflation is eternal, the probability doesn't matter as long as it is not precisely zero: once it starts, inflation perpetuates itself and quickly dominates the Universe.[4][108]:223–225 However, Albrecht and Lorenzo Sorbo argued that the probability of an inflationary cosmos, consistent with today's observations, emerging by a random fluctuation from some pre-existent state is much higher than that of a non-inflationary cosmos. This is because the "seed" amount of non-gravitational energy required for the inflationary cosmos is so much less than that for a non-inflationary alternative, which outweighs any entropic considerations.[109]

Another problem that has occasionally been mentioned is the trans-Planckian problem or trans-Planckian effects.[110] Since the energy scale of inflation and the Planck scale are relatively close, some of the quantum fluctuations that have made up the structure in our universe were smaller than the Planck length before inflation. Therefore, there ought to be corrections from Planck-scale physics, in particular the unknown quantum theory of gravity. Some disagreement remains about the magnitude of this effect: about whether it is just on the threshold of detectability or completely undetectable.[111]

47.5.4   Hybrid inflation

Another kind of inflation, called *hybrid inflation*, is an extension of new inflation. It introduces additional scalar fields, so that while one of the scalar fields is responsible for normal slow roll inflation, another triggers the end of inflation: when inflation has continued for sufficiently long, it becomes favorable to the second field to decay into a much lower energy state.[112]

In hybrid inflation, one scalar field is responsible for most of the energy density (thus determining the rate of expansion), while another is responsible for the slow roll (thus determining the period of inflation and its termination). Thus fluctuations in the former inflaton would not affect inflation termination, while fluctuations in the latter would not affect the rate of expansion. Therefore, hybrid inflation is not eternal.[113][114] When the second (slow-rolling) inflaton reaches the bottom of its potential, it changes the location of the minimum of the first inflaton's potential, which leads to a fast roll of the inflaton down its potential, leading to termination of inflation.

47.5.5   Inflation and string cosmology

The discovery of flux compactifications opened the way for reconciling inflation and string theory.[115] *Brane inflation* suggests that inflation arises from the motion of D-branes[116] in the compactified geometry, usually towards a stack of anti-D-branes. This theory, governed by the *Dirac-Born-Infeld action*, is different from ordinary inflation. The dynamics are not completely understood. It appears that special conditions are necessary since inflation occurs in tunneling between two vacua in the string landscape. The process of tunneling between two vacua is a form of old inflation, but new inflation must then occur by some other mechanism.

47.5.6 Inflation and loop quantum gravity

When investigating the effects the theory of loop quantum gravity would have on cosmology, a loop quantum cosmology model has evolved that provides a possible mechanism for cosmological inflation. Loop quantum gravity assumes a quantized spacetime. If the energy density is larger than can be held by the quantized spacetime, it is thought to bounce back.[117]

47.6 Alternatives

Other models explain some of the observations explained by inflation. However none of these "alternatives" has the same breadth of explanation and still require inflation for a more complete fit with observation. They should therefore be regarded as adjuncts to inflation, rather than as alternatives.

47.6.1 Big bounce

The big bounce hypothesis attempts to replace the cosmic singularity with a cosmic contraction and bounce, thereby explaining the initial conditions that led to the big bang.[118] The flatness and horizon problems are naturally solved in the Einstein-Cartan-Sciama-Kibble theory of gravity, without needing an exotic form of matter or free parameters.[119][120] This theory extends general relativity by removing a constraint of the symmetry of the affine connection and regarding its antisymmetric part, the torsion tensor, as a dynamical variable. The minimal coupling between torsion and Dirac spinors generates a spin-spin interaction that is significant in fermionic matter at extremely high densities. Such an interaction averts the unphysical Big Bang singularity, replacing it with a cusp-like bounce at a finite minimum scale factor, before which the Universe was contracting. The rapid expansion immediately after the Big Bounce explains why the present Universe at largest scales appears spatially flat, homogeneous and isotropic. As the density of the Universe decreases, the effects of torsion weaken and the Universe smoothly enters the radiation-dominated era.

47.6.2 String theory

String theory requires that, in addition to the three observable spatial dimensions, additional dimensions exist that are curled up or compactified (see also Kaluza–Klein theory). Extra dimensions appear as a frequent component of supergravity models and other approaches to quantum gravity. This raised the contingent question of why four space-time dimensions became large and the rest became unobservably small. An attempt to address this question, called *string gas cosmology*, was proposed by Robert Brandenberger and Cumrun Vafa.[121] This model focuses on the dynamics of the early universe considered as a hot gas of strings. Brandenberger and Vafa show that a dimension of spacetime can only expand if the strings that wind around it can efficiently annihilate each other. Each string is a one-dimensional object, and the largest number of dimensions in which two strings will generically intersect (and, presumably, annihilate) is three. Therefore, the most likely number of non-compact (large) spatial dimensions is three. Current work on this model centers on whether it can succeed in stabilizing the size of the compactified dimensions and produce the correct spectrum of primordial density perturbations.[122] Supporters admit that their model "does not solve the entropy and flatness problems of standard cosmology ..... and we can provide no explanation for why the current universe is so close to being spatially flat".[123]

47.6.3 Ekpyrotic and cyclic models

The ekpyrotic and cyclic models are also considered adjuncts to inflation. These models solve the horizon problem through an expanding epoch well *before* the Big Bang, and then generate the required spectrum of primordial density perturbations during a contracting phase leading to a Big Crunch. The Universe passes through the Big Crunch and emerges in a hot Big Bang phase. In this sense they are reminiscent of Richard Chace Tolman's oscillatory universe; in Tolman's model, however, the total age of the Universe is necessarily finite, while in these models this is not necessarily so. Whether the correct spectrum of density fluctuations can be produced, and whether the Universe can successfully navigate the Big Bang/Big Crunch transition, remains a topic of controversy and current research. Ekpyrotic models avoid the magnetic monopole problem as long as the temperature at the Big Crunch/Big Bang transition remains below the Grand Unified Scale, as this is the temperature required to produce magnetic monopoles in the first place. As things stand, there is no evidence of any 'slowing down' of the expansion, but this is not surprising as each cycle is expected to last on the order of a trillion years.

47.6.4 Varying C

Another adjunct, the varying speed of light model was offered by Jean-Pierre Petit in 1988, John Moffat in 1992 as well Albrecht and João Magueijo in 1999, instead of superluminal expansion the speed of light was 60 orders of magnitude faster than its current value solving the horizon and homogeneity problems in the early universe.

47.7 Criticisms

Since its introduction by Alan Guth in 1980, the inflationary paradigm has become widely accepted. Nevertheless, many physicists, mathematicians, and philosophers of science have voiced criticisms, claiming untestable predictions and a lack of serious empirical support.[4] In 1999, John Earman and Jesús Mosterín published a thorough critical review of inflationary cosmology, concluding, "we do not think that there are, as yet, good grounds for admitting any of the models of inflation into the standard core of cosmology."[5]

In order to work, and as pointed out by Roger Penrose from 1986 on, inflation requires extremely specific initial conditions of its own, so that the problem (or pseudo-problem) of initial conditions is not solved: "There is something fundamentally misconceived about trying to explain the uniformity of the early universe as resulting from a thermalization process. [...] For, if the thermalization is actually doing anything [...] then it represents a definite increasing of the entropy. Thus, the universe would have been even more special before the thermalization than after."[124] The problem of specific or "fine-tuned" initial conditions would not have been solved; it would have gotten worse. At a conference in 2015, Penrose said that "inflation isn't falsifiable, it's falsified. [...] BICEP did a wonderful service by bringing all the Inflation-ists out of their shell, and giving them a black eye."[6]

A recurrent criticism of inflation is that the invoked inflation field does not correspond to any known physical field, and that its potential energy curve seems to be an ad hoc contrivance to accommodate almost any data obtainable. Paul Steinhardt, one of the founding fathers of inflationary cosmology, has recently become one of its sharpest critics. He calls 'bad inflation' a period of accelerated expansion whose outcome conflicts with observations, and 'good inflation' one compatible with them: "Not only is bad inflation more likely than good inflation, but no inflation is more likely than either.... Roger Penrose considered all the possible configurations of the inflaton and gravitational fields. Some of these configurations lead to inflation ... Other configurations lead to a uniform, flat universe directly – without inflation. Obtaining a flat universe is unlikely overall. Penrose's shocking conclusion, though, was that obtaining a flat universe without inflation is much more likely than with inflation – by a factor of 10 to the googol (10 to the 100) power!"[4][108] Together with Anna Ijjas and Abraham Loeb, he wrote articles claiming that the inflationary paradigm is in trouble in view of the data from the Planck satellite.[125][126] Counter-arguments were presented by Alan Guth, David Kaiser, and Yasunori Nomura[127] and by Andrei Linde,[128] saying that "cosmic inflation is on a stronger footing than ever before".[127]

47.8 See also

- Brane cosmology
- Conservation of angular momentum
- Cosmology
- Dark flow
- Doughnut theory of the universe
- Hubble's law
- Non-minimally coupled inflation
- Nonlinear optics
- Varying speed of light
- Warm inflation

47.9 Notes

[1] "First Second of the Big Bang". *How The Universe Works 3*. 2014. Discovery Science.

[2] Tyson, Neil deGrasse and Donald Goldsmith (2004), *Origins: Fourteen Billion Years of Cosmic Evolution*, W. W. Norton & Co., pp. 84–5.

[3] Tsujikawa, Shinji (28 Apr 2003). "Introductory review of cosmic inflation": 4257. arXiv:hep-ph/0304257. Bibcode:2003hep.ph....4257T. In fact temperature anisotropies observed by the COBE satellite in 1992 exhibit nearly scale-invariant spectra as predicted by the inflationary paradigm. Recent observations of WMAP also show strong evidence for inflation.

[4] Steinhardt, Paul J. (2011). "The inflation debate: Is the theory at the heart of modern cosmology deeply flawed?" (*Scientific American*, April; pp. 18-25).

[5] Earman, John; Mosterín, Jesús (March 1999). "A Critical Look at Inflationary Cosmology". *Philosophy of Science* **66**: 1–49. doi:10.2307/188736 (inactive 2015-01-14). JSTOR 188736.

[6] Hložek, Renée (12 June 2015). "CMB@50 day three". Retrieved 15 July 2015.
This is a collation of remarks from the third day of the "Cosmic Microwave Background @50" conference held at Princeton, 10–12 June 2015.

[7] Guth, Alan H. (1997). *The Inflationary Universe: The Quest for a New Theory of Cosmic Origins*. Basic Books. pp. 233–234. ISBN 0201328402.

[8] "The Medallists: A list of past Dirac Medallists". *ictp.it*.

[9] Staff (17 March 2014). "BICEP2 2014 Results Release". *National Science Foundation*. Retrieved 18 March 2014.

[10] Clavin, Whitney (17 March 2014). "NASA Technology Views Birth of the Universe". *NASA*. Retrieved 17 March 2014.

[11] Overbye, Dennis (17 March 2014). "Space Ripples Reveal Big Bang's Smoking Gun". *The New York Times*. Retrieved 17 March 2014.

[12] Using Tiny Particles To Answer Giant Questions. Science Friday, 3 April 2009.

[13] See also Faster than light#Universal expansion.

[14] Spergel, D.N. (2006). "Three-year Wilkinson Microwave Anisotropy Probe (WMAP) observations: Implications for cosmology". WMAP... confirms the basic tenets of the inflationary paradigm...

[15] "Our Baby Universe Likely Expanded Rapidly, Study Suggests". *Space.com*.

[16] Melia, Fulvio (2007). "The Cosmic Horizon". *Monthly Notices of the Royal Astronomical Society* **382** (4): 1917–1921. arXiv:0711.4181. Bibcode:2007MNRAS.382.1917M. doi:10.1111/j.1365-2966.2007.12499.x.

[17] Melia, Fulvio; et al. (2009). "The Cosmological Spacetime". *International Journal of Modern Physics D* **18** (12): 1889–1901. arXiv:0907.5394. Bibcode:2009IJMPD..18.1889M. doi:10.1142/s0218271809015746.

[18] Kolb and Turner (1988).

[19] Barbara Sue Ryden (2003). *Introduction to cosmology*. Addison-Wesley. ISBN 978-0-8053-8912-8. Not only is inflation very effective at driving down the number density of magnetic monopoles, it is also effective at driving down the number density of every other type of particle, including photons.[202–207]

[20] This is usually quoted as 60 e-folds of expansion, where $e^{60} \approx 10^{26}$. It is equal to the amount of expansion since reheating, which is roughly $E_{\text{infl}}^{\text{ation}}/T_0$, where $T_0 = 2.7$ K is the temperature of the cosmic microwave background today. See, *e.g.* Kolb and Turner (1998) or Liddle and Lyth (2000).

[21] Guth, *Phase transitions in the very early universe*, in *The Very Early Universe*, ISBN 0-521-31677-4 eds Hawking, Gibbon & Siklos

[22] See Kolb and Turner (1988) or Mukhanov (2005).

[23] Kofman, Lev; Linde, Andrei; Starobinsky, Alexei (1994). "Reheating after inflation". *Physical Review Letters* **73** (5): 3195–3198. arXiv:hep-th/9405187. Bibcode:1986CQGra...3..811K. doi:10.1088/0264-9381/3/5/011.

[24] Much of the historical context is explained in chapters 15–17 of Peebles (1993).

[25] Misner, Charles W.; Coley, A A; Ellis, G F R; Hancock, M (1968). "The isotropy of the universe". *Astrophysical Journal* **151** (2): 431. Bibcode:1998CQGra..15..331W. doi:10.1088/0264-9381/15/2/008.

[26] Misner, Charles; Thorne, Kip S. and Wheeler, John Archibald (1973). *Gravitation*. San Francisco: W. H. Freeman. pp. 489–490, 525–526. ISBN 0-7167-0344-0.

[27] Weinberg, Steven (1971). *Gravitation and Cosmology*. John Wiley. pp. 740, 815. ISBN 0-471-92567-5.

[28] Lemaître, Georges (1933). "The expanding universe". *Annales de la Société Scientifique de Bruxelles* **47A**: 49., English in *Gen. Rel. Grav.* **29**:641–680, 1997.

[29] R. C. Tolman (1934). *Relativity, Thermodynamics, and Cosmology*. Oxford: Clarendon Press. ISBN 0-486-65383-8. LCCN 34032023. Reissued (1987) New York: Dover ISBN 0-486-65383-8.

[30] Misner, Charles W.; Leach, P G L (1969). "Mixmaster universe". *Physical Review Letters* **22** (15): 1071–74. Bibcode:2008JPhA...41o5201A. doi:10.1088/1751-8113/41/15/155201.

[31] Dicke, Robert H. (1970). *Gravitation and the Universe*. Philadelphia: American Philosopical Society.

[32] Dicke, Robert H.; P. J. E. Peebles (1979). "The big bang cosmology – enigmas and nostrums". In ed. S. W. Hawking and W. Israel. *General Relativity: an Einstein Centenary Survey*. Cambridge University Press.

[33] Alan P. Lightman (1 January 1993). *Ancient Light: Our Changing View of the Universe*. Harvard University Press. ISBN 978-0-674-03363-4.

[34] "WMAP- Content of the Universe". *nasa.gov*.

[35] Since supersymmetric Grand Unified Theory is built into string theory, it is still a triumph for inflation that it is able to deal with these magnetic relics. See, *e.g.* Kolb and Turner (1988) and Raby, Stuart (2006). ed. Bruce Hoeneisen, ed. "Grand Unified Theories". arXiv:hep-ph/0608183.

[36] 't Hooft, Gerard (1974). "Magnetic monopoles in Unified Gauge Theories". *Nuclear Physics B* **79** (2): 276–84. Bibcode:1974NuPhB..79..276T. doi:10.1016/0550-3213(74)90486-6.

[37] Polyakov, Alexander M. (1974). "Particle spectrum in quantum field theory". *JETP Letters* **20**: 194–5. Bibcode:1974JETPL..20..194P.

[38] Guth, Alan; Tye, S. (1980). "Phase Transitions and Magnetic Monopole Production in the Very Early Universe". *Physical Review Letters* **44** (10): 631–635; Erratum *ibid.*,**44**:963, 1980. Bibcode:1980PhRvL..44..631G. doi:10.1103/PhysRevLett.44.631.

[39] Einhorn, Martin B; Stein, D. L.; Toussaint, Doug (1980). "Are Grand Unified Theories Compatible with Standard Cosmology?". *Physical Review D* **21** (12): 3295–3298. Bibcode:1980PhRvD..21.3295E. doi:10.1103/PhysRevD.21.3295.

[40] Zel'dovich, Ya.; Khlopov, M. Yu. (1978). "On the concentration of relic monopoles in the universe". *Physics Letters B* **79** (3): 239–41. Bibcode:1978PhLB...79..239Z. doi:10.1016/0370-2693(78)90232-0.

[41] Preskill, John (1979). "Cosmological production of superheavy magnetic monopoles". *Physical Review Letters* **43** (19): 1365–1368. Bibcode:1979PhRvL..43.1365P. doi:10.1103/PhysRevLett.43.1365.

[42] See, *e.g.* Yao, W.–M.; Amsler, C.; Asner, D.; Barnett, R. M.; Beringer, J.; Burchat, P. R.; Carone, C. D.; Caso, C.; Dahl, O.; d'Ambrosio, G.; De Gouvea, A.; Doser, M.; Eidelman, S.; Feng, J. L.; Gherghetta, T.; Goodman, M.; Grab, C.; Groom, D. E.; Gurtu, A.; Hagiwara, K.; Hayes, K. G.; Hernández-Rey, J. J.; Hikasa, K.; Jawahery, H.; Kolda, C.; Kwon, Y.; Mangano, M. L.; Manohar, A. V.; Masoni, A.; et al. (2006). "Review of Particle Physics". *J. Phys. G* **33** (1): 1–1232. arXiv:astro-ph/0601168. Bibcode:2006JPhG...33....1Y. doi:10.1088/0954-3899/33/1/001.

[43] Rees, Martin. (1998). *Before the Beginning* (New York: Basic Books) p. 185 ISBN 0-201-15142-1

[44] de Sitter, Willem (1917). "Einstein's theory of gravitation and its astronomical consequences. Third paper". *Monthly Notices of the Royal Astronomical Society* **78**: 3–28. Bibcode:1917MNRAS..78....3D. doi:10.1093/mnras/78.1.3.

[45] Starobinsky, A. A. (December 1979). "Spectrum Of Relict Gravitational Radiation And The Early State Of The Universe". *Journal of Experimental and Theoretical Physics Letters* **30**: 682. Bibcode:1979JETPL..30..682S.; Starobinskii, A. A. (December 1979). "Spectrum of relict gravitational radiation and the early state of the universe". *Pisma Zh. Eksp. Teor. Fiz. (Soviet Journal of Experimental and Theoretical Physics Letters)* **30**: 719. Bibcode:1979ZhPmR..30..719S.

[46] SLAC seminar, "10^{-35} seconds after the Big Bang", 23 January 1980. see Guth (1997), pg 186

[47] Guth, Alan H. (1981). "Inflationary universe: A possible solution to the horizon and flatness problems" (PDF). *Physical Review D* **23** (2): 347–356. Bibcode:1981PhRvD..23..347G. doi:10.1103/PhysRevD.23.347.

[48] Chapter 17 of Peebles (1993).

[49] Starobinsky, Alexei A. (1980). "A new type of isotropic cosmological models without singularity". *Physics Letters B* **91**: 99–102. Bibcode:1980PhLB...91...99S. doi:10.1016/0370-2693(80)90670-X.

[50] Kazanas, D. (1980). "Dynamics of the universe and spontaneous symmetry breaking". *Astrophysical Journal* **241**: L59–63. Bibcode:1980ApJ...241L..59K. doi:10.1086/183361.

[51] Sato, K. (1981). "Cosmological baryon number domain structure and the first order phase transition of a vacuum". *Physics Letters B* **33**: 66–70. Bibcode:1981PhLB...99...66S. doi:10.1016/0370-2693(81)90805-4.

[52] Einhorn, Martin B; Sato, Katsuhiko (1981). "Monopole Production In The Very Early Universe In A First Order Phase Transition". *Nuclear Physics B* **180** (3): 385–404. Bibcode:1981NuPhB.180..385E. doi:10.1016/0550-3213(81)90057-2.

[53] Linde, A (1982). "A new inflationary universe scenario: A possible solution of the horizon, flatness, homogeneity, isotropy and primordial monopole problems". *Physics Letters B* **108** (6): 389–393. Bibcode:1982PhLB..108..389L. doi:10.1016/0370-2693(82)91219-9.

[54] Albrecht, Andreas; Steinhardt, Paul (1982). "Cosmology for Grand Unified Theories with Radiatively Induced Symmetry Breaking" (PDF). *Physical Review Letters* **48** (17): 1220–1223. Bibcode:1982PhRvL..48.1220A. doi:10.1103/PhysRevLett.48.1220.

[55] J.B. Hartle (2003). *Gravity: An Introduction to Einstein's General Relativity* (1st ed.). Addison Wesley. p. 411. ISBN 0-8053-8662-9

[56] See Linde (1990) and Mukhanov (2005).

[57] Chibisov, Viatcheslav F.; Chibisov, G. V. (1981). "Quantum fluctuation and "nonsingular" universe". *JETP Letters* **33**: 532–5. Bibcode:1981JETPL..33..532M.

[58] Mukhanov, Viatcheslav F. (1982). "The vacuum energy and large scale structure of the universe". *Soviet Physics JETP* **56**: 258–65.

[59] See Guth (1997) for a popular description of the workshop, or *The Very Early Universe*, ISBN 0-521-31677-4 eds Hawking, Gibbon & Siklos for a more detailed report

[60] Hawking, S.W. (1982). "The development of irregularities in a single bubble inflationary universe". *Physics Letters B* **115** (4): 295–297. Bibcode:1982PhLB..115..295H. doi:10.1016/0370-2693(82)90373-2.

[61] Starobinsky, Alexei A. (1982). "Dynamics of phase transition in the new inflationary universe scenario and generation of perturbations". *Physics Letters B* **117** (3–4): 175–8. Bibcode:1982PhLB..117..175S. doi:10.1016/0370-2693(82)90541-X.

[62] Guth, A.H. (1982). "Fluctuations in the new inflationary universe". *Physical Review Letters* **49** (15): 1110–3. Bibcode:1982PhRvL..49.1110G. doi:10.1103/PhysRevLett.49.1110.

[63] Bardeen, James M.; Steinhardt, Paul J.; Turner, Michael S. (1983). "Spontaneous creation Of almost scale-free density perturbations in an inflationary universe". *Physical Review D* **28** (4): 679–693. Bibcode:1983PhRvD..28..679B. doi:10.1103/PhysRevD.28.679.

[64] Perturbations can be represented by Fourier modes of a wavelength. Each Fourier mode is normally distributed (usually called Gaussian) with mean zero. Different Fourier components are uncorrelated. The variance of a mode depends only on its wavelength in such a way that within any given volume each wavelength contributes an equal amount of power to the spectrum of perturbations. Since the Fourier transform is in three dimensions, this means that the variance of a mode goes as k^{-3} to compensate for the fact that within any volume, the number of modes with a given wavenumber k goes as k^3.

[65] Tegmark, M.; Eisenstein, Daniel J.; Strauss, Michael A.; Weinberg, David H.; Blanton, Michael R.; Frieman, Joshua A.; Fukugita, Masataka; Gunn, James E.; et al. (August 2006). "Cosmological constraints from the SDSS luminous red galaxies". *Physical Review D* **74** (12). arXiv:astro-ph/0608632. Bibcode:2006PhRvD..74l3507T. doi:10.1103/PhysRevD.74.123507.

[66] Steinhardt, Paul J. (2004). "Cosmological perturbations: Myths and facts". *Modern Physics Letters A* **19** (13 & 16): 967–82. Bibcode:2004MPLA...19..967S. doi:10.1142/S0217732304014252.

[67] Boyle, Latham A.; Steinhardt, PJ; Turok, N (2006). "Inflationary predictions for scalar and tensor fluctuations reconsidered". *Physical Review Letters* **96** (11): 111301. arXiv:astro-ph/0507455. Bibcode:2006PhRvL..96k1301B. doi:10.1103/PhysRevLett.96.111301. PMID 16605810.

[68] Tegmark, Max (2005). "What does inflation really predict?". *JCAP* **0504** (4): 001. arXiv:astro-ph/0410281. Bibcode:2005JCAP...04..001T. doi:10.1088/1475-7516/2005/04/001.

[69] This is known as a "red" spectrum, in analogy to redshift, because the spectrum has more power at longer wavelengths.

[70] Komatsu, E.; Smith, K. M.; Dunkley, J.; Bennett, C. L.; Gold, B.; Hinshaw, G.; Jarosik, N.; Larson, D.; et al. (January 2010). "Seven-Year Wilkinson Microwave Anisotropy Probe (WMAP) Observations: Cosmological Interpretation". *The Astrophysical Journal Supplement Series* **192** (2): 18. arXiv:1001.4538. Bibcode:2011ApJS..192...18K. doi:10.1088/0067-0049/192/2/18.

[71] Spergel, D. N.; Verde, L.; Peiris, H. V.; Komatsu, E.; Nolta, M. R.; Bennett, C. L.; Halpern, M.; Hinshaw, G.; et al. (2003). "First year Wilkinson Microwave Anisotropy Probe (WMAP) observations: determination of cosmological parameters". *Astrophysical Journal Supplement Series* **148** (1): 175–194. arXiv:astro-ph/0302209. Bibcode:2003ApJS..148..175S. doi:10.1086/377226.

[72] See cosmic microwave background#Low multipoles for details and references.

[73] Overbye, Dennis (24 March 2014). "Ripples From the Big Bang". *New York Times*. Retrieved 24 March 2014.

[74] Ade, P.A.R. (BICEP2 Collaboration); et al. (19 June 2014). "Detection of B-Mode Polarization at Degree Angular Scales by BICEP2". *Physical Review Letters* **112** (24): 241101. arXiv:1403.3985. Bibcode:2014PhRvL.112x1101A. doi:10.1103/PhysRevLett.112.241101. PMID 24996078.

[75] Woit, Peter (13 May 2014). "BICEP2 News". *Not Even Wrong*. Columbia University. Retrieved 19 January 2014.

[76] Overbye, Dennis (19 June 2014). "Astronomers Hedge on Big Bang Detection Claim". *New York Times*. Retrieved 20 June 2014.

[77] Amos, Jonathan (19 June 2014). "Cosmic inflation: Confidence lowered for Big Bang signal". *BBC News*. Retrieved 20 June 2014.

[78] Planck Collaboration Team (19 September 2014). "Planck intermediate results. XXX. The angular power spectrum of polarized dust emission at intermediate and high Galactic latitudes". *ArXiv*. arXiv:1409.5738. Bibcode:2014arXiv1409.5738P. Retrieved 22 September 2014.

[79] Overbye, Dennis (22 September 2014). "Study Confirms Criticism of Big Bang Finding". *New York Times*. Retrieved 22 September 2014.

[80] Clavin, Whitney (30 January 2015). "Gravitational Waves from Early Universe Remain Elusive". *NASA*. Retrieved 30 January 2015.

[81] Overbye, Dennis (30 January 2015). "Speck of Interstellar Dust Obscures Glimpse of Big Bang". *New York Times*. Retrieved 31 January 2015.

[82] Rosset, C.; PLANCK-HFI collaboration (2005). "Systematic effects in CMB polarization measurements". *Exploring the universe: Contents and structures of the universe (XXXIXth Rencontres de Moriond)*.

[83] Loeb, A.; Zaldarriaga, M (2004). "Measuring the small-scale power spectrum of cosmic density fluctuations through 21 cm tomography prior to the epoch of structure formation". *Physical Review Letters* **92** (21): 211301. arXiv:astro-ph/0312134. Bibcode:2004PhRvL..92u1301L. doi:10.1103/PhysRevLett.92.211301. PMID 15245272.

[84] Guth, Alan (1997). *The Inflationary Universe*. Addison–Wesley. ISBN 0-201-14942-7.

[85] Choi, Charles (Jun 29, 2012). "Could the Large Hadron Collider Discover the Particle Underlying Both Mass and Cosmic Inflation?". Scientific American. Retrieved Jun 25, 2014."The virtue of so-called Higgs inflation models

is that they might explain inflation within the current Standard Model of particle physics, which successfully describes how most known particles and forces behave. Interest in the Higgs is running hot this summer because CERN, the lab in Geneva, Switzerland, that runs the LHC, has said it will announce highly anticipated findings regarding the particle in early July."

[86] Salvio, Alberto (2013-08-09). "Higgs Inflation at NNLO after the Boson Discovery". *Phys.Lett.* *B727 (2013) 234-239* **727**: 234–239. arXiv:1308.2244. Bibcode:2013PhLB..727..234S. doi:10.1016/j.physletb.2013.10.042.

[87] Technically, these conditions are that the logarithmic derivative of the potential, $\epsilon = (1/2)(V'/V)^2$ and second derivative $\eta = V''/V$ are small, where V is the potential and the equations are written in reduced Planck units. See, *e.g.* Liddle and Lyth (2000), pg 42-43.

[88] Salvio, Strumia (2014-03-17). "Agravity". *JHEP* *1406* *(2014)* *080* **2014**. arXiv:1403.4226. Bibcode:2014JHEP...06..080S. doi:10.1007/JHEP06(2014)080.

[89] Linde, Andrei D. (1983). "Chaotic inflation". *Physics Letters B* **129** (3): 171–81. Bibcode:1983PhLB..129..177L. doi:10.1016/0370-2693(83)90837-7.

[90] Technically, this is because the inflaton potential is expressed as a Taylor series in $\varphi/m P_l$, where φ is the inflaton and $m P_l$ is the Planck mass. While for a single term, such as the mass term $m_\varphi{}^4(\varphi/m P_l)^2$, the slow roll conditions can be satisfied for φ much greater than $m P_l$, this is precisely the situation in effective field theory in which higher order terms would be expected to contribute and destroy the conditions for inflation. The absence of these higher order corrections can be seen as another sort of fine tuning. See *e.g.* Alabidi, Laila; Lyth, David H (2006). "Inflation models and observation". *JCAP* **0605** (5): 016. arXiv:astro-ph/0510441. Bibcode:2006JCAP...05..016A. doi:10.1088/1475-7516/2006/05/016.

[91] See, *e.g.* Lyth, David H. (1997). "What would we learn by detecting a gravitational wave signal in the cosmic microwave background anisotropy?". *Physical Review Letters* **78** (10): 1861–3. arXiv:hep-ph/9606387. Bibcode:1997PhRvL..78.1861L. doi:10.1103/PhysRevLett.78.1861.

[92] Brandenberger, Robert H. (November 2004). "Challenges for inflationary cosmology". arXiv:astro-ph/0411671.

[93] Vilenkin, Alexander (1983). "The birth of inflationary universes". *Physical Review D* **27** (12): 2848–2855. Bibcode:1983PhRvD..27.2848V. doi:10.1103/PhysRevD.27.2848.

[94] A. Linde (1986). "Eternal chaotic inflation". *Modern Physics Letters A* **1** (2): 81–85. Bibcode:1986MPLA....1...81L. doi:10.1142/S0217732386000129. A. Linde (1986).

"Eternally existing self-reproducing chaotic inflationary universe" (PDF). *Physics Letters B* **175** (4): 395–400. Bibcode:1986PhLB..175..395L. doi:10.1016/0370-2693(86)90611-8.

[95] A. Borde, A. Guth and A. Vilenkin (2003). "Inflationary space-times are incomplete in past directions". *Physical Review Letters* **90** (15): 151301. arXiv:gr-qc/0110012. Bibcode:2003PhRvL..90o1301B. doi:10.1103/PhysRevLett.90.151301. PMID 12732026.

[96] A. Borde (1994). "Open and closed universes, initial singularities and inflation". *Physical Review D* **50** (6): 3692–702. arXiv:gr-qc/9403049. Bibcode:1994PhRvD..50.3692B. doi:10.1103/PhysRevD.50.3692.

[97] A. Borde and A. Vilenkin (1994). "Eternal inflation and the initial singularity". *Physical Review Letters* **72** (21): 3305–9. arXiv:gr-qc/9312022. Bibcode:1994PhRvL..72.3305B. doi:10.1103/PhysRevLett.72.3305.

[98] Linde (2005, §V).

[99] Carroll, Sean M.; Chen, Jennifer (2005). "Does inflation provide natural initial conditions for the universe?". *Gen. Rel. Grav.* **37** (10): 1671–4. arXiv:gr-qc/0505037. Bibcode:2005GReGr..37.1671C. doi:10.1007/s10714-005-0148-2.

[100] Carroll, Sean M.; Jennifer Chen (2004). "Spontaneous inflation and the origin of the arrow of time". arXiv:hep-th/0410270.

[101] Aguirre, Anthony; Gratton, Steven (2003). "Inflation without a beginning: A null boundary proposal". *Physical Review D* **67** (8): 083515. arXiv:gr-qc/0301042. Bibcode:2003PhRvD..67h3515A. doi:10.1103/PhysRevD.67.083515.

[102] Aguirre, Anthony; Gratton, Steven (2002). "Steady-State Eternal Inflation". *Physical Review D* **65** (8): 083507. arXiv:astro-ph/0111191. Bibcode:2002PhRvD..65h3507A. doi:10.1103/PhysRevD.65.083507.

[103] Hartle, J.; Hawking, S. (1983). "Wave function of the universe". *Physical Review D* **28** (12): 2960–2975. Bibcode:1983PhRvD..28.2960H. doi:10.1103/PhysRevD.28.2960.; See also Hawking (1998).

[104] Hawking (1998), p. 129.

[105] Wikiquote

[106] Page, Don N. (1983). "Inflation does not explain time asymmetry". *Nature* **304** (5921): 39–41. Bibcode:1983Natur.304...39P. doi:10.1038/304039a0.; see also Roger Penrose's book The Road to Reality: A Complete Guide to the Laws of the Universe.

[107] Hawking, S. W.; Page, Don N. (1988). "How probable is inflation?". *Nuclear Physics B* **298** (4): 789–809. Bibcode:1988NuPhB.298..789H. doi:10.1016/0550-3213(88)90008-9.

[108] Paul J. Steinhardt; Neil Turok (2007). *Endless Universe: Beyond the Big Bang*. Broadway Books. ISBN 978-0-7679-1501-4.

[109] Albrecht, Andreas; Sorbo, Lorenzo (2004). "Can the universe afford inflation?". *Physical Review D* **70** (6): 063528. arXiv:hep-th/0405270. Bibcode:2004PhRvD..70f3528A. doi:10.1103/PhysRevD.70.063528.

[110] Martin, Jerome; Brandenberger, Robert (2001). "The trans-Planckian problem of inflationary cosmology". *Physical Review D* **63** (12): 123501. arXiv:hep-th/0005209. Bibcode:2001PhRvD..63l3501M. doi:10.1103/PhysRevD.63.123501.

[111] Martin, Jerome; Ringeval, Christophe (2004). "Superimposed Oscillations in the WMAP Data?". *Physical Review D* **69** (8): 083515. arXiv:astro-ph/0310382. Bibcode:2004PhRvD..69h3515M. doi:10.1103/PhysRevD.69.083515.

[112] Robert H. Brandenberger, "A Status Review of Inflationary Cosmology", proceedings Journal-ref: BROWN-HET-1256 (2001), (available from arXiv:hep-ph/0101119v1 11 January 2001)

[113] Andrei Linde, "Prospects of Inflation", *Physica Scripta Online* (2004) (available from arXiv:hep-th/0402051)

[114] Blanco-Pillado et al., "Racetrack inflation", (2004) (available from arXiv:hep-th/0406230)

[115] Kachru, Shamit; Kallosh, Renata; Linde, Andrei; Maldacena, Juan; McAllister, Liam; Trivedi, Sandip P (2003). "Towards inflation in string theory". *JCAP* **0310** (10): 013. arXiv:hep-th/0308055. Bibcode:2003JCAP...10..013K. doi:10.1088/1475-7516/2003/10/013.

[116] G. R. Dvali, S. H. Henry Tye, *Brane inflation, Phys.Lett.* **B450**, 72-82 (1999), arXiv:hep-ph/9812483.

[117] Bojowald, Martin (October 2008). "Big Bang or Big Bounce?: New Theory on the Universe's Birth". Retrieved 2015-08-31.

[118] Itzhak Bars; Paul Steinhardt; Neil Turok (November 20, 2013). "Sailing through the big crunch-big bang transition". arXiv:1312.0739v2. In the standard big bang inflationary model, the cosmic singularity problem is left unresolved and the cosmology is geodesically incomplete. Consequently, the origin of space and time and the peculiar, exponentially fine-tuned initial conditions required to begin inflation are not explained. In a recent series of papers, we have shown how to construct the complete set of homogeneous classical cosmological solutions of the standard model coupled to gravity, in which the cosmic singularity is replaced by a bounce: the smooth transition from contraction and big crunch to big bang and expansion.

[119] Poplawski, N. J. (2010). "Cosmology with torsion: An alternative to cosmic inflation". *Physics Letters B* **694** (3): 181–185. arXiv:1007.0587. Bibcode:2010PhLB..694..181P. doi:10.1016/j.physletb.2010.09.056.

[120] Poplawski, N. (2012). "Nonsingular, big-bounce cosmology from spinor-torsion coupling". *Physical Review D* **85** (10): 107502. arXiv:1111.4595. Bibcode:2012PhRvD..85j7502P. doi:10.1103/PhysRevD.85.107502.

[121] Brandenberger, R; Vafa, C. (1989). "Superstrings in the early universe". *Nuclear Physics B* **316** (2): 391–410. Bibcode:1989NuPhB.316..391B. doi:10.1016/0550-3213(89)90037-0.

[122] Battefeld, Thorsten; Watson, Scott (2006). "String Gas Cosmology". *Reviews Modern Physics* **78** (2): 435–454. arXiv:hep-th/0510022. Bibcode:2006RvMP...78..435B. doi:10.1103/RevModPhys.78.435.

[123] Brandenberger, Robert H.; Nayeri, ALI; Patil, Subodh P.; Vafa, Cumrun (2007). "String Gas Cosmology and Structure Formation". *International Journal of Modern Physics A* **22** (21): 3621–3642. arXiv:hep-th/0608121. Bibcode:2007IJMPA..22.3621B. doi:10.1142/S0217751X07037159.

[124] Penrose, Roger (2004). *The Road to Reality: A Complete Guide to the Laws of the Universe*. London: Vintage Books, p. 755. See also Penrose, Roger (1989). "Difficulties with Inflationary Cosmology". *Annals of the New York Academy of Sciences* **271**: 249–264. Bibcode:1989NYASA.571..249P. doi:10.1111/j.1749-6632.1989.tb50513.x.

[125] Ijjas, Anna; Steinhardt, Paul J.; Loeb, Abraham. "Inflationary paradigm in trouble after Planck2013". *Physics Letters* **B723**: 261–266. arXiv:1304.2785. Bibcode:2013PhLB..723..261I. doi:10.1016/j.physletb.2013.05.023.

[126] Ijjas, Anna; Steinhardt, Paul J.; Loeb, Abraham. "Inflationary schism after Planck2013". *Physics Letters* **B736**: 142–146. Bibcode:2014PhLB..736..142I. doi:10.1016/j.physletb.2014.07.012.

[127] Guth, Alan H.; Kaiser, David I.; Nomura, Yasunori. "Inflationary paradigm after Planck 2013". *Physics Letters* **B733**: 112–119. arXiv:1312.7619. Bibcode:2014PhLB..733..112G. doi:10.1016/j.physletb.2014.03.020.

[128] Linde, Andrei. "Inflationary cosmology after Planck 2013". arXiv:1402.0526. Bibcode:2014arXiv1402.0526L.

47.10 References

- Guth, Alan (1997). *The Inflationary Universe: The Quest for a New Theory of Cosmic Origins*. Perseus. ISBN 0-201-32840-2.

- Hawking, Stephen (1998). *A Brief History of Time*. Bantam. ISBN 0-553-38016-8.

- Hawking, Stephen; Gary Gibbons (1983). *The Very Early Universe*. Cambridge University Press. ISBN 0-521-31677-4.

- Kolb, Edward; Michael Turner (1988). *The Early Universe*. Addison-Wesley. ISBN 0-201-11604-9.

- Linde, Andrei (1990). *Particle Physics and Inflationary Cosmology*. Chur, Switzerland: Harwood. arXiv:hep-th/0503203. ISBN 3-7186-0490-6.

- Linde, Andrei (2005) "Inflation and String Cosmology", *eConf* **C040802** (2004) L024; *J. Phys. Conf. Ser.* **24** (2005) 151–60; arXiv:hep-th/0503195 v1 2005-03-24.

- Liddle, Andrew; David Lyth (2000). *Cosmological Inflation and Large-Scale Structure*. Cambridge. ISBN 0-521-57598-2.

- Lyth, David H.; Riotto, Antonio (1999). "Particle physics models of inflation and the cosmological density perturbation". *Phys. Rept.* **314** (1–2): 1–146. arXiv:hep-ph/9807278. Bibcode:1999PhR...314....1L. doi:10.1016/S0370-1573(98)00128-8.

- Mukhanov, Viatcheslav (2005). *Physical Foundations of Cosmology*. Cambridge University Press. ISBN 0-521-56398-4.

- Vilenkin, Alex (2006). *Many Worlds in One: The Search for Other Universes*. Hill and Wang. ISBN 0-8090-9523-8.

- Peebles, P. J. E. (1993). *Principles of Physical Cosmology*. Princeton University Press. ISBN 0-691-01933-9.

47.11   External links

- Was Cosmic Inflation The 'Bang' Of The Big Bang?, by Alan Guth, 1997

- An Introduction to Cosmological Inflation by Andrew Liddle, 1999

- update 2004 by Andrew Liddle

- hep-ph/0309238 Laura Covi: Status of observational cosmology and inflation

- hep-th/0311040 David H. Lyth: Which is the best inflation model?

- The Growth of Inflation *Symmetry*, December 2004

- Guth's logbook showing the original idea

- WMAP Bolsters Case for Cosmic Inflation, March 2006

- NASA March 2006 WMAP press release

- Max Tegmark's *Our Mathematical Universe* (2014), "Chapter 5: Inflation"

47.12 Text and image sources, contributors, and licenses

47.12.1 Text

- **Introduction to M-theory** *Source:* https://en.wikipedia.org/wiki/Introduction_to_M-theory?oldid=717221473 *Contributors:* AxelBoldt, Sodium, BF, Michael Hardy, Dcljr, Sannse, Dgrant, Looxix~enwiki, Brinticus, Ojs, Schneelocke, Gutza, Morven, Tea2min, Ds13, Kocio, Sjö, Bhny, Jason.nussbaum, Aftermath, Shawnc, SmackBot, Famouslongago, Bduke, Silly rabbit, Scwlong, Aleenf1, Hu12, JRSpriggs, Jac16888, Gmusser, Headbomb, Peter Gulutzan, Julia Rossi, Knotwork, Rothorpe, Mclay1, Chris G, Rajpaj, Falcor84, R'n'B, Maurice Carbonaro, It Is Me Here, Gwen Gale, TXiKiBoT, PhysPhD, Accounting4Taste, Dlm slovakia, Dstebbins, ClueBot, MelonBot, Rankiri, Saeed.Veradi, Addbot, FactCheckerPrime, Morning277, Tide rolls, Luckas-bot, AnomieBOT, Gitman4, Jim1138, Materialscientist, Xqbot, Omnipaedista, Tom.Reding, NqpZ, EmausBot, Slightsmile, Hhhippo, Smiwi, Superzeal, ClueBot NG, Rgwkenyon, Gilderien, Helpful Pixie Bot, Ramaksoud2000, TheDude2011, LightandDark2000, Dave Bowman - Discovery Won, Reatlas, WorldWideJuan, Jamesmcmahon0, Comp.arch, Dimension10, Mfb, Tetra quark, Baking Soda and Anonymous: 79

- **History of string theory** *Source:* https://en.wikipedia.org/wiki/History_of_string_theory?oldid=703828599 *Contributors:* Bogdangiusca, Andrewman327, Wjhonson, Tea2min, Lethe, Tom harrison, Moyogo, Alison, Eep², Iamunknown, Danski14, Xaphan9966, Angr, Joriki, Wackyvorlon, Mpatel, Rjwilmsi, Nightscream, Ground Zero, DavideAndrea, Bhny, Netrapt, SmackBot, Colonies Chris, Lambiam, Owlbuster, Zejames, Dan Gluck, Michael C Price, Alaibot, Opheicus, Headbomb, Appraiser, AlphaEta, Hans Dunkelberg, HowardFrampton, Red Act, Likebox, WikiLaurent, Sabri76, Ssaco, Truthnlove, Addbot, Tassedethe, OlEnglish, HerculeBot, Citation bot, Omnipaedista, Edited0001, Patchy1, Knowandgive, Citation bot 1, Tom.Reding, Foobarnix, ClueBot NG, Frietjes, Helpful Pixie Bot, Bibcode Bot, AHusain314, Rmlkcl, Polytope24 and Anonymous: 27

- **String theory** *Source:* https://en.wikipedia.org/wiki/String_theory?oldid=717496071 *Contributors:* AxelBoldt, Sodium, Mav, Bryan Derksen, Zundark, The Anome, Tarquin, Taw, Eean, Malcolm Farmer, Hephaestos, Olivier, Drseudo, Stevertigo, Spiff~enwiki, Edward, PhilipMW, Michael Hardy, Bewildebeast, Dante Alighieri, Gabbe, Graue, Tgeorgescu, Mcarling, CesarB, Looxix~enwiki, Ahoerstemeier, Theresa knott, Suisui, Angela, Den fjättrade ankan~enwiki, Jdforrester, Julesd, Salsa Shark, Schneelocke, Charles Matthews, Timwi, Bemoeial, Jitse Niesen, 4lex, Greenrd, ErikStewart, Furrykef, Saltine, Phys, Omegatron, Bevo, Topbanana, Trent, Nufy8, Robbot, Craig Stuntz, Fredrik, Chris 73, R3m0t, COGDEN, Mirv, Wjhonson, Sverdrup, Academic Challenger, DHN, Hadal, Khlo, ElBenevolente, HaeB, Xanzzibar, Tea2min, Giftlite, DocWatson42, Christopher Parham, Awolf002, Mporter, Amorim Parga, Mikez, Harp, Kim Bruning, Tom harrison, Ferkelparade, Leflyman, Fropuff, No Guru, Anville, Moyogo, Curps, Pashute, Nomad~enwiki, Mboverload, Solipsist, SWAdair, DemonThing, Wmahan, Btphelps, MSTCrow, Decoy, Chowbok, Gadfium, Steuard, Pgan002, Quadell, Carandol~enwiki, Antandrus, Beland, JoJan, Khaosworks, Tothebarricades.tk, Thincat, Tomruen, Shidobu, Icairns, Lumidek, NoPetrol, Avihu, Fanghong~enwiki, Trevor MacInnis, Lacrimosus, Zro, Mike Rosoft, D6, Urvabara, Felix Wan, Jkl, Discospinster, ElTyrant, Rich Farmbrough, Rhobite, Pjacobi, Alien life form, Vapour, Silence, Kzzl, LindsayH, Mani1, Pavel Vozenilek, Paul August, Bender235, Kjoonlee, Mashford, Kelvinc, Perlman10s, Panu~enwiki, Brian0918, Dpotter, Livajo, El C, Laurascudder, Shanes, Zegoma beach, RoyBoy, Causa sui, Bobo192, Directorstratton, Janna Isabot, Smalljim, John Vandenberg, Flxmghvgvk, I9Q79oL78KiL0QTFHgyc, Physicistjedi, Bongoo, 4v4l0n42, Merope, Geschichte, Linuxlad, Phils, Merenta, Alansohn, Gary, JYolkowski, Enirac Sum, Ryanmcdaniel, Arthena, Borisblue, Rd232, Plumbago, Axl, R Calvete, Lightdarkness, Kocio, Bart133, Wtmitchell, Isaac, Tycho, Cal 1234, Fadereu, CloudNine, Sciurinæ, Computerjoe, Kusma, DV8 2XL, Pwqn, Gene Nygaard, Ringbang, Ceyockey, Falcorian, Bobrayner, Joriki, Mel Etitis, Linas, BillC, Jacobolus, HFarmer, Before My Ken, Netdragon, MONGO, GeorgeOrr, Mpatel, Bbatsell, GregorB, 马马马马马, Joke137, Christopher Thomas, Dysepsion, GSlicer, Jan.bannister, Graham87, Magister Mathematicae, Hillbrand, BD2412, Elvey, Galwhaa, Raymond Hill, JIP, RxS, Athelwulf, Edison, Sjakkalle, Rjwilmsi, Xgamer4, Jake Wartenberg, Arabani, MarSch, TheRingess, Jmcc150, Aero66, Crazynas, Juan Marquez, R.e.b., Bubba73, DoubleBlue, Zelos, AlisonW, Asafavi, Lionelbrits, Conorific, Zunz, Mathbot, Crazycomputers, RexNL, Gurch, Algri, TeaDrinker, Zifnabxar, XAXISx, Erik4, Phoenix2~enwiki, Antimatter15, Ggb667, Chobot, Visor, DVdm, Mhking, VolatileChemical, Bgwhite, Algebraist, Ben Tibbetts, YurikBot, Ugha, Wavelength, Borgx, NuclearFusion~enwiki, Angus Lepper, Hairy Dude, Jimp, Hillman, Cyferx, Wolfmankurd, Pip2andahalf, RussBot, Moronoman, Crazytales, Pippo2001, Bhny, Pigman, SpuriousQ, Branman515, Stephenb, Gaius Cornelius, Eleassar, Rsrikanth05, Bovineone, Cheesus, Shanel, NawlinWiki, Tong~enwiki, Mike18xx, SCZenz, Cleared as filed, Bdiah, Pym98, SColombo, Haemo, FF2010, Closedmouth, Reyk, Brina700, Chris Brennan, Vicarious, Brianlucas, Geoffrey.landis, Hitchhiker89, Spliffy, Pred, ArielGold, Roy Fultun, Ilmari Karonen, Katieh5584, Pentasyllabic, Lunch, DVD R W, WikiFew, That Guy, From That Show!, Street Scholar, AndrewWTaylor, QSquared, Sardanaphalus, Vanka5, MacsBug, Hvitlys, SmackBot, Kurochka, Zazaban, Tom Lougheed, Prodego, KnowledgeOfSelf, Hydrogen Iodide, Melchoir, Vald, Skrewtape, Atomota, Canthusus, GaeusOctavius, Cool3, Andyvn22, Gilliam, Skizzik, RobertM525, Dauto, Bluebot, SSJ 5, Keegan, Aidan Croft, Thumperward, Oli Filth, Silly rabbit, Timneu22, SchfiftyThree, Moshe Constantine Hassan Al-Silverburg, Complexica, Rediahs, RayAYang, Aero77, Adamstevenson, Ikiroid, Epastore, Baronnet, Ned Scott, Sbharris, Colonies Chris, Konstable, Sct72, Scwlong, Can't sleep, clown will eat me, Timothy Clemans, Onorem, Neilanderson, EvelinaB, TKD, KerathFreeman, Addshore, UU, The tooth, Pepsidrinka, Somebody2292, --=The Doctor=--, Fuhghettaboutit, Cybercobra, Irish Souffle, Nakon, Jdlambert, James McNally, MichaelBillington, Lostart, Insineratehymn, Drphilharmonic, SpiderJon, DMacks, Ihatetoregister, Where, Michael IFA, Yevgeny Kats, Vasiliy Faronov, Byelf2007, Angela26, Visium, Rory096, Zymurgy, Harryboyles, Mdl53711, T-dot, Titus III, Ergative rlt, MagnaMopus, UberCryxic, Vgy7ujm, Lazylaces, Linnell, Mgiganteus1, Nonsuch, IronGargoyle, Ckatz, DoItAgain, AstroGod, Kirbytime, Jimbo Mahoney, FredrickS, Invisifan, Ryulong, Ryanjunk, MathStuf, Mike Doughney, Norm mit, Hindol, Dan Gluck, Huntscorpio, Iridescent, K, Sunoco, You? Me? Us?, CzarB, Rabinzkaman, JoeBot, Lottamiata, Tony Fox, Vrkaul, Torrazzo, Gil Gamesh, Areldyb, Courcelles, Tawkerbot2, Gebrah, Shamvil, Fdssdf, DKqwerty, Lbr123, Harold f, Heqs, Devourer09, Duduong, Sarvagnya, Dewayne76, JForget, Cg-realms, InvisibleK, CRGreathouse, CmdrObot, Earthlyreason, Van helsing, Olaf Davis, CBM, Rawling, Jibal, Witten Is God, Nunquam Dormio, Giko, KnightLago, Thubsch, Leujohn, SlashDot, TheTito, Karenjc, Myasuda, Emarv, Cydebot, Gmusser, Gogo Dodo, Jkokavec, Kahananite, Quajafrie, Michael C Price, Doug Weller, DumbBOT, Narayanese, AlphaNumeric, SRoughsedge, Vanished User jdksfajlasd, Woland37, Zalgo, Daniel Olsen, UberScienceNerd, Bkazaz, DJBullfish, Thijs!bot, Epbr123, Rwmnau, Babemachine, Pimpin101, Mbell, O, Faigl.ladislav, Ucanlookitup, Andyjsmith, Headbomb, Tcturner2002, Marek69, Brahmajnani, Arthurcprado~enwiki, Y.t., D3gtrd, Babemonkey, Dark dude, Duncan McB, EdJohnston, MichaelMaggs, Ancientanubis, Natalie Erin, Hempfel, Jomoal99, Mmortal03, Mentifisto, Geekdom04, AntiVandalBot, Luna Santin, Seaphoto, Ed270791, Opelio, Doc Tropics, David136a, NithinBekal, Dotdotdotdash, Helicoptor, Poshzombie, MontanNito, Dylan Lake, Maximilian77, Shlomi Hillel, Db63376, SamIAmNot, Knotwork, Res2216firestar, Superior IQ Genius, MER-C, Andonic, Sitethief, 100110100, TallulahBelle, Nestamachine, Savant13, Daynightrader, Goldenglove, Charibdis, Acroterion, Ophion, Aigisthos, Editmyhandman, Aruben537,

Magioladitis, WolfmanSF, Bongwarrior, VoABot II, Yandman, JamesBWatson, باسم, Qutt, Jespinos, Kevinmon, Aka042, Froid, DAGwyn, Catgut, Panser Born, Ensign beedrill, Perspectival, JJ Harrison, Dirac66, Justanother, Aziz1005, Cpl Syx, ChazBeckett, Teardrop onthefire, WLU, Stephen shenker, Robin S, SkepticVK, Joshua Davis, Mkroh, B9 hummingbird hovering, S3000, Hdt83, MartinBot, FlieGerFaUstMe262, Ytomem, Shimwell, Arjun01, KrishSundaresan, Anaxial, Jay Litman, Alexcalamaro, Andrej.westermann, Smokizzy, LedgendGamer, Cyrus Andiron, Peteryoung144, Tgeairn, Artaxiad, HEL, AlphaEta, J.delanoy, AstroHurricane001, Maurice Carbonaro, Yonidebot, Morris729, M C Y 1008, 69gangsta420, It Is Me Here, Shawn in Montreal, Janus Shadowsong, Bailo26, Fredsie, Madagaskar07, Duchesserin, AntiSpamBot, CHIAGEHYANG, Chiswick Chap, Watsup1313, Belovedfreak, HaloInverse, NewEnglandYankee, Scott1329m, Thesis4Eva, Policron, Jrcla2, KylieTastic, WJBscribe, Rnricklefs, Jamesofur, Eyelidlessness, Jonnyk aus, Kvdveer, JavierMC, Izno, Xiahou, CardinalDan, Sheliak, HamatoKameko, Malik Shabazz, Concertmusic, JohnBlackburne, JustinHagstrom, Fences and windows, Wooba doob, Philip Trueman, DoorsAjar, HowardFrampton, TXiKiBoT, Zidonuke, Red Act, Kriak, Calwiki, Technopat, Hqb, Andrius.v, Anonymous Dissident, Crohnie, AlysTarr, Qxz, Vanished user ikijeirw34iuaeolaseriffic, Impunv, Seraphim, Martin451, Don4of4, ABigGreenHippo, Huperphuff, LeaveSleaves, Kaenneth, StringyGuy, Maxim, Erth64net, Meters, Lamro, Rickstauduhar, Enviroboy, Turgan, Anna512, PhysPhD, Northfox, NPguy, Matthew Sanders, Luke Walkerson, Newbyguesses, MissMJ, SieBot, Escher26, J.A.Ireland, BA (IHPST), 4wajzkd02, Robdunst, Dreamafter, Pallab1234, Dbelange, MTHarden, Lemonflash, Kylemew, Yintan, GlassCobra, Discrete, Bentogoa, Likebox, Flyer22 Reborn, Exert, ProGeek314, Arbor to SJ, Babawhitemoose, Caidh, Dhatfield, Audree, Oxymoron83, Pretty Green, Weaselstomp, Manway, Alex.muller, Taco Manipulator, Tschach, Manheat84, Anchor Link Bot, Mikebernstein, ImperialismGo, Nergaal, Ionfield, Ayleuss, Sh4wz0r, Naturespace, ImageRemovalBot, Martarius, Phyte, ClueBot, The Thing That Should Not Be, String4d, Illusion96, Polyamorph, Mpd1989, Alexdeburca18, Wiggl3sLimited, Excirial, Kjramesh, Jusdafax, Resoru, WikiZorro, Eeekster, Verum~enwiki, Tamaratrouts, Gtstricky, Humanino, Brews ohare, NuclearWarfare, Cenarium, Arjayay, Razorflame, Scoobey, BOTarate, Sideswiper, Thingg, Capudo, BVBede, Versus22, Introductory adverb clause, MelonBot, SoxBot III, Egmontaz, Notpayingthepsychiatrist, DumZiBoT, BahTab, TimothyRias, Aj00200, Reaperfromhell, Dunkaroo207, XLinkBot, AlexGWU, Impshum, Saeed.Veradi, Little Mountain 5, Guy392, David424, Truthnlove, Qweeveen, Tayste, Addbot, Steven66s, Denali134, Elemented9, Varrey280303, Eric Drexler, Some jerk on the Internet, Fizzycyst, Uruk2008, DOI bot, Jojhutton, AngryBacon, Non-dropframe, Captain-tucker, Auspex1729, Kongr43gpen, Fgnievinski, Rhetoric Of A Sophist, Ronhjones, CanadianLinuxUser, Cst17, Download, Glane23, Bassbonerocks, Chzz, Favonian, Kronix35, LinkFA-Bot, Udugunit, Aktsu, Tassedethe, Numbo3-bot, Anpecota, Tide rolls, HerpesVirus, SDJ, OlEnglish, Scourge of God, Davidmedlar, Couldbenoway66, Yobot, Maxdamantus, Terrisknickers, Kartano, TaBOT-zerem, Julia W, Unique and proud of it, FireMouseHQ, Terrifictriffid, ArchonMagnus, CinchBug, Synchronism, AnomieBOT, Cleeseheb, 1exec1, Charlesvi, Bigdaddy4x4, Gitman4, Jim1138, IRP, Mintrick, Drweetmola, Ornamentalone, M00npirate, Gautam10, Csigabi, Poli-Psy, Materialscientist, 90 Auto, Citation bot, Teleprinter Sleuth, Vuerqex, Twri, Frankenpuppy, Fuzzy Bob Saget, DirlBot, Georgepowell2008, Heidisql, Cureden, Ekwos, Capricorn42, Gensanders, NFD9001, Anna Frodesiak, Tomwsulcer, A23649, Pra1998, Coretheapple, RadiX, Jagbag2, Vandalism destroyer, Ab1, Omnipaedista, Bandit5005, Shirik, RibotBOT, Waleswatcher, Saalstin, Amaury, Aaron35510, Caz34, Doulos Christos, Sewblon, Born Gay, Capricorn24, SchnitzelMannGreek, A. di M., SpacePyjamas, Kierkkadon, A.amitkumar, Dougofborg, StringLove, Nobelprizewinner, Astiburg, FrescoBot, Fortdj33, Paine Ellsworth, Goodbye Galaxy, HJ Mitchell, Steve Quinn, Vhann, Kwiki, Xhaoz, Citation bot 1, Batong, Gil987, Pinethicket, I dream of horses, Tallboyhoops1991, Three887, Steveo27five, RedBot, Sardinita, Serols, Vhsatheeshkumar, Swisstingle, DeletionUK, Reconsider the static, IVAN3MAN, Remingtonhill1, Orenburg1, Coltonhs, Angus Guilherme, Smamaret, Bethovenn, Dinamik-bot, Dc987, Oswaldo Zapata, Egemont, Syebo, Alaithiran, Reaper Eternal, Seahorseruler, Ybungalobill, Quaker phil, Specs112, Dr. Aakash Patel, Tbhotch, StormbringerUK, Minimac, Mathgenius3141592, Keegscee, Omgwaffels, Mick le pick, Solancel, Aznhero3793, Dwielark, Afteread, Enauspeaker, EmausBot, MaooaM, Immunize, Az29, Milkocookie, Faolin42, Fotoni, RA0808, RenamedUser01302013, 8digits, Yukiseaside, Slightsmile, Tommy2010, Winner 42, Wikipelli, Dcirovic, JonezyKiDx, Joe Gazz84, ZéroBot, Timeitsways, John Cline, Cogiati, Quaqa, Chrispaps2413, Nasulikid, Vollrath2323, Benjamin1414141414141414, Arbnos, Green Lane, A930913, Bamyers99, Azeraphale, H3llBot, Encyclopadia, Danga1988, Ollainen, PoisonGM, Wayne Slam, OnePt618, Knome335, L Kensington, Lulzprotuns, Kranix, Rpcappello, Maschen, Vastly~enwiki, Donner60, CatFiggy, CountMacula, Orange Suede Sofa, Etov, M1k3 101, Bill william compton, Wakabaloola, TERBAFAN, Nickslspride34, NeuralLotus, Isocliff, Brechbill123, Xanchester, ClueBot NG, Martti Muukkonen, KagakuKyouju, Jeff Song, This lousy T-shirt, Satellizer, Name Omitted, Marcdean123, Wiki incorp, Frietjes, O.Koslowski, Alexdamaino9, Dream of Nyx, Blackhall616, Widr, Sashhere, WikiPuppies, Stu181, T00g00d96, Pluma, Storm.sarup, Helpful Pixie Bot, Manzeet, Waffleboy36, HMSSolent, Mikeshelton1, Bibcode Bot, 2001:db8, Phillip.phillipson, Hoaxinator, Lowercase sigmabot, Thor cherubim, BG19bot, Mrshabam, Nishch, Flowerhat15, AvocatoBot, Housegeek224, MahRanch, Benzband, Altaïr, Benhenchdickthomas, Shreyakstring, Sweaty maori sphincter, DaFalk, Dsabo74, Ratanmaitra, MM4EVAH, Steven.w.kowalski, Minsbot, JGallardo2600, Dylanlatham, Myfriendganesha, OCCullens, Likeaboss189, Sean271293, LinusE8, BattyBot, Several Pending, Aldrich2122, CommanderMoka, Cyberbot II, The Illusive Man, ChrisGualtieri, KoalamaN2, Trevorkid45, Catsloveit07, Alex Modzz, Rustyjamsen, Goh ryangoh, Dexbot, Exolius, Hilander316, Alman1234321, SuperCalzer, LightandDark2000, MeekMelange, BQND, Cdarrai1, Kephir, TheMonkeyboy524, Michael Anon, TwoTwoHello, Mattfat8, Lugia2453, Anruy, Rachel weld, Jamesx12345, AHusain314, BossEditors, Hillbillyholiday, Joeinwiki, Mattninja, Theshadow444, Asaa82, Jakemarz197, Kzhang1025, Epicgenius, Spongbob456789, ꧁, TestMaster, Ianreisterariola, GrapperJ, Makeitnasty, Moemajdi, I am One of Many, NualaIvy, BAZINGASS, St3fanPC, Eyesnore, Isaac grozd, Jordanissexyaf1999, Baruch6525, Mosbruckercj, Ihatedirac2k13, Jonamithy121314, 123physicsquantum, Jt198, RaphaelQS, HeyJude70, AParker628, DimReg, A.k.blaze1, Joshuk, Zenibus, Nianoobasik, Ihelpapplen, Gamo To Apoel, SacredLabyrinth, Ginsuloft, Vampre1122, Dimension10, Howard Wolowitz, AddWittyNameHere, Polytope24, Elysion, Tutun12$, Longerboats5, SimonWombat8, Konveyor Belt, Vtank54, Micheal545, Hck24, Caliae19, Hexafish, Simpick, TheRealTheKoi, Bballbro62, Monkbot, ArmyPath, Gabero.88, TheQ Editor, Jtsmith098, Joshmiller1, Hanseer360, XXvPIEvXx, Dbennett 24, Ghikpenos, Nick65633, Saundra03, Thehippothatknows, Sewwgers, Teelaskeletor, Cirksena, Balockaye1234, PloppyDoo, Yesufu29, Lumpy2k14, Podayeruma, Abstract92, Sbenfiel, Monkman2k4, Swegwegdgfyetkfoffkkfkfkv, John95541234, Poopman224, ScrapIronIV, Tetra quark, GeneralizationsAreBad, Shivansh2014n, KasparBot, SHUCKYLUCKY, Fabiotheoto, FartGoblin, Joca potato, Joshcool246, Theoretical Physisist4444, JanetTom55, Reg7d88, CHANDLER MERRILL, Baking Soda, FklfjDKFd bfl, Rajputclann, Entranced98, Jahziahk, Mjhog, Strong81, WikiTikiDude007, ILoveShukli, Qwerty2345ß, A1D1A2D2 and Anonymous: 1594

- **Superstring theory** *Source:* https://en.wikipedia.org/wiki/Superstring_theory?oldid=718279717 *Contributors:* Mav, Bryan Derksen, Stevertigo, Michael Hardy, Erik Zachte, Minesweeper, Looix~enwiki, Ahoerstemeier, JWSchmidt, Cyan, Palfrey, Evercat, Schneelocke, Hashar, Charles Matthews, Tpbradbury, Motor, David Shay, Omegatron, Bevo, Bcorr, Robbot, Fredrik, Hadal, Vuara, Tea2min, Giftlite, Barbara Shack, Herbee, Fropuff, Anville, Maarten van Vliet, WalkinDownThirtyThree, Christopherlin, Steuard, Karol Langner, Lumidek, Prestonmarkstone, Rich Farmbrough, Igorivanov~enwiki, Autiger, Pavel Vozenilek, El C, Rgdboer, Shadow demon, Causa sui, Billymac00, BM, Gary, Pion, Tycho, Cal 1234, Redvers, Postrach, Supercool Dude, Mindmatrix, Mpatel, Joke137, Mandarax, Bill37212, Yamamoto Ichiro, Bubbleboys,

Chobot, Ben Tibbetts, Wavelength, RussBot, Chris Capoccia, Chensiyuan, Cate, Chaos, NawlinWiki, Astral, Voidxor, TheMadBaron, Zerodamage, Allens, SmackBot, Android 93, Kurochka, McGeddon, Kintetsubuffalo, Cesoid, Silly rabbit, Stevage, Baronnet, Colonies Chris, Scwlong, Mesons, Kurrupt3d, Bjankuloski06en~enwiki, Makyen, MathStuf, Hu12, Iridescent, Kahalachan, Gatortpk, Mattbr, Neelix, Gregbard, Nauticashades, Cydebot, Davidanzaldua, ChKa, Headbomb, Escarbot, Jj137, Shlomi Hillel, Dougher, JAnDbot, 100110100, 28421u2232nfenfcenc, Aziz1005, Connor Behan, Rickard Vogelberg, Jean-Pierre Petit~enwiki, Hans Dunkelberg, Maurice Carbonaro, Bot-Schafter, SmilesALot, Student7, Cmichael, Sheliak, JayCo777, Calwiki, Andrius.v, Molinogi, Billinghurst, Lamro, Enviroboy, Seraphita~enwiki, Drschawrz, Henry Delforn (old), Lightmouse, Altzinn, Gratedparmesan, ClueBot, Arakunem, Vergil 577, Drmies, Frdayeen, Vizzini101, Niceguyedc, Neverquick, Resoru, Mastertek, Kakofonous, Princess Janay, Alex123irish123, Madeinmexico567, Oldnoah, Madeinmexico566, Truthnlove, YeAaMsLtA, Addbot, Physicman123, CWatchman, Cuaxdon, Semdino, AnnaFrance, LinkFA-Bot, TaBOT-zerem, Evans1982, Gerixau, Eric-Wester, Magog the Ogre, AnomieBOT, DemocraticLuntz, ^musaz, Josh Guffin, Jim1138, Materialscientist, Citation bot, Renaissancee, BLP-outrageous move logs, Omnipaedista, RibotBOT, Paine Ellsworth, Steve Quinn, Tom.Reding, Klavesin, Tkachyk, Dinamik-bot, Bj norge, Orphan Wiki, Idh0854, Arbnos, Vramasub, L Kensington, Particle hep, Isocliff, ClueBot NG, Widr, Adminium, Delivernews, Bibcode Bot, Khanduras, Quarkgluonsoup, Flowerhat15, MythosMagic, OCCullens, Aldrich2122, Graphium, AHusain314, Jochen Burghardt, WorldWideJuan, Jakec, Liquidityinsta, E8xE8, Polytope24, FlaviusCorcoata, Cirksena, BakedLikaBiscuit, Crito10, KasparBot, Rantonels, Baking Soda and Anonymous: 184

- **String duality** *Source:* https://en.wikipedia.org/wiki/String_duality?oldid=703639606 *Contributors:* Kingturtle, Fropuff, Mpatel, SmackBot, Dan Gluck, Thadius856, PhysPhD, Addbot, Bte99, Lightbot, Xqbot, Omnipaedista, Sa'y, CaroleHenson, ChrisGualtieri, Yikkayaya and Anonymous: 3

- **S-duality** *Source:* https://en.wikipedia.org/wiki/S-duality?oldid=709314225 *Contributors:* CYD, The Anome, Michael Hardy, Angela, Charles Matthews, Phys, Lumidek, Gauge, Linas, Mpatel, Rjwilmsi, CJLL Wright, Shell Kinney, Sardanaphalus, SmackBot, Reedy, Colonies Chris, Scwlong, QFT, Pjoef, MenoBot, AnonyScientist, Addbot, Yobot, JackieBot, Citation bot, Omnipaedista, Erik9bot, Tom.Reding, Vhsatheeshkumar, Lightlowemon, ZéroBot, Maschen, ClueBot NG, Bibcode Bot, BG19bot, Enyokoyama, AHusain314, Polytope24 and Anonymous: 7

- **T-duality** *Source:* https://en.wikipedia.org/wiki/T-duality?oldid=709314218 *Contributors:* TakuyaMurata, Charles Matthews, Reddi, David Gerard, Alison, Lumidek, Lysdexia, BD2412, Roboto de Ajvol, WAS 4.250, Sardanaphalus, Unyoyega, MalafayaBot, Scwlong, QFT, Dan Gluck, Outriggr (2006-2009), Headbomb, RogueTeddy, Eujin16, Truthnlove, Addbot, MagnusA.Bot, Yobot, Citation bot, LilHelpa, Omnipaedista, HissingFauna, Patchy1, Tom.Reding, EmausBot, ZéroBot, Bibcode Bot, BG19bot, Solomon7968, AHusain314, Polytope24, Anrnusna and Anonymous: 10

- **D-brane** *Source:* https://en.wikipedia.org/wiki/D-brane?oldid=696680667 *Contributors:* Zundark, TakuyaMurata, Karada, JWSchmidt, AugPi, Smack, Schneelocke, Gandalf61, Michael Snow, Fropuff, Anville, Just Another Dan, Phe, Lumidek, Rgrg, H0riz0n, El C, Constantine, I9Q79oL78KiL0QTFHgyc, FlaBot, Bhny, Nick, Zwobot, Sardanaphalus, KnightRider~enwiki, Teemu Ruskeepää, Colonies Chris, Scwlong, QFT, Eric Olson, Fuhghettaboutit, Vampus, JarahE, Twyder, Eewild, 345Kai, Cydebot, Headbomb, J. W. Love, Nick Number, Magioladitis, Jpod2, STBot, HEL, VolkovBot, TXiKiBoT, PhysPhD, Jonathanrcoxhead, Excirial, Alexbot, ResidueOfDesign, Addbot, LaaknorBot, Tassedethe, Lightbot, Luckas-bot, Yobot, Amirobot, Azcolvin429, Royote, Citation bot, Twri, ChristopherKingChemist, Omnipaedista, Galaktiker, Mentibot, Wakabaloola, Petrb, Frietjes, Luizpuodzius, OCCullens, Polytope24 and Anonymous: 30

- **Brane** *Source:* https://en.wikipedia.org/wiki/Brane?oldid=714337937 *Contributors:* Bth, Michael Hardy, DIG~enwiki, Samuelsen, JWSchmidt, Silverfish, Wetman, Bcorr, Blainster, Fropuff, Just Another Dan, D3, Lumidek, Yuriz, Rhobite, H0riz0n, Ben Standeven, RoyBoy, Mairi, Constantine, GatesPlusPlus, Kocio, Agguarx, Mpatel, Liface, BD2412, Quiddity, R.e.b., Mathbot, BradBeattie, Metropolitan90, YurikBot, Wavelength, NawlinWiki, Wknight94, Closedmouth, SmackBot, Kurochka, Jwestbrook, Autarch, Seanor32, Silly rabbit, Colonies Chris, Nsmith4658, Mesons, Monotonehell, TheVikingRaider, Yevgeny Kats, Spiritia, PaddyM, Czoller, Calmargulis, BeenAroundAWhile, Adailton, Julius M-D, J. W. Love, Julia Rossi, Chrisjj3, MER-C, Steveprutz, Just H, N.Nahber, Urco, Alexrussell101, Cyborg Ninja, Idioma-bot, VolkovBot, Anonymous Dissident, Michael Frind, Paucabot, Drschawrz, SieBot, Tresiden, OKBot, ClueBot, The Thing That Should Not Be, SilvonenBot, NonvocalScream, Addbot, Jujutsuka, Royote, LilHelpa, Patmethenyfan, Omnipaedista, Nagualdesign, Kgrad, Tbhotch, Tesseract2, EmausBot, MathMaven, ClueBot NG, Mikeflem, Gilderien, Baseball Watcher, Frietjes, DBigXray, Lowercase sigmabot, BG19bot, Solomon7968, OCCullens, BattyBot, Brirush, E8xE8, Dimension10, Polytope24, Ihsanturk, Eno Lirpa and Anonymous: 50

- **Fundamental interaction** *Source:* https://en.wikipedia.org/wiki/Fundamental_interaction?oldid=713122554 *Contributors:* AxelBoldt, Zundark, The Anome, Tarquin, AstroNomer, William Avery, Roadrunner, Ellmist, Robert Foley, Heron, Isis~enwiki, Stevertigo, Patrick, Michael Hardy, Gdarin, CesarB, Looxix~enwiki, Cyp, William M. Connolley, Theresa knott, Mxn, Bemoeial, Reddi, Zoicon5, Finlay McWalter, Robbot, Lowellian, Brjaga, Roscoe x, Seth Ilys, Ancheta Wis, Giftlite, Christopher Parham, Herbee, Monedula, Xerxes314, Alison, Pcarbonn, Beland, Melikamp, Karol Langner, AmarChandra, Mike Rosoft, Jørgen Friis Bak, JimJast, Discospinster, Guanabot, FT2, Harriv, Quietly, GoldenRing, Clement Cherlin, El C, Lycurgus, Joanjoc~enwiki, Alereon, Euyyn, Kanzure, Army1987, Rbj, Haham hanuka, Nsaa, Jumbuck, Foant, Dachannien, Kdau, ReyBrujo, Reaverdrop, BDD, Someoneinmyheadbutit'snotme, DV8 2XL, Kazvorpal, Woohookitty, Linas, Mindmatrix, Sabejias, StradivariusTV, Mpatel, Miss Madeline, Isnow, Elvey, Chun-hian, Koavf, Strait, Jmcc150, RE, Gadha, FlaBot, DClement, ZoneSeek, Alfred Centauri, Lmatt, Rell Canis, Mstroeck, Chobot, Subtractive, Visor, GangofOne, Mysekurity, YurikBot, Ashleyisachild, Bambaiah, Lucinos~enwiki, Wavesmikey, Chaos, FFLaguna, Dbfirs, Trigger hippie77, Enormousdude, Shimei, RG2, Bweenie, Phr en, GrinBot~enwiki, SmackBot, Unyoyega, Andy M. Wang, Vvarkey, Jjalexand, Mithaca, Acipsen, DHN-bot~enwiki, Colonies Chris, Andy120290, Addshore, SundarBot, Jgwacker, LeoNomis, Sadi Carnot, TTE, SashatoBot, FrozenMan, Philosophus, A. Parrot, Fangfufu, GDallimore, Avanishsharma, CRGreathouse, Green caterpillar, McVities, MaxEnt, A. Exeunt, Scott.medling, LouisBB, Thijs!bot, Martin Hogbin, Mojo Hand, Headbomb, Dfrg.msc, Dodecahedron~enwiki, HolyT, JAnDbot, The penfool, Fordskydog, MER-C, TheEditrix2, Fabrictramp, Leyo, Trusilver, Joshuaali, Idioma-bot, VolkovBot, TXiKiBoT, Anonymous Dissident, MackSalmon, Praveen pillay, BotKung, Gnomon13, Lamro, RMW42, EmxBot, Neparis, SieBot, WereSpielChequers, ToePeu.bot, Avargasm, RadicalOne, Dhatfield, SuperSpy00bob, Sbowers3, BartekChom, Beast of traal, Lightmouse, Nskillen, Sunrise, OKBot, Bpeps, C0nanPayne, StewartMH, Sfan00 IMG, ClueBot, MichaelVernonDavis, SuperHamster, Djr32, Sadiqsaleem09, PixelBot, Eeekster, Zamis45, Yonskii, 1ForTheMoney, Noctibus, Truthnlove, Addbot, Mabdul, LinkFA-Bot, F Notebook, Lightbot, Legobot, Clay Juicer, Luckas-bot, Yobot, II MusLiM HyBRiD II, Rifter0x0000, AnomieBOT, Glen Dillon, Girl Scout cookie, Cleroth, JackieBot, Piano non troppo, Flewis, AthenaO, Xqbot, Gap9551, Omnipaedista, RibotBOT, Alvin Seville, A. di M., Ironboy11, Goodbye Galaxy, Jmbenham, Unkownkid2400, Rameshngbot, Jschnur, RedBot, Σ, Frankjohnson123, IVAN3MAN, Right-wing genius, Lokentaren, Setsuna29, EngineerFromVega, RjwilmsiBot, Anuandraj, Beyond My Ken, Deadlyops, Carbo1200, Kbasford, ClueBot NG, Greedohun, MelbourneStar, Grannis3, Kasirbot, Kaos Magician, Einsteiner900, Widr, Shelbylv, CasualVisitor, BG19bot, Hz.tiang, JimmyMachineGunHand,

Cengime, BattyBot, Prof. Squirrel, Dexbot, Mogism, Makecat-bot, Ryan.laff, Jamesx12345, Ttitts, CsDix, Kenanwang, GregRos, Prokaryotes, Robertpb97, Occurring, Basedrawnz, Learnerktm, Julietvbarbara, Barbarousbunch815, Brendapallister, Nicholaspurcellstudio, Pickleslover, Tetra quark, Isambard Kingdom, Claudio.nahmad.arcaraz, Trumpet21, KasparBot, The oracle 2015, Mitzionne, CAPTAIN RAJU, Ram9bo, Huritisho, G-dac, Okamialvis, Incendiary Iconoclasm and Anonymous: 261

- **Theory of everything** *Source:* https://en.wikipedia.org/wiki/Theory_of_everything?oldid=716784654 *Contributors:* AxelBoldt, Paul Drye, CYD, The Anome, Eclecticology, Toby Bartels, Roadrunner, Zippy, Stevertigo, Lorenzarius, Michael Hardy, Rojclague, Nixdorf, Takuya-Murata, Karada, Skysmith, Kosebamse, CesarB, Anders Feder, Angela, Julesd, Salsa Shark, Ugen64, Poor Yorick, Evercat, Schneelocke, Feed-mecereal, Timwi, Dcoetzee, Dysprosia, Jitse Niesen, Wik, Jakenelson, Omegatron, Raul654, Nnh, Kevin M C Harkess, UninvitedCompany, Fredrik, Altenmann, Nurg, Naddy, Gandalf61, Mirv, Academic Challenger, Rursus, Blainster, Caknuck, Wereon, Diberri, Pengo, Tea2min, Hooloovoo, Ancheta Wis, Dbenbenn, Mporter, Jabra, Ferkelparade, Bfinn, Xerxes314, Curps, Alison, FeloniousMonk, McGravin, Behnam, Gzornenplatz, JRR Trollkien, Steuard, Andycjp, Sonjaaa, Antandrus, Kim54, Tomruen, Lumidek, Gscshoyru, WpZurp, TJSwoboda, Zondor, Mike Rosoft, JimJast, Discospinster, Rich Farmbrough, H0riz0n, Pjacobi, Vsmith, Pluke, Autiger, Mal~enwiki, Pavel Vozenilek, Floorsheim, El C, Lycurgus, Sourcecode, Oldsoul, PhilHibbs, Sietse Snel, Jpgordon, Atraxani, Smalljim, Slicky, LostLeviathan, Matpitka, Juesch, Dan-ski14, Alansohn, Gary, DariuszT, ShardPhoenix, Kocio, Pion, Hdeasy, Bart133, Schaefer, BanyanTree, ClockworkSoul, Tycho, Suruena, Count Iblis, DV8 2XL, Gene Nygaard, Euphrosyne, Squidwina, Ott, Siafu, Roylee, Woohookitty, Mindmatrix, RHaworth, TigerShark, Savantnavas, MrDarcy, Mpatel, GregorB, Athletec64, Christopher Thomas, Aarghdvaark, Ashmoo, BD2412, Drbogdan, Rjwilmsi, Kinu, Strait, Lordsatri, Dennis Estenson II, HappyCamper, LjL, Bubba73, The wub, Yamamoto Ichiro, JohnDBuell, FayssalF, ColinJF, Wragge, Windchaser, Musical Linguist, Mindloss, RexNL, Gurch, Pete.Hurd, Lmatt, Diza, Zayani, Spencerk, Chobot, Sharkface217, DVdm, Hmonroe, Bgwhite, Ptah~enwiki, Ugha, Wavelength, Hillman, StuffOfInterest, Phantomsteve, Arado, John Smith's, Zigamorph, SpuriousQ, Jobe457, Stephenb, Cambridge-BayWeather, Rsrikanth05, Vibritannia, Neilbeach, Salsb, Big Brother 1984, Anomalocaris, NawlinWiki, Joncolvin, ErkDemon, Trovatore, ETTan, Schrei, THB, Syrthiss, Wknight94, Richardcavell, FF2010, CWenger, Kevin, Caco de vidro, Katieh5584, Banus, Sbyrnes321, Nark-straws, SmackBot, R.E. Freak, Kurochka, DuoDeathscyther 02, Bayardo, McGeddon, Delldot, Kintetsubuffalo, Portillo, Rmosler2100, Bluebot, Jjalexand, 7777777s, Silly rabbit, George Church, Colonies Chris, A. B., Calc rulz, Nicknitro71, Zsinj, TallyJoe, John Hyams, Jamse, Scott3, Jefffire, Serenity-Fr, Bilgrau, Avb, Rrburke, Addshore, Justin Stafford, DrL, Mr.LMNOP, Rassisi, Spanyard, Byelf2007, Nishkid64, Giovanni33, Soap, Cronholm144, Loadmaster, Stupid Corn, Benjaminlobato, FredrickS, SirFozzie, Waggers, Alexander Gieg, Gcavep, JMK, Abel Cavaşi, Newone, Courcelles, Tubezone, Esn, Dave Runger, Valoem, JRSpriggs, Kurtan~enwiki, 0-8, Duduong, Friendly Neighbour, CRGreathouse, Geremia, Tkoeppe, Ken Gallager, DepartedUser2, Cydebot, Vanished user 2340rujowierfj08234irjwfw4, Ninguém, Steel, Peterdjones, He-brides, David edwards, Michael C Price, Raoul NK, Wortzman, Ulnevets, Konradek, Mojo Hand, Raymond Feilner, Headbomb, Marek69, Inve40, Twcjr, Duncan McB, KrakatoaKatie, Luna Santin, Gdo01, Byrgenwulf, Myanw, Knotwork, Len Raymond, JAnDbot, Barek, MER-C, Txomin, Inks.LWC, Matthew Fennell, Instinct, MoralMajority, Promking, Bongwarrior, VoABot II, JamesBWatson, JBKramer, DAGwyn, Theroadislong, Lenschulwitz, 28421u2232nfenfcenc, Peatbog, Allstarecho, Fang 23, Spellmaster, Philg88, Peter J Schoen, Denis tarasov, Mart-inBot, R'n'B, JCarlos, J.delanoy, Pharaoh of the Wizards, Maurice Carbonaro, LordAnubisBOT, Pyrospirit, AntiSpamBot, NewEnglandYankee, DadaNeem, Cometstyles, WJBscribe, Foofighter20x, Econofire, Squids and Chips, Germanium, Reelrt, ChaosCon343, Danwills, RingtailedFox, Jeff G., TXiKiBoT, Nxavar, Rei-bot, Vishal144, IllaZilla, Pouya sh, Corvus cornix, Michael H 34, Martin451, Cheffoxx, Betanon, BotKung, Ev-erything counts, Popopp, MrMelonhead, Stephenmolesey, Andrewaskew, James McBride, Deanlsinclair, Pageman~enwiki, Monty845, Logan, Kpa4941, PaddyLeahy, Dogah, SieBot, Tiddly Tom, Robdunst, Wing gundam, Gammanon, Bentogoa, Likebox, Tiptoety, SteakNShake, Momo san, Freeman501, BartekChom, Monkeyspangler, Lightmouse, Anakin101, Divinestuff, Carbogen, Ayleuss, Soporaeternus, ArepoEn, ClueBot, LAX, Cliff, Ian the Aussie, Monomath1, Boing! said Zebedee, Heldbacktheband, LonelyBeacon, Neverquick, Excirial, WikiZorro, Tamara-trouts, Wndl42, Brews ohare, PhySusie, Morel, Mastertek, Mikaey, 7, Crowsnest, Thinking Stone, TimothyRias, PatDunphey, JKeck, XLinkBot, Bvssvni, Ougner, Truthnlove, YeAaMsLtA, Thatguyflint, Tayste, Balungifrancis, Addbot, Proofreader77, Some jerk on the Internet, Uruk2008, DOI bot, Couchie, Johnchang6868, Discrepancy, Mjamja, Bobtron5000, Fluffernutter, KaityJoe, MrOllie, Favonian, Barak Sh, F Notebook, Tide rolls, Scientryst, WikiDreamer Bot, Meisam, Blah28948, Yobot, Finiter, Ptbotgourou, Ezequiels.90, Jgmoxness, Amble, Mirandamir, RDemelo, AnomieBOT, ^musaz, Girl Scout cookie, 9258fahsflkh917fas, Theunify, Anxfisa, Kanat Abildinov, Materialscientist, Citation bot, Subhajit Ganguly, Fleaman5000, Amareto2, Addihockey10, Smk65536, Mlpearc, GrouchoBot, Rwmeo, Omnipaedista, Shirik, RibotBOT, Fa.alt3r3g0, Fsdjfsdfk, Chaseroads, 卡卡, FrescoBot, Paine Ellsworth, Ribashka, Krj373, Steven Avraham Rosten, PhysicsExplorer, Ottokar~enwiki, Tank hasmukh Khimjibhai, Tank theorist of everything, Hasmukh Khimjibhai Tank, DivineAlpha, Citation bot 1, Gil987, Three887, Tom.Reding, A8UDI, NarSakSasLee, Casimir9999, AndrewGrieder, Aknochel, IVAN3MAN, SchreyP, Noel Edward, Natwatchmaker, Weedwhacker128, Suffusion of Yellow, Koozedine, RjwilmsiBot, Specal ops, Afteread, DASHBot, Golumbo, EmausBot, Ikerus, Katherine, Dewritech, RA0808, K6ka, Zero939, Thecheesykid, Hhhippo, CanonLawJunkie, Traxs7, Arbnos, SporkBot, AManWithNoPlan, DanielBurnstein, FinalRapture, Aatu Koskensilta, Staszek Lem, Sridattadev, M00se1989, Wiggles007, Andrushkkutza, Maschen, Vedoder, Donner60, GIAN PHIL, Davi-daedwards, WHF Christie, Terra Novus, Matevz91, Isocliff, Sanno89, Cgt, Will Beback Auto, ClueBot NG, Stein Sivertsen, ClaudeDes, Lord God Almighty, Hindustanilanguage, Helpful Pixie Bot, Nightingale.zj, B21O303V3941W42371, Bibcode Bot, Wiki13, Akashankitjain, Neu-tral current, Aranea Mortem, Stimulieconomy, Steven.w.kowalski, MathewTownsend, Flyerbri, GroupT, Megajakeroo, La marts boys, Zofo, LightandDark2000, Josepht404, Nickhwee, Davidyevgeny, Kingcircle, Vith Nix, Illuusio, Davidyevgenyroven, QuantumNico, Vladimir Leonov, Friek555, HesterShaw, Sol1, Phaedrx, Jwratner1, Jmassion, HeymynamesJon, Bigfootrobert, Elitousson, Mdsheraj, Kdmeaney, JaconaFrere, Somecdnguy4, 22merlin, Monkbot, LollyBear12, StacyPoyPie, Mujii loving, Mayojohns, Gronk Oz, Yoyosami, Hakan tomaşoğlu, Pfpguy, Cirk-sena, Svm sudhan, 39Debangshu, Quantalogos, GeneralizationsAreBad, KasparBot, Christos Theopoulos, Asterixf2, Patrickmantonio, Quack-riot, Sir Cumference, Soaring, Squido609, Simple abundance 1945, Umesh Mitra, YukinonKanade, Pitadvocate1178, Gaurdain8, Gaurdian8, Anonymous-Pres and Anonymous: 526

- **An Exceptionally Simple Theory of Everything** *Source:* https://en.wikipedia.org/wiki/An_Exceptionally_Simple_Theory_of_Everything? oldid=699423130 *Contributors:* Michael Hardy, TakuyaMurata, Karada, Ciphergoth, Mxn, Disdero, Dragons flight, Carlossuarez46, Nurg, Rur-sus, David Gerard, Giftlite, Mporter, Beefman, Oberiko, Alison, Mateuszica, Jossi, Tomruen, Urvabara, RossPatterson, Rich Farmbrough, R6144, David Schaich, D-Notice, Pavel Vozenilek, Oldsoul, Calton, Jheald, SteinbDJ, Eric Herboso, Firsfron, Mindmatrix, Mpatel, GregorB, Christopher Thomas, Bikeable, Rjwilmsi, Salix alba, Bubba73, Wragge, Ground Zero, John Baez, Fragglet, Diza, Chobot, Wavelength, JWB, StuffOfInterest, Markhoney, Bhny, SCZenz, Mccready, BOT-Superzerocool, Dv82matt, Georgewilliamherbert, Smurrayinchester, Giovan-nino~enwiki, Jdhedden, Luk, SmackBot, Bayardo, FlashSheridan, Silly rabbit, Colonies Chris, John Hyams, Writtenright, Addshore, Kendrick7, Ged UK, TenPoundHammer, Saluton~enwiki, Brent williams, Special-T, FredrickS, PRRfan, Alexander Gieg, Norm mit, Benplowman, Dave

Runger, JRSpriggs, Merzbow, Gmusser, Aglisi, Xxanthippe, Tdvance, DumbBOT, Adailton, Koeplinger, Qwyrxian, Headbomb, Itsmejudith, Igodard, PeterStJohn, SalvNaut, David Eppstein, WLU, TimidGuy, J.delanoy, Opaqueice, Christopher norton, Cuzkatzimhut, VolkovBot, John Darrow, JohnBlackburne, Plclark, Michael H 34, Everything counts, Antixt, AmigoNico, Gnocchi, Revent, Le Pied-bot~enwiki, Kildor, Navy.enthusiast, GiveItSomeThought, Zeyn1, MmmmJoel, ClueBot, EGetzler, Alexbot, Muhandes, Charles.O.Wilson, SockPuppetForTomruen, Mlaffs, 1776historybuffer, Deeds4u, XLinkBot, Addbot, DOI bot, Riyuky, Discrepancy, Debresser, Cesiumfrog, Scientryst, Yobot, Jgmoxness, AnomieBOT, ThaddeusB, USConsLib, Citation bot, The sock that should not be, Omnipaedista, Dstup, Waleswatcher, A Quest For Knowledge, WaysToEscape, Citation bot 1, Tom282f3~enwiki, Jonesey95, Tom.Reding, Standardfact, Miles1228, NerdyScienceDude, Afteread, DASHBot, Verbapple, Golumbo, EmausBot, Detogain, QuotScheme, ZéroBot, Totalbr, H3llBot, Quondum, Wiggles007, Helpful Pixie Bot, Eldora2440, Physicrocks, Dilaton, Nickhwee, Comp.arch, Spiral161, Monkbot, Chemistry1111 and Anonymous: 152

- **Mathematical universe hypothesis** *Source:* https://en.wikipedia.org/wiki/Mathematical_universe_hypothesis?oldid=717263046 *Contributors:* Zundark, Fubar Obfusco, William Avery, Michael Hardy, Sverdrup, Dbachmann, Bender235, Jhertel, DanielVallstrom, BRW, Woohookitty, GregorB, Rjwilmsi, MarSch, Dianelos, Jfraatz, Hillman, Salsb, Joncolvin, Banus, Bluebot, Lambiam, Dark Formal, John H, Morgan, Billgunn, George100, Mbell, Headbomb, Parsiferon, Tim333, Magioladitis, NerdyNSK, Maxim, Lamro, Martarius, Silent Key, XLinkBot, MystBot, Addbot, Discrepancy, Favonian, Luckas-bot, Yobot, AnomieBOT, Omnipaedista, IO Device, Machine Elf 1735, Randalliser, Kartasto, Bibcode Bot, BG19bot, Erik.Bjareholt, John Aiello, Comp.arch, Sol1, Jwratner1, 22merlin, Monkbot, Frettled Gruntbuggly, HouseOfChange, Quantalogos, Baking Soda, Hyperforin and Anonymous: 44

- **Compactification (physics)** *Source:* https://en.wikipedia.org/wiki/Compactification_(physics)?oldid=696134739 *Contributors:* The Anome, Michael Hardy, Charles Matthews, Mpatel, Eyu100, Eubot, Salsb, SmackBot, Ben Jos, Noah Salzman, JarahE, Dan Gluck, Headbomb, Isilanes, AlleborgoBot, Ozooxo, AnonyScientist, AlexGWU, Addbot, Luckas-bot, Wireader, Dogbert66, EmausBot, ZéroBot, Chemistry1111 and Anonymous: 9

- **Extra dimensions** *Source:* https://en.wikipedia.org/wiki/Extra_dimensions?oldid=705556627 *Contributors:* Michael Hardy, Lumidek, Jon Awbrey, A876, Mojo Hand, R'n'B, Mild Bill Hiccup, Addbot, Davdde, 🔲🔲, Dadonene89, ZéroBot, D.Lazard, Invadibot, SoledadKabocha, CAPTAIN RAJU and Anonymous: 2

- **Quantum gravity** *Source:* https://en.wikipedia.org/wiki/Quantum_gravity?oldid=715870969 *Contributors:* AstroNomer, Matusz, Miguel~enwiki, Roadrunner, Stevertigo, Ubiquity, Bobby D. Bryant, Mcarling, NuclearWinner, Anders Feder, Susurrus, Coren, Charles Matthews, Timwi, Reddi, Tpbradbury, Phys, Bevo, Raul654, BenRG, Frazzydee, Jeffq, Sdedeo, Rholton, Wereon, Ilya (usurped), Seth Ilys, Ancheta Wis, Giftlite, Herbee, Fropuff, Endlessnameless, Malyctenar, Jason Quinn, Finn-Zoltan, YapaTi~enwiki, Lumidek, Marcus2, Joyous!, TJSwoboda, Vitaleyes, Davidclifford, JimJast, Rich Farmbrough, Guanabot, FT2, Masudr, Pjacobi, Pie4all88, David Schaich, Bender235, Clement Cherlin, El C, PhilHibbs, Army1987, Apyule, VBGFscJUn3, PWilkinson, Daniel Arteaga~enwiki, Keenan Pepper, Cjthellama, DonJStevens, Velella, Dabbler, Tycho, Cal 1234, RJFJR, Count Iblis, ThomasWinwood, Anarchimede, Scarykitty, Woohookitty, Igny, ToddFincannon, Mpatel, GregorB, Joke137, Christopher Thomas, Marudubshinki, Graham87, Yurik, Kroggz, Rjwilmsi, Eoghanacht, Jrasowsky, JHMM13, Smithfarm, Ems57fcva, FayssalF, Itinerant1, Lmatt, Chobot, Hmonroe, YurikBot, Hillman, ErkDemon, JocK, SCZenz, Roy Brumback, Bota47, Zunaid, JonathanD, 2over0, Arthur Rubin, Modify, LeonardoRob0t, Caco de vidro, RG2, KasugaHuang, Resolute, SmackBot, Samdutton, Vald, Eskimbot, Hbackman, Onebravemonkey, Gilliam, Chris the speller, Ben.c.roberts, Cthuljew, Silly rabbit, Complexica, Colonies Chris, QFT, Soosed, Theanphibian, Shushruth, Ck lostsword, Yevgeny Kats, DJIndica, Lambiam, Mike1901, Vampus, Vincenzo.romano, Jaganath, JorisvS, Bjankuloski06, RoboDick~enwiki, IronGargoyle, Dicklyon, SirFozzie, Treyp, Twunchy, Iridescent, Piccor, Kurtan~enwiki, Harold f, CalebNoble, Duduong, Paulmlieberman, TVC 15, Phatom87, UncleBubba, TAz69x, Sam Staton, ST47, B, Patrick O'Leary, Epbr123, Koeplinger, Klasovsky, Markus Pössel, Keraunos, Headbomb, Marek69, MichaelMaggs, Tim Shuba, MER-C, ParadiZio, Clementvidal, Perlygatekeeper, VoABot II, Alvatros~enwiki, Bdalevin, SHCarter, Jpod2, DAGwyn, Nucleophilic, LorenzoB, Rickard Vogelberg, DancingPenguin, Rettetast, Victor Blacus, AstroHurricane001, Yonidebot, Acalamari, Mstuomel, Fullmetal2887, NewEnglandYankee, DorganBot, CardinalDan, Idioma-bot, Sheliak, VolkovBot, Pleasantville, Seattle Skier, AlnoktaBOT, TXiKiBoT, Dllahr, Rdekleer, Saibod, Cyberchip, Wikiwikimoore, Carlorovelli, LoreMiles, StevenJohnston, SieBot, LeadSongDog, Bentogoa, Coldcreation, ReluctantPhilosopher, StaticG, GarbagEcol, ClueBot, The Thing That Should Not Be, EoGuy, Polyamorph, Andwor9, Notburnt, Tms9, Alexbot, Resoru, Eeekster, Tamaratrouts, Brews ohare, SchreiberBike, Askahrc, BOTarate, Lambtron, DumZiBoT, XLinkBot, Rror, Facts707, SilvonenBot, Theonlydavewilliams, Mhsb, Truthnlove, Ttimespan, Trifonov~enwiki, Addbot, Mortense, Grayfell, Eric Drexler, Gravitophoton, DOI bot, AkhtaBot, CanadianLinuxUser, Frosty726, LaaknorBot, Delaszk, Tassedethe, Tide rolls, Taketa, Titan1129, Krano, Luckas-bot, Yobot, WikiDan61, Pigetrational, Wireader, Allowgolf~enwiki, Wiki Roxor, Jim1138, IRP, Sz-iwbot, Quantity, Materialscientist, Citation bot, ArthurBot, LilHelpa, Amareto2, Ekwos, KrisBogdanov, Rolfguthmann, StealthCopyEditor, 🔲🔲, Dan6hell66, Rabsmith, Hep thinker, Paine Ellsworth, DrArthurRubinPHD, Lagelspeil, Nunc aut numquam, Vacuunaut, Van Speijk, Knowandgive, Craig Pemberton, Udifuchs, Citation bot 2, Citation bot 1, Citation bot 4, Jonesey95, Hirvenkürpa, Tom.Reding, Pmokeefe, Serols, Casimir9999, Dac04, Dude1818, Valeriy Pischenko, Follyland, TrueTeargem, N0814444, Earthandmoon, Korepin, DARTH SIDIOUS 2, Musictime4me, RjwilmsiBot, EmausBot, Francophile124, Octaazacubane, Fotoni, Slightsmile, Garfield Salazar, Hhhippo, JSquish, John Cline, Fæ, LostAlone, Brazmyth, Throwmeaway, Arbnos, Ebrambot, Kusername, DanielBurnstein, TonyMath, L Kensington, Maschen, Donner60, Parusaro, Apratim07, Terra Novus, Isocliff, Googledin!, ClueBot NG, SpikeTorontoRCP, Science writer, Preon, Raidr, Jhmmok, 336, Widr, Helpful Pixie Bot, Bibcode Bot, BG19bot, Bardsley Rides a Segway, Apelikedawg, FiveColourMap, Trevayne08, Mr.viktor.stepanov, Brainssturm, BattyBot, Jimw338, Ryanr666, Kryomaxim, Garuda0001, CuriousMind01, Saehry, TwoTwoHello, Sanathdevalapurkar, Andyhowlett, GabeIglesia, Sanathlab, Roiwallace, Spencer.mccormick, Spencerfjase, MrShlongNo1, Marc D. Garrett, D00d00ballz, Gigantmozg, Susan.grayeff, Polytope24, Frinthruit, Anrnusna, Dfyytj, Monkbot, Umut Alihan Dikel, Negative24, Amortias, Klj1234, Pfpguy, Egarcitenre, KasparBot, Christos Theopoulos, Schidan, Jespergrimstrup, DiscreteEditor, Peter SamFan, Xerxeese and Anonymous: 306

- **String (physics)** *Source:* https://en.wikipedia.org/wiki/String_(physics)?oldid=717215160 *Contributors:* Andres, Wereon, Fropuff, El C, Mpatel, Gwernol, Roboto de Ajvol, KnightRider~enwiki, Scwlong, Fiziker, EPM, ServAce85, Drewbarfield, TriTertButoxy, Astrobradley, Dan Gluck, Markjoseph125, Epbr123, Headbomb, The Radio Star, Hempfel, B-80, Qwerty Binary, Maurice Carbonaro, Andrius.v, PhysPhD, Anchor Link Bot, ClueBot, Addbot, Allowgolf~enwiki, Pcb95, FrescoBot, Fisuaq, Mathewmathewmeixnermeixner, Frietjes, Widr, Hansan29, Davida98, Polytope24 and Anonymous: 26

- **Type I string theory** *Source:* https://en.wikipedia.org/wiki/Type_I_string_theory?oldid=629001015 *Contributors:* DJ Clayworth, Fropuff, Moyogo, Lumidek, Brian0918, Velella, Mpatel, EricCHill, Reyk, Sardanaphalus, SmackBot, Unyoyega, Colonies Chris, Jmnbatista, LordAnubisBOT, MarkJefferys, Idioma-bot, Sheliak, Thomas.W, Sagnotti, Addbot, Lightbot, Materialscientist, Omnipaedista, Erik9bot, Calmer Waters, EmausBot, ClueBot NG, BG19bot, Frze, Hmainsbot1, Zacht.carnevale, Dimension10, Polytope24 and Anonymous: 6

- **Type II string theory** *Source:* https://en.wikipedia.org/wiki/Type_II_string_theory?oldid=629086741 *Contributors:* Stevertigo, Fropuff, Christopherlin, Lumidek, Jag123, MarSch, Mgnbar, Reyk, Sardanaphalus, SmackBot, Unyoyega, Colonies Chris, Jmnbatista, Joshua Davis, LordAnubisBOT, Idioma-bot, Sheliak, OKBot, Jovianeye, MystBot, Addbot, AkhtaBot, Yobot, Omnipaedista, Erik9bot, FrescoBot, Emaus-Bot, 336, Luizpuodzius, Dimension10, Polytope24, Luca.agozzino and Anonymous: 5

- **Heterotic string theory** *Source:* https://en.wikipedia.org/wiki/Heterotic_string_theory?oldid=640076972 *Contributors:* Charles Matthews, Denni, Hugo~enwiki, Giftlite, Fropuff, Lumidek, Rich Farmbrough, Pearle, NTK, Mpatel, Chobot, YurikBot, Bhny, Hwasungmars, 2over0, Sardanaphalus, KnightRider~enwiki, SmackBot, Schmiteye, Fplay, QFT, Jmnbatista, Headbomb, Lamontacranston, STBot, MarkJefferys, She-liak, JohnBlackburne, Spiral5800, Legoktm, TimothyRias, MystBot, Addbot, Debresser, Tassedethe, Lightbot, OlEnglish, Jack who built the house, AnomieBOT, Citation bot, Omnipaedista, Erik9bot, Steve Quinn, Orenburg1, EmausBot, WaterfordPBR, Staszek Lem, ClueBot NG, BG19bot, AHusain314, I am One of Many, Dimension10, Polytope24 and Anonymous: 31

- **Orthogonal group** *Source:* https://en.wikipedia.org/wiki/Orthogonal_group?oldid=709475729 *Contributors:* AxelBoldt, Zundark, The Anome, Patrick, Michael Hardy, TakuyaMurata, Looxix~enwiki, Stevan White, Loren Rosen, Charles Matthews, Phys, MathMartin, Softcafe, Weialawaga~enwiki, Giftlite, BenFrantzDale, Fropuff, Dratman, Chadernook, Vivacissamamente, Gauge, Pt, Kwamikagami, Iyerkri, Msh210, Oleg Alexandrov, Guardian of Light, BD2412, Salix alba, HappyCamper, R.e.b., Masnevets, Algebraist, YurikBot, Wavelength, Shell Kinney, Crasshopper, Paul D. Anderson, KnightRider~enwiki, SmackBot, Incnis Mrsi, Unyoyega, Renamed user 1, Bluebot, Nbarth, Tamfang, Lambiam, Ulner, Jim.belk, TooMuchMath, Technohead1980, Bruno321, Tawkerbot2, CmdrObot, Thijs!bot, Marek69, Nick Number, Magioladitis, Schm-loof, Policron, JohnBlackburne, Wolfrock, Drschawrz, Paolo.dL, Thehotelambush, JerroldPease-Atlanta, JackSchmidt, Anchor Link Bot, Mr. Stradivarius, XLinkBot, Addbot, Morriswa, Zorrobot, Ettrig, Yobot, Niout, Mathiscool, Noideta, Xqbot, Citation bot 1, DreamingInRed~enwiki, Woona, John of Reading, Superlaser1, Slawekb, Somethingcompletelydifferent, Quondum, D.Lazard, Zephyrus Tavvier, Jjenkins5123, Help-ful Pixie Bot, BG19bot, Solomon7968, Deltahedron, Mark L MacDonald, Makecat-bot, Spectral sequence, Umberto Lupo, CsDix, Cyrapas, Promise her a definition, Zimboras, Mathphysman, Dhm4444, WillemienH and Anonymous: 57

- **E8 (mathematics)** *Source:* https://en.wikipedia.org/wiki/E8_(mathematics)?oldid=709000616 *Contributors:* Zundark, Frecklefoot, Michael Hardy, Gabbe, Angela, Charles Matthews, Phys, Jeffq, Gwrede, Nurg, Mattflaschen, Tea2min, Giftlite, Cobaltbluetony, Fropuff, Gro-Tsen, Ja-son Quinn, Tagishsimon, Tomruen, Pmanderson, Lumidek, Rich Farmbrough, Bender235, Oldsoul, RoyBoy, John Vandenberg, Scott Ritchie, Reubot, Anthony Appleyard, RJFJR, Oleg Alexandrov, Jeffrey O. Gustafson, Ae-a, Lhademmor, Rjwilmsi, Salix alba, Gareth McCaughan, HappyCamper, R.e.b., Marcol, Bubba73, John Baez, Mathbot, Ayla, Bgwhite, JWB, Dmharvey, Michael Slone, AVM, KSmrq, Wimt, JocK, Ravedave, Romanempire, Crasshopper, A bit iffy, SmackBot, Croberts, MerlinMM, Master of Puppets, Nbarth, Jahiegel, Davidarichter, Van-ished User 0001, Nexxuz, Cronholm144, Fuzzy510, Trounce, CUTKD, Dan Gluck, Strazys, JRSpriggs, CRGreathouse, Aussiepete, Michael C Price, Adailton, Headbomb, Marek69, OrenBochman, Sherbrooke, Escarbot, Yellowdesk, Bundas, Emax0, Tengfred, Vudicarus, Exceptg, David Eppstein, STBot, Rettetast, Rocchini, TomyDuby, Mstuomel, Joshua Issac, DavidCBryant, Remember the dot, Phoenix1304, GregWoodhouse, Wilmot1, Morenooso, JohnBlackburne, Ambrose H. Field, Fences and windows, Cashmundy, Leafyplant, ZELBOG~enwiki, PaulTanenbaum, Enigmaman, Lamro, Poltair, 7hobo, Arcfrk, Canavalia, Drschawrz, JackSchmidt, Navy.enthusiast, Jruderman, Sfan00 IMG, Jlrodri, Nilradi-cal, SoxBot, Revotfel, Farnite1, XLinkBot, Jimmyknauer, Addbot, Discrepancy, Ersik, SpBot, Yobot, Niout, Amirobot, Jgmoxness, Ewsnow, Citation bot, Twri, Junidhar, Maxwell helper, Ben.renaud, Double sharp, Trappist the monk, Tition1, Chronulator, Mean as custard, The tree stump, Afteread, ChuispastonBot, Mikhail Ryazanov, ClueBot NG, MelbourneStar, Helpful Pixie Bot, Dilaton, Tonyxc600, Tupags, Mark L MacDonald, Jochen Burghardt, Hooman khodayari, Slomo Shapiro, CsDix, Master Lenman, Furjo, Expendable123, Monkbot, Jcmckeown, Jerry08baddog, Chemistry1111 and Anonymous: 110

- **Spacetime** *Source:* https://en.wikipedia.org/wiki/Spacetime?oldid=717992586 *Contributors:* Paul Drye, The Cunctator, Dreamyshade, Bryan Derksen, Malcolm Farmer, Josh Grosse, XJaM, Karl Palmen, Stevertigo, Patrick, Infrogmation, Smelialichu, Michael Hardy, Wshun, Pit~enwiki, Dcljr, Karada, Mcarling, Looxix~enwiki, William M. Connolley, Snoyes, Kingturtle, Glenn, Loren Rosen, HolIgor, Adam Bishop, Dcoetzee, Reddi, Jay, E23~enwiki, Omegatron, Fvw, Robbot, Kristof vt, Goethean, Ashley Y, Sverdrup, Blainster, DHN, Papadopc, Tea2min, Fin-lander, Matt Gies, Giftlite, ByteCoder, Wolfkeeper, Herbee, Tom Radulovich, Everyking, Snowdog, Michael Devore, Niteowlneils, Yekrats, Eequor, Utcursch, Beland, Karol Langner, Wikimol, JimWae, Karl-Henner, Adashiel, ELApro, Chris Howard, Juan Ponderas, Discospinster, Rich Farmbrough, Cacycle, Ascánder, Dolda2000, Bender235, Ben Standeven, El C, Rgdboer, Lankiveil, Shoujun, Teorth, Che090572, Rbj, Tobacman, I9Q79oL78KiL0QTFHgyc, Como, Obradovic Goran, Free Bear, Keenan Pepper, Sourcer66~enwiki, Riana, Geoff-codes, Rey-Brujo, Arag0rn, DonQuixote, Eddie Dealtry, DominicC13, H2g2bob, Loxley~enwiki, Camw, StradivariusTV, TheNightFly, Pkeck, ^demon, Doran, Jeff3000, Mpatel, GregorB, Palica, Graham87, Deltabeignet, Li-sung, Mkn1234, MekaD, Rjwilmsi, KYPark, Kinu, Vary, MarSch, FayssalF, Lebha, Mathbot, Alexjohnc3, Jrtayloriv, Exelban, Pete.Hurd, Tardis, Chobot, Tene, DVdm, VolatileChemical, YurikBot, Wavelength, Splintercellguy, Wolfmankurd, CanadianCaesar, Yamara, NawlinWiki, Mipadi, Trovatore, Schlafly, JocK, Crasshopper, Tony1, T, Zythe, Gad-get850, Sahands, Light current, Zzuuzz, StuRat, KGasso, JoanneB, Heathhunnicutt, Anclation~enwiki, RG2, Teply, Mejor Los Indios, Qero, Eigenlambda, Sardanaphalus, SmackBot, RDBury, Formativ, Maksim-e~enwiki, Forteller~enwiki, RaulMiller, Ashill, Kurochka, Lestrade, In-verseHypercube, KnowledgeOfSelf, C.Fred, AndreasJS, Jaytan, Alex earlier account, JeffieAlex, Yamaguchi⬚⬚, Gilliam, NickGarvey, JMiall, Oli Filth, TheScurvyEye, Silly rabbit, Complexica, Dabigkid, Jerome Charles Potts, Nbarth, Sbharris, Bryan Truitt, Can't sleep, clown will eat me, Tamfang, Chlewbot, Rrburke, Celarnor, Tsop, CanDo, Dylanrush, RaCha'ar, Mtmelendez, Looris, Richard001, Hammer1980, Roman-ski, Sayden, Kuru, MagnaMopus, Hernoor, Homan2006, 16@r, Loadmaster, Lampman, Hypnosifl, Ace Frahm, Inquisitus, FVP, Shoeofdeath, Newone, Yourstruly, Andrew Hampe, Lxl, Aeons, Xammer, Paolodm, CalebNoble, Robinhw, JForget, Twipie, Blve23, Jsd, Jnoa, WeggeBot, Myasuda, Azakreski, Joshua BishopRoby, Cydebot, AniMate, Kanags, Fl, MC10, Llort, Eu.stefan, Palindromica, Manfroze, DarkLink, Amelio-rate!, DBaba, TarquiniusWikipedius, Kylewriter, Raoul NK, Letranova, Thijs!bot, Wikid77, Jedibob5, HappyInGeneral, Gamer007, Headbomb, Vertium, RolanGaros, Pigalle, Washingtonlerias, Ubuthustra, D.H, Nick Number, Klausness, Sam42, DarthNemesis, Northumbrian, Escarbot, WikiSlasher, AntiVandalBot, Seaphoto, Maxibons, Tim Shuba, Braindrain0000, Tempest115, Jrw@pobox.com, Narssarssuaq, Husond, MER-C, Andrewdolby, RogierBrussee, Bongwarrior, VoABot II, Bakken, Appraiser, Faizhaider, Cuardin, Stijn Vermeeren, Trebor1, Catgut, Cardamon, NMarkRoberts, IkonicDeath, MetsBot, Mwasim1, JaGa, GuelphGryphon98, NatureA16, FisherQueen, Flowanda, MartinBot, TechnoFaye, Wikeepeedier, Player 03, Tgeairn, HEL, J.delanoy, Bobvinson, Maurice Carbonaro, Foober, 3halfinchfloppy, Lantonov, NewEnglandYankee, LeighvsOptimvsMaximvs, KylieTastic, Ja 62, Vinsfan368, Izno, Idioma-bot, Makewater, 28bytes, VolkovBot, XCelam, JohnBlackburne, Al-

ClueBot, Seervoitek, Rodhullandemu, Jorisverbiest, Feebas factor, ChandlerMapBot, Nilradical, Wikeepedian, Stephen Poppitt, Addbot, Vectorboson, Luckas-bot, Yobot, Planlips, Dickdock, AnomieBOT, Icalanise, Materialscientist, Xqbot, Br77rino, Balaonair, ⁇⁇, Paine Ellsworth, Blackoutjack, Kikeku, Rameshngbot, Tom.Reding, RedBot, Alarichus, Michael9422, Silicon-28, TjBot, EmausBot, WikitanvirBot, Quazar121, Solomonfromfinland, JSquish, Fimin, Quondum, AManWithNoPlan, EdoBot, ClueBot NG, PBot1, EthanChant, Bibcode Bot, BG19bot, Petermahlzahn, KingKhan85, ChrisGualtieri, BoethiusUK, DerekWinters, Tentinator, JNrgbKLM, Mohit rajpal, KasparBot, Jiswin1992, Even This Is Taken, Wulframm, Chemistry1111 and Anonymous: 121

- **Kaluza–Klein theory** *Source:* https://en.wikipedia.org/wiki/Kaluza%E2%80%93Klein_theory?oldid=717951430 *Contributors:* Sodium, The Anome, XJaM, Roadrunner, Rlee0001, Stevertigo, JohnOwens, Michael Hardy, Looxix~enwiki, Ahoerstemeier, Susurrus, Smack, Charles Matthews, Timwi, Reddi, Wik, Phys, Carbuncle, Ancheta Wis, Kim Bruning, Fropuff, Mdob, Iantresman, Rauyran, Brianhe, Rich Farmbrough, Roo72, Ponder, Paul August, Bender235, Szquirrel, John Vandenberg, I9Q79oL78KiL0QTFHgyc, Geschichte, RJFJR, Notjim, Linas, Lgallindo, Trapolator, Mpatel, Joke137, Rjwilmsi, MarSch, GünniX, YurikBot, Rt66lt, Hillman, Geologician, Buster79, Gillis, Kewp, Petri Krohn, Pred, Caco de vidro, KasugaHuang, Jodarom, Kurochka, Nihonjoe, Zazaban, Unyoyega, Hmains, Chris the speller, MalafayaBot, Colonies Chris, QFT, Legaleagle86, Jon Awbrey, Beetstra, DabMachine, Rschwieb, FrEd 00, Tawkerbot2, CmdrObot, Mattbr, Wfdavis, Moyerjax, MaxEnt, Epbr123, Mojo Hand, Headbomb, JustAGal, Isilanes, Golf Bravo, Magioladitis, JoseAntonioOrtegaRuiz, Jpod2, Mbc362, Maliz, Cpiral, TomyDuby, Quantling, Lseixas, Sheliak, Cuzkatzimhut, Red Act, Impunv, Thomas.schick, AlleborgoBot, YohanN7, ArdClose, EoGuy, Mild Bill Hiccup, Masterpiece2000, Canis Lupus, EverettYou, AnonyScientist, Albambot, Addbot, Gravitophoton, Protonk, Lightbot, Matěj Grabovský, Yobot, Turul2, Jo3sampl, Citation bot, Omnipaedista, RibotBOT, Paine Ellsworth, Quiden711, Tom.Reding, RockSolidCosmo, Crabhiggins, Bj norge, David.c.stone, Arbnos, Quondum, TonyMath, Helpful Pixie Bot, Bibcode Bot, BG19bot, Zerothat, Ownedroad9, Metsfreak2121, MSUGRA, Mogism, Lianatajo, MuonRay, Orderofmagnitudeapproximation, Frinthruit, Monkbot, ManitouLance, Claudio Orzalesi and Anonymous: 86

- **N = 4 supersymmetric Yang–Mills theory** *Source:* https://en.wikipedia.org/wiki/N_%3D_4_supersymmetric_Yang%E2%80%93Mills_theory?oldid=718152767 *Contributors:* The Anome, Michael Hardy, Bearcat, Tomruen, Hydrox, David Schaich, Jonathanischoice, Rjwilmsi, Bgwhite, Malcolma, Myasuda, Yellowdesk, Tonyfaull, Connor Behan, Drschawrz, Yobot, AnomieBOT, Omnipaedista, FrescoBot, Banej, Maschen, Zfeinst, Knook, Bibcode Bot, BattyBot, Khazar2, Enyokoyama, AHusain314, Randrian, Polytope24, Monkbot and Anonymous: 14

- **Coupling constant** *Source:* https://en.wikipedia.org/wiki/Coupling_constant?oldid=705485117 *Contributors:* CYD, The Anome, Michael Hardy, SebastianHelm, Phys, Wereon, Giftlite, Xerxes314, Alison, Macrakis, Wmahan, Doshell, Karol Langner, Lumidek, Dmr2, Aranel, Cmdrjameson, Jag123, Lonesoldier, Ricky81682, Oleg Alexandrov, SeventyThree, Zbxgscqf, TheRingess, R.e.b., Erkcan, Goudzovski, Ahpook, Bambaiah, Archelon, Cheesus, Spike Wilbury, SmackBot, Eskimbot, Bluebot, Ardm2, QFT, TriTertButoxy, Lambiam, JorisvS, 137 0, Whatever1111, Headbomb, Shambolic Entity, Kxxchen, Buurma, .anacondabot, Grimlock, Jpod2, Connor Behan, STBotD, TXiKiBoT, Venny85, Robsalmond, DragonBot, Truthnlove, Addbot, Qmark42, Luckas-bot, Yobot, AnomieBOT, GrouchoBot, Omnipaedista, Miracle Pen, EmausBot, Maschen, Zueignung, Chris.vonnegut, Dhm4444, Feon, Dhm44444 and Anonymous: 30

- **Montonen–Olive duality** *Source:* https://en.wikipedia.org/wiki/Montonen%E2%80%93Olive_duality?oldid=639115740 *Contributors:* Zundark, Michael Hardy, Charles Matthews, Phys, Giftlite, Fropuff, Lumidek, RussBot, Conscious, Mgary, David Cherney, Legobot, Yobot, Omnipaedista and Anonymous: 3

- **Supermembranes** *Source:* https://en.wikipedia.org/wiki/Supermembranes?oldid=712555963 *Contributors:* Berek, RomanSpa, Headbomb, Nick Number, Drschawrz, LilHelpa, FrescoBot and Anonymous: 4

- **Matrix theory (physics)** *Source:* https://en.wikipedia.org/wiki/Matrix_theory_(physics)?oldid=709314232 *Contributors:* Mav, RadicalBender, Ancheta Wis, Fropuff, Lumidek, Adashiel, FlaBot, QFT, Radagast83, Alphachimpbot, Stephen Shenker, Bibcode Bot, BG19bot, Polytope24 and Anonymous: 7

- **Noncommutative geometry** *Source:* https://en.wikipedia.org/wiki/Noncommutative_geometry?oldid=715533318 *Contributors:* Youssefsan, Michael Hardy, TakuyaMurata, GTBacchus, William M. Connolley, Charles Matthews, Phys, Tea2min, Giftlite, Lupin, Jrdioko, Tristanreid, DefLog~enwiki, CSTAR, Chris Howard, JonL, Paul August, Gauge, Count Iblis, Ceyockey, Oleg Alexandrov, Japanese Searobin, Linas, -Ril-, Mpatel, Triddle, GregorB, Rjwilmsi, R.e.b., John Z, Bmicomp, Siddhant, RussBot, Michael Slone, Shell Kinney, Yserarau, Crasshopper, JonathanD, Caco de vidro, SmackBot, Unyoyega, Zoran.skoda, Scwlong, CRGreathouse, Ntsimp, Thijs!bot, Konradek, Headbomb, Magioladitis, Taborgate, Lantonov, Cuzkatzimhut, Schucker, JohnBlackburne, LokiClock, Henry Delforn (old), Sphilbrick, Alteration×10, Mild Bill Hiccup, DragonBot, ProfessorTarantoga, Pqnelson, Addbot, Krampma, Yobot, Wireader, Felipe Gonçalves Assis, 9258fahsflkh917fas, Xqbot, Omnipaedista, Gsard, Tonyxty, Stephan Spahn, ZéroBot, Booqorm, Anselrill, Davidaedwards, Helpful Pixie Bot, Bibcode Bot, BG19bot, Solomon7968, Brad7777, Solmoss, Qtom.masters, Minimalrho, Deltahedron, Enyokoyama, Mark viking, Foredit, Yonathanyeremy and Anonymous: 41

- **AdS/CFT correspondence** *Source:* https://en.wikipedia.org/wiki/AdS/CFT_correspondence?oldid=704753554 *Contributors:* Bryan Derksen, The Anome, Stevertigo, Michael Hardy, Charles Matthews, Grendelkhan, Phys, COGDEN, Giftlite, Mporter, LeYaYa, Fropuff, Michael Devore, Eequor, Aside, Quadell, Tomruen, Lumidek, Rich Farmbrough, Cacycle, Pjacobi, Schuetzm, David Schaich, Dmr2, Jag123, Physicistjedi, Merlinme, Mpatel, Jugger90, BD2412, Rjwilmsi, FlaBot, Bgwhite, YurikBot, Salsb, Grafen, Hwasungmars, Psu256, Knotnic, Sardanaphalus, Honza Záruba, GaeusOctavius, Silly rabbit, Colonies Chris, QFT, John, JorisvS, Mr Stephen, Dan Gluck, Metre01, Bobamnertiopsis, Gebrah, Thijs!bot, Mbell, Headbomb, Landscape~enwiki, JAnDbot, Magioladitis, WolfmanSF, CommonsDelinker, TheSeven, Sigmundur, TXiKiBoT, Hqb, Piperh, Lamro, Drschawrz, CodeTalker, Elassint, Ozooxo, Piledhigheranddeeper, Kville105125, Mastertek, Another Believer, AnonyScientist, Addbot, Tassedethe, Lightbot, SPat, AnomieBOT, Citation bot, PowerUserPCDude, Pra1998, Eugene-elgato, Cannolis, Vhsatheeshkumar, Meier99, Nilock, RjwilmsiBot, GoingBatty, Outriggr, Serketan, Werieth, StringTheory11, SporkBot, AManWithNoPlan, YtivarG, Isocliff, ClueBot NG, Raidr, Bibcode Bot, BG19bot, QuarkyPi, Hmainsbot1, AHusain314, BronzeRatio, Mark viking, VoxelBot, Liquidityinsta, TFA Protector Bot, Dimension10, Polytope24, DThung, Username123454321, DvtheDv, Sileby10, Barbcash, Kismit33, Janaabul and Anonymous: 66

- **Quasiparticle** *Source:* https://en.wikipedia.org/wiki/Quasiparticle?oldid=715462380 *Contributors:* CYD, Glenn, Charles Matthews, Phys, Topbanana, Lumos3, Donarreiskoffer, Giftlite, Dratman, Jason Quinn, Sysin, Lumidek, Vsmith, RoyBoy, Lysdexia, Gene Nygaard, Wafry, Rjwilmsi, Arnero, Rune.welsh, Srleffler, Chobot, YurikBot, Wavelength, Shaddack, Sbyrnes321, SmackBot, Stepa, Chris the speller, Complexica, UNV, Vanished user 9i39j3, Euchiasmus, JorisvS, Brienanni, Iridescent, WilliamDParker, CmdrObot, Cydebot, Mbell, Hazmat2, Headbomb, BehnamFarid, Widefox, James Slezak, Yellowdesk, Michael.j.sykora, David Eppstein, Stevvers, Connor Behan, Gernewvic, Venny85, Antixt, AlleborgoBot, Kbrose, Goeie, ClueBot, Mr Accountable, Alexbot, Addbot, Mathieu Perrin, SPat, Luckas-bot, Yobot, Genius.scholar, LilHelpa, ChristopherKingChemist, Doraemonpaul, Freddy78, RedAcer, Citation bot 1, Trappist the monk, Logiolgeirss, ZéroBot, StringTheory11, Diego Grez

Bot, Tls60, RockMagnetist, ClueBot NG, PBot1, Asi013, Helpful Pixie Bot, Bibcode Bot, Rolancito, F=q(E+v^B), Shaginyan, Mark viking, LeoKadanoff, Susan.grayeff and Anonymous: 50

- **Vacuum solution** *Source:* https://en.wikipedia.org/wiki/Vacuum_solution?oldid=679362162 *Contributors:* Charles Matthews, Mpatel, Rjwilmsi, Hillman, Sandstein, Colonies Chris, Headbomb, Henry Delforn (old), Biggerj1, Addbot, Citation bot 1, Skyerise, Bibcode Bot and Luizpuodzius

- **Einstein field equations** *Source:* https://en.wikipedia.org/wiki/Einstein_field_equations?oldid=715836363 *Contributors:* XJaM, Boud, Michael Hardy, Ahoerstemeier, Mxn, Charles Matthews, Tpbradbury, Phys, Wtrmute, Raul654, Nnh, Chris 73, Altenmann, Lowellian, Wereon, Jheise, Cyrius, Ancheta Wis, Giftlite, Christopher Parham, Lethe, Ausir, Fropuff, DefLog~enwiki, Beland, Karol Langner, Anythingy-ouwant, Spoirier~enwiki, AmarChandra, D6, Hydrox, ThomasK, Bender235, Pt, I9Q79oL78KiL0QTFHgyc, Guiltyspark, Larryv, Krellis, Daniel Arteaga~enwiki, Alansohn, Bootstoots, Wtmitchell, Allen McC.~enwiki, Gene Nygaard, Nightstallion, Oleg Alexandrov, Simetrical, Woohookitty, Oliphaunt, Mpatel, Joke137, Abd, Rjwilmsi, Zbxgscqf, Ems57fcva, Sango123, Lionelbrits, Itinerant1, Meeve, BradBeattie, Chobot, DVdm, Roboto de Ajvol, YurikBot, Kungfuadam, KasugaHuang, Sardanaphalus, SmackBot, Incnis Mrsi, Nickst, Jab843, Gilliam, Skizzik, Silly rabbit, DHN-bot~enwiki, Modest Genius, GeorgeMoney, Fuhghettaboutit, Yevgeny Kats, JorisvS, Quaeler, MOBle, JRSpriggs, Arthurlo, Icek~enwiki, Xxanthippe, Michael C Price, Headbomb, AntiVandalBot, Luna Santin, Prolog, Cinnamon42, Gökhan, YK Times, Pervect, Magioladitis, Bongwarrior, DAGwyn, Lantonov, NewEnglandYankee, Policron, Sjwk, Sheliak, Oshwah, Alan Rockefeller, Keller dude4, Andy Dingley, Gabrielsleitao, Ddagit, Ggygreg, Ehud Lesar, Drschawrz, YohanN7, Yintan, Flyer22 Reborn, JerroldPease-Atlanta, BenoniBot~enwiki, Coldcreation, PerryTachett, Denisarona, ClueBot, Blanchardb, Itzguru, Estirabot, SchreiberBike, Jpsfitz, TimothyRias, Addbot, Gravitophoton, DOI bot, Landon1980, MrOllie, Download, Favonian, Vysotsky, Lightbot, OlEnglish, Zorrobot, Luckas-bot, Yobot, AnomieBOT, Ulric1313, Hunnjazal, Citation bot, ArthurBot, LilHelpa, Xqbot, NOrbeck, J04n, Armbrust, Omnipaedista, Crowsmen, Fres-coBot, Paine Ellsworth, Sławomir Biały, Citation bot 1, Wandering-teacher, RedBot, Full-date unlinking bot, Fredkinfollower, Trappist the monk, Le Docteur, Earthandmoon, Virtakuono, EmausBot, Netheril96, TuHan-Bot, Solomonfromfinland, Hhhippo, Cogiati, Thewhyman, H3llBot, Quondum, SporkBot, FinalRapture, Whatsthatcomingoverthehill, Maschen, Donner60, Xanchester, ClueBot NG, Jj1236, Rezabot, Widr, Euty, Helpful Pixie Bot, Bibcode Bot, Jwchong, Frze, F=q(E+v^B), Brad7777, BattyBot, David.moreno72, Twhitguy14, CuriousMind01, 1imking, Hublolly, Razibot, Rfassbind, I am One of Many, Melonkelon, Frinthruit, Endercreeper01, QueenFan, Wishva de Silva, DanielBenjaminH, BU Rob13, Jman0631, Efstabro, דוקטורגלי and Anonymous: 163

- **Anti-de Sitter space** *Source:* https://en.wikipedia.org/wiki/Anti-de_Sitter_space?oldid=713588067 *Contributors:* Edward, Patrick, Michael Hardy, Karada, Looxix~enwiki, Docu, Charles Matthews, Saltine, Phys, Pierre Boreal, BenRG, Giftlite, Lethe, Eequor, Steuard, Lumidek, Cacycle, Ben Standeven, Kevin Lamoreau, PWilkinson, Hooperbloob, Jérôme, Alai, Ceyockey, Linas, Mpatel, BD2412, Zbxgscqf, MarSch, R.e.b., Mathbot, BradBeattie, Chobot, Roboto de Ajvol, YurikBot, Hillman, Ohwilleke, Salsb, CharlesHBennett, SDS, SmackBot, Chris the speller, Bluebot, Moky~enwiki, Acgetchell, Yevgeny Kats, Ulner, Ekjon Lok, JorisvS, Hetar, Vanisaac, Harold f, CmdrObot, Myasuda, Meno25, Headbomb, Shambolic Entity, Roice3, Mickwilson20, R'n'B, Ross Fraser, Schucker, Kyle the bot, Wing gundam, Wmpearl, Alexbot, 1ForThe-Money, AnonyScientist, Addbot, AlexPenson, AnomieBOT, Point-set topologist, Sławomir Biały, I dream of horses, Thinking of England, Rausch, Lightlowemon, Jordgette, Nilock, WikitanvirBot, Slawekb, Quondum, Maschen, Posteinstein, Johnuio, T.seppelt, Jimw338, Khazar2, Alexander Chekmenev, Verdana Bold, Absolutelypuremilk and Anonymous: 41

- **Minkowski space** *Source:* https://en.wikipedia.org/wiki/Minkowski_space?oldid=709121334 *Contributors:* XJaM, William Avery, Stever-tigo, Patrick, PhilipMW, Michael Hardy, Tim Starling, Karada, Stupidmoron, Hawthorn, Charles Matthews, Kbk, Zoicon5, LMB, Phys, Fvw, Josh Cherry, Jheise, Marc Venot, Decumanus, Giftlite, Gene Ward Smith, BenFrantzDale, Lethe, MathKnight, Fropuff, Dratman, Anythingy-ouwant, Frau Hitt~enwiki, Hidaspal, Bender235, Syp, Rgdboer, I9Q79oL78KiL0QTFHgyc, Phils, Anthony Appleyard, NukWik, Egg, Ring-bang, Japanese Searobin, Stemonitis, Linas, StradivariusTV, Mpatel, Tlroche, Rjwilmsi, Zbxgscqf, R.e.b., Lionelbrits, Mathbot, Chobot, Dylan Thurston, DVdm, Hmonroe, YurikBot, Hairy Dude, Rhythm, Nick, Arthur Rubin, KasugaHuang, That Guy, From That Show!, Sardanaphalus, KnightRider~enwiki, SmackBot, Turbos10~enwiki, Vald, ZerodEgo, Xie Xiaolei, Bluebot, Alexwagner, Complexica, Waprap, Jbergquist, An-drei Stroe, Gregapan, Lambiam, Eliyak, Jim.belk, NongBot~enwiki, Bosons, Xenure, Loadmaster, Dan Gluck, JRSpriggs, CRGreathouse, Cyde-bot, Michael C Price, Thijs!bot, Martin Hogbin, Headbomb, D.H, Noclevername, Gökhan, JAnDbot, Jyotirmoyb, Sullivan.t.j, First Harmonic, Maurice Carbonaro, TomyDuby, Goutui, WaiteDavid137, Fylwind, Equazcion, XCelam, JohnBlackburne, Red Act, Nxavar, DWP17, PaulTa-nenbaum, Geometry guy, Zain Ebrahim111, StevenJohnston, YohanN7, Juanmantoya, Paradoctor, Happysailor, DaveBeal, Henry Delforn (old), Udirock, Mr. Stradivarius, Renata500, Martarius, ClueBot, RODERICKMOLASAR, Bastien Sens-Méyé~enwiki, Sun Creator, Carriearch-dale, Forbes72, Whizmd, Addbot, Mortense, Kwvan, Gregz08, 84user, Tide rolls, Cesiumfrog, Luckas-bot, Yobot, Amirobot, AnomieBOT, Sfaefaol, Illegal604, Corwin323, ArthurBot, Nanog, NOrbeck, RibotBOT, Ashi009, MeDrewNotYou, Paine Ellsworth, Sławomir Biały, Gui-tarstud101, Tsester, Serols, Stephen Henry Davies, Tcnuk, Trappist the monk, Retired user 0001, Dinamik-bot, John of Reading, WikitanvirBot, AlexUT, Netheril96, Hhhippo, Quondum, Thine Antique Pen, Maschen, JFB80, ClueBot NG, Jack Greenmaven, CaroleHenson, Helpful Pixie Bot, BG19bot, F=q(E+v^B), Aisteco, ChrisGualtieri, Deltahedron, Makecat-bot, Twhitguy14, TwoTwoHello, CsDix, Frinthruit, JaconaFrere, Monkbot, Biblioworm, WillemienH, KasparBot, Arnaud Dorthe, Baking Soda and Anonymous: 98

- **6D (2,0) superconformal field theory** *Source:* https://en.wikipedia.org/wiki/6D_(2%2C0)_superconformal_field_theory?oldid=717311251 *Contributors:* Rjwilmsi, Bibcode Bot, Polytope24, AHusain3141 and BD2412bot

- **ABJM superconformal field theory** *Source:* https://en.wikipedia.org/wiki/ABJM_superconformal_field_theory?oldid=616994359 *Contribu-tors:* Polytope24

- **Chern–Simons theory** *Source:* https://en.wikipedia.org/wiki/Chern%E2%80%93Simons_theory?oldid=698383625 *Contributors:* Michael Hardy, Charles Matthews, Phys, Giftlite, LeYaYa, Fropuff, Dmr2, Bender235, Joke137, BD2412, Wavelength, Gaius Cornelius, K.C. Tang, Doncram, Deville, JamieVicary, JarahE, JoeBot, Headbomb, Escarbot, Y Angora, Hamsterlopithecus, STBot, DrKay, Kevin Hickerson, Poli-cron, Sheliak, Kyle the bot, Temurjin, StevenJohnston, General Epitaph, Mild Bill Hiccup, SchreiberBike, Pichpich, Koumz, Addbot, TStein, Luckas-bot, Yobot, AnomieBOT, IRP, Xqbot, Sionus, False vacuum, FrescoBot, Quondum, Nemo47, Maschen, Aerthis, Raidr, KLBot2, Bib-code Bot, YFdyh-bot, Enyokoyama and Anonymous: 30

- **String cosmology** *Source:* https://en.wikipedia.org/wiki/String_cosmology?oldid=666337214 *Contributors:* Stevertigo, Phys, Nurg, Thincat, Lumidek, RJHall, I9Q79oL78KiL0QTFHgyc, Rjwilmsi, Iamfscked, SmackBot, Colonies Chris, Lambiam, Headbomb, Peter Gulutzan, Nono le petit robot~enwiki, Addbot, AkhtaBot, OlEnglish, Yobot, Citation bot, Pra1998, Citation bot 1, Tom.Reding, Bibcode Bot, Polytope24, Tetra quark and Anonymous: 5

- **String theory landscape** *Source:* https://en.wikipedia.org/wiki/String_theory_landscape?oldid=707702367 *Contributors:* Maury Markowitz, Sho Uemura, Fropuff, Eequor, I9Q79oL78KiL0QTFHgyc, Kusma, DV8 2XL, Ringbang, GregorB, BlaiseFEgan, Joke137, YurikBot, Bhny, Archelon, IAMTHEEGGMAN, Gadget850, SmackBot, Kurochka, Colonies Chris, CmdrObot, Mbell, Headbomb, Peter Gulutzan, Landscape~enwiki, Tim Shuba, Glen, Dr. Morbius, IgorSF, Jrcla2, Gbawden, Martarius, WurmWoode, Niceguyedc, Addbot, DOI bot, Cuaxdon, SeniorInt, AnomieBOT, Macquaire, Omnipaedista, Citation bot 2, Citation bot 1, Tom.Reding, Xaviertan, Ale And Quail, RA0808, ThePowerofX, Isocliff, Bibcode Bot, Polytope24, Tetra quark and Anonymous: 30

- **Calabi–Yau manifold** *Source:* https://en.wikipedia.org/wiki/Calabi%E2%80%93Yau_manifold?oldid=709351077 *Contributors:* Bryan Derksen, The Anome, Zoe, Hephaestos, Michael Hardy, TakuyaMurata, Docu, Disdero, Charles Matthews, The Anomebot, Phys, Owen, Pfortuny, Justo, Hadal, Tosha, Giftlite, Gtrmp, Lethe, MSGJ, Fropuff, Anville, Daibhid C, BesigedB, Quadell, Lumidek, Pjacobi, Dmr2, Bender235, Gauge, El C, Art LaPella, Euyyn, Deryck Chan, Ceyockey, Joriki, Isnow, Rjwilmsi, MarSch, SMC, R.e.b., Lionelbrits, John Baez, Mathbot, YurikBot, KSmrq, Giro720, Aaron Meyers, LeonardoRob0t, Ilmari Karonen, Lunch, SmackBot, Kurochka, Unyoyega, PrimeHunter, Nbarth, Colonies Chris, Chlewbot, Pissant, Acdx, Zchenyu, Jim.belk, Ddirac, JarahE, Dan Gluck, Mreplatinum, Olaf Davis, CBM, WLior, Juhachi, Thubsch, SlashDot, Headbomb, Mordent~enwiki, Steelpillow, Felix116, Jakob.scholbach, Migp, Zsejk, Silentaria, Cander0000, CommonsDelinker, Paulnwatts, FANSTARbot, Spiegelprime, Dorftrottel, Pleasantville, Enderminh, Anonymous Dissident, Geometry guy, Arcfrk, AlleborgoBot, JerrySteal, ReluctantPhilosopher, ClueBot, DragonBot, SpikeToronto, Toyotacorona, Addbot, Uruk2008, DOI bot, AkhtaBot, Zahd, Ronhjones, Tassedethe, Lightbot, Luckas-bot, Yobot, Tr!stan, AnomieBOT, Materialscientist, LilHelpa, MauritsBot, Xqbot, Jbourjai, Omnipaedista, Point-set topologist, Patchy1, Citation bot 1, Rausch, Gluino, ZéroBot, Aerthis, RockMagnetist, ClueBot NG, Zeugding, NabilMiri, Wilson868, Bibcode Bot, Blaspie55, Brad7777, Volons, Jeremy112233, Jimw338, Enyokoyama, Twhitguy14, Indiscipline046 and Anonymous: 70

- **Brane cosmology** *Source:* https://en.wikipedia.org/wiki/Brane_cosmology?oldid=694290677 *Contributors:* AxelBoldt, Camembert, Bth, Twilsonb, DIG~enwiki, AugPi, Pedant17, Rursus, David Gerard, Barbara Shack, Herbee, LeYaYa, Tomothy, H0riz0n, Pjacobi, El C, Constantine, I9Q79oL78KiL0QTFHgyc, Oliphaunt, Joke137, Strait, Algri, Eric B, YurikBot, Hairy Dude, Gaius Cornelius, Salsb, Bucketsofg, BOT-Superzerocool, Bmju, Closedmouth, Tim314, SmackBot, Kurochka, SSJemmett, TimBentley, Jprg1966, Dugodugo, Colonies Chris, Stevenmitchell, Savidan, Ligulembot, Yevgeny Kats, Quaeler, CapitalR, George100, Verdy p, Michael C Price, Headbomb, Escarbot, VictorAnyakin, Skylights76, Robin S, FlieGerFaUstMe262, HiEv, Afluent Rider, Vendrov, PlanetStar, James Banogon, Fjados, Masterpiece2000, MelonBot, Thinking Stone, Mixen Dixon, The Thin Man Who Never Leaves, Yobot, Materialscientist, Almabot, Louperibot, DrilBot, Full-date unlinking bot, Tkachyk, Solomonfromfinland, MerlIwBot, Bibcode Bot, Davidiad, Ownedroad9, PamFromMD2, Harsh 2580, Twhitguy15, E8xE8, Monkbot, BradNorton1979, Tetra quark, Srednuas Lenoroc and Anonymous: 29

- **Inflation (cosmology)** *Source:* https://en.wikipedia.org/wiki/Inflation_(cosmology)?oldid=713771276 *Contributors:* Bryan Derksen, The Anome, Diatarn iv~enwiki, Roadrunner, David spector, Hephaestos, Stevertigo, Edward, Nealmcb, Boud, Michael Hardy, Tim Starling, Dcljr, Cyde, Ellywa, William M. Connolley, Theresa knott, Jeff Relf, Mxn, Timwi, Rednblu, Bartosz, Pierre Boreal, Raul654, Chuunen Baka, Robbot, Gandalf61, Rursus, Ancheta Wis, Giftlite, Barbara Shack, Mikez, Lethe, Dratman, Curps, Jcobb, Gracefool, Just Another Dan, Andycjp, HorsePunchKid, Beland, Elroch, JDoolin, Burschik, Shadypalm88, Eep², Mike Rosoft, DanielCD, Noisy, Rich Farmbrough, FT2, Pjacobi, Luxdormiens, Dbachmann, Bender235, AdamSolomon, Pt, Worldtraveller, Art LaPella, Orlady, Drhex, Guettarda, I9Q79oL78KiL0QTFHgyc, Jeodesic, Rsholmes, Anthony Appleyard, Plumbago, JHG, Schaefer, EmmetCaulfield, Cgmusselman, Dirac1933, Oleg Alexandrov, Matevzk, Yeastbeast, StradivariusTV, BillC, Bluemoose, Wdanwatts, Joke137, Rnt20, Malangthon, Ketiltrout, Drbogdan, Rjwilmsi, Zbxgscqf, Mattmartin, Strait, Eyu100, Jehochman, Ems57fcva, Bubba73, FlaBot, Nihiltres, Itinerant1, Phoenix2~enwiki, Chobot, Hermitage, Bgwhite, YurikBot, Wavelength, Supasheep, Ytrottier, Gaius Cornelius, Anomalocaris, NawlinWiki, LiamE, Davemck, JonathanD, Enormousdude, 2over0, Arthur Rubin, Argo Navis, Physicsdavid, Profero, Luk, SmackBot, Haza-w, KnowledgeOfSelf, Lawrencekhoo, Onsly, Jdthood, Salmar, Jefffire, Hve, QFT, Vanished User 0001, Stevenmitchell, BIL, Lostart, Ligulembot, Yevgeny Kats, Byelf2007, Lambiam, J 1982, Rcapone, JorisvS, Heliogabulus, Dan Gluck, Spebudmak, JoeBot, UncleDouggie, Fsotrain09, Oshah, JRSpriggs, Chetvorno, Friendly Neighbour, Drinibot, Vanished user 2345, Brownlee, SuperMidget, Cydebot, BobQQ, Mortus Est, Cyhawk, Ttiotsw, Julian Mendez, Dr.enh, Michael C Price, Kozuch, LilDice, Thijs!bot, Headbomb, Z10x, Jklumker, Alfredr, Dawnseeker2000, Pollira, Rico402, Lfstevens, Gmarsden, JAnDbot, Olaf, LinkinPark, GurchBot, Magioladitis, Jpod2, Vanished user ty12kl89jq10, Rickard Vogelberg, Dr. Morbius, Bhenderson, TomS TDotO, Tarotcards, Wesino, Student7, Potatoswatter, Ollie 9045, Ja 62, Useight, Idioma-bot, Sheliak, Tokenhost, VolkovBot, ABF, ColdCase, Philip Trueman, TXiKiBoT, Calwiki, Thrawn562, Gobofro, SwordSmurf, Northfox, PaddyLeahy, SieBot, Wing gundam, OpenLoop, Likebox, Flyer22 Reborn, Mimihitam, Hockeyboi34, Lightmouse, Sunrise, Southtown, Hamiltondaniel, Epistemion, ClueBot, Niceguyedc, ChandlerMapBot, Jusdafax, ResidueOfDesign, Ploft, Scog, SchreiberBike, TimothyRias, Katsushi, MidwestGeek, Addbot, Roentgenium111, DOI bot, Blethering Scot, Ronhjones, Glane23, Deamon138, TStein, Barak Sh, Tassedethe, Zorrobot, Ben Ben, Legobot, Yinweichen, Luckas-bot, Yobot, Amirobot, Aldebaran66, Amble, Isotelesis, Magog the Ogre, AnomieBOT, Pyrrhon8, Rubinbot, Piano non troppo, Collieuk, Ulric1313, Citation bot, Xqbot, Plastadity, Capricorn42, P14nic997, False vacuum, Waleswatcher, Ignoranteconomist, Bigger digger, Chatul, 🐱🐱, CES1596, FrescoBot, Mesterhd, Paine Ellsworth, Schnufflus, Charles Edwin Shipp, Bbhustles, Ahnoneemoos, Pinethicket, Tom.Reding, Ganondolf, Σ, Aknochel, Mercy11, Trappist the monk, Jordgette, Wdanbae, Aabaakawad, Michael9422, CobraBot, Deathflyer, Mathewsyriac, EmausBot, Thucyd, GoingBatty, Wikipelli, Kiatdd, Italia2006, Werieth, ZéroBot, Chasrob, Wackywace, Chharvey, Bamyers99, Suslindisambiguator, AManWithNoPlan, RaptureBot, Maschen, HCPotter, Crux007, RockMagnetist, Whoop whoop pull up, ClueBot NG, J kay831, Law of Entropy, Supermint, Helpful Pixie Bot, Bibcode Bot, Lowercase sigmabot, BG19bot, Negativecharge, MSgtpotter, Badon, BML0309, Zedshort, Hamish59, Minsbot, BattyBot, SupernovaExplosion, ChrisGualtieri, JYBot, Rfassbind, Ikjyotsingh, Astroali, Lepton01, Pkanella, Chwon, Rolf h nelson, Comp.arch, Kogge, Hilmer B, Anrnusna, Stamptrader, Dodi 8238, Epaminondas of Thebes, Man of Steel 85, Abitslow, Monkbot, Accnln, BradNorton1979, YeOldeGentleman, Waters.Justin, Tetra quark, Isambard Kingdom, Sleepy Geek, Anand2202, Jmc76, Quasiopinionated, EnigmaLord515, Phseek, Trekkiepanda and Anonymous: 212

47.12.2 Images

- **File:AdS3.svg** *Source:* https://upload.wikimedia.org/wikipedia/commons/4/47/AdS3.svg *License:* CC BY-SA 3.0 *Contributors:* This file was derived from: AdS3 (new).png
Original artist:

- **File:Standard_Model_of_Elementary_Particles.svg** *Source:* https://upload.wikimedia.org/wikipedia/commons/0/00/Standard_Model_of_Elementary_Particles.svg *License:* CC BY 3.0 *Contributors:* Own work by uploader, PBS NOVA [1], Fermilab, Office of Science, United States Department of Energy, Particle Data Group *Original artist:* MissMJ

- **File:Standard_conf.png** *Source:* https://upload.wikimedia.org/wikipedia/commons/9/90/Standard_conf.png *License:* Public domain *Contributors:* Own work (Original text: *self-made*) *Original artist:* Gerd Kortemeyer (talk)

- **File:Star_polygon_9_4.png** *Source:* https://upload.wikimedia.org/wikipedia/commons/8/80/Star_polygon_9_4.png *License:* Public domain *Contributors:* http://upload.wikimedia.org/wikipedia/en/8/80/Star_polygon_9_4.png *Original artist:* Tomruen

- **File:Stephen_Hawking.StarChild.jpg** *Source:* https://upload.wikimedia.org/wikipedia/commons/e/eb/Stephen_Hawking.StarChild.jpg *License:* Public domain *Contributors:* Original. Source (StarChild Learning Center). Directory listing. *Original artist:* NASA

- **File:StringTheoryDualities.svg** *Source:* https://upload.wikimedia.org/wikipedia/commons/8/8a/StringTheoryDualities.svg *License:* CC BY-SA 3.0 *Contributors:*

- StringTheoryDualities.jpg *Original artist:*

- derivative work: Alex Dunkel (Maky)

- **File:Stylised_Lithium_Atom.svg** *Source:* https://upload.wikimedia.org/wikipedia/commons/e/e1/Stylised_Lithium_Atom.svg *License:* CC-BY-SA-3.0 *Contributors:* based off of Image:Stylised Lithium Atom.png by Halfdan. *Original artist:* SVG by Indolences. Recoloring and ironing out some glitches done by Rainer Klute.

- **File:Sub-atomic_particles.png** *Source:* https://upload.wikimedia.org/wikipedia/commons/9/9c/Sub-atomic_particles.png *License:* CC BY-SA 4.0 *Contributors:* Own work *Original artist:* Cjean42

- **File:Swiss-Commemorative-Coin-1979b-CHF-5-obverse.png** *Source:* https://upload.wikimedia.org/wikipedia/commons/f/fe/Swiss-Commemorative-Coin-1979b-CHF-5-obverse.png *License:* Public domain *Contributors:* http://www.swissmint.ch/upload/_pdf/produkte/d/GM_BILDERd07-CuNi.pdf *Original artist:* Kurt Wirth, Bern

- **File:Symmetricwave2.png** *Source:* https://upload.wikimedia.org/wikipedia/commons/1/1d/Symmetricwave2.png *License:* CC BY 3.0 *Contributors:* Own work *Original artist:* TimothyRias

- **File:Tesseract.gif** *Source:* https://upload.wikimedia.org/wikipedia/commons/5/55/Tesseract.gif *License:* Public domain *Contributors:* Own work *Original artist:* Jason Hise at English Wikipedia

- **File:Text_document_with_red_question_mark.svg** *Source:* https://upload.wikimedia.org/wikipedia/commons/a/a4/Text_document_with_red_question_mark.svg *License:* Public domain *Contributors:* Created by bdesham with Inkscape; based upon Text-x-generic.svg from the Tango project. *Original artist:* Benjamin D. Esham (bdesham)

- **File:Torus_cycles.png** *Source:* https://upload.wikimedia.org/wikipedia/commons/5/54/Torus_cycles.png *License:* CC-BY-SA-3.0 *Contributors:* ? *Original artist:* ?

- **File:Uniform_tiling_433-t0.png** *Source:* https://upload.wikimedia.org/wikipedia/commons/1/13/Uniform_tiling_433-t0.png *License:* Public domain *Contributors:* Transferred from en.wikipedia *Original artist:* Tomruen at en.wikipedia

- **File:Uniform_tiling_433-t0_(formatted).svg** *Source:* https://upload.wikimedia.org/wikipedia/commons/2/21/Uniform_tiling_433-t0_%28formatted%29.svg *License:* CC BY-SA 4.0 *Contributors:* This file was derived from: Uniform tiling 433-t0.png *Original artist:*

- derivative work: Polytope24

- **File:Wikibooks-logo-en-noslogan.svg** *Source:* https://upload.wikimedia.org/wikipedia/commons/d/df/Wikibooks-logo-en-noslogan.svg *License:* CC BY-SA 3.0 *Contributors:* Own work *Original artist:* User:Bastique, User:Ramac et al.

- **File:Wikiquote-logo.svg** *Source:* https://upload.wikimedia.org/wikipedia/commons/f/fa/Wikiquote-logo.svg *License:* Public domain *Contributors:* ? *Original artist:* ?

- **File:Wikisource-logo.svg** *Source:* https://upload.wikimedia.org/wikipedia/commons/4/4c/Wikisource-logo.svg *License:* CC BY-SA 3.0 *Contributors:* Rei-artur *Original artist:* Nicholas Moreau

- **File:Wikiversity-logo.svg** *Source:* https://upload.wikimedia.org/wikipedia/commons/9/91/Wikiversity-logo.svg *License:* CC BY-SA 3.0 *Contributors:* Snorky (optimized and cleaned up by verdy_p) *Original artist:* Snorky (optimized and cleaned up by verdy_p)

- **File:Wiktionary-logo-en.svg** *Source:* https://upload.wikimedia.org/wikipedia/commons/f/f8/Wiktionary-logo-en.svg *License:* Public domain *Contributors:* Vector version of Image:Wiktionary-logo-en.png. *Original artist:* Vectorized by Fvasconcellos (talk · contribs), based on original logo tossed together by Brion Vibber

- **File:Winding_Number_-1.svg** *Source:* https://upload.wikimedia.org/wikipedia/commons/1/18/Winding_Number_-1.svg *License:* Public domain *Contributors:* Own work *Original artist:* Jim.belk

- **File:Winding_Number_-2.svg** *Source:* https://upload.wikimedia.org/wikipedia/commons/9/92/Winding_Number_-2.svg *License:* Public domain *Contributors:* Own work *Original artist:* Jim.belk

- **File:Winding_Number_0.svg** *Source:* https://upload.wikimedia.org/wikipedia/commons/6/64/Winding_Number_0.svg *License:* Public domain *Contributors:* Own work *Original artist:* Jim.belk

- **File:Winding_Number_1.svg** *Source:* https://upload.wikimedia.org/wikipedia/commons/0/0e/Winding_Number_1.svg *License:* Public domain *Contributors:* Own work *Original artist:* Jim.belk

- **File:Winding_Number_2.svg** *Source:* https://upload.wikimedia.org/wikipedia/commons/a/aa/Winding_Number_2.svg *License:* Public domain *Contributors:* Own work *Original artist:* Jim.belk

- **File:Winding_Number_3.svg** *Source:* https://upload.wikimedia.org/wikipedia/commons/5/54/Winding_Number_3.svg *License:* Public domain *Contributors:* Own work *Original artist:* Jim.belk
- **File:Winding_Number_Animation_Small.gif** *Source:* https://upload.wikimedia.org/wikipedia/commons/a/ac/Winding_Number_Animation_Small.gif *License:* Public domain *Contributors:* Own work *Original artist:* Jim.belk
- **File:Winding_Number_Around_Point.svg** *Source:* https://upload.wikimedia.org/wikipedia/commons/8/8f/Winding_Number_Around_Point.svg *License:* Public domain *Contributors:* Own work *Original artist:* Jim.belk
- **File:World_line.svg** *Source:* https://upload.wikimedia.org/wikipedia/commons/1/16/World_line.svg *License:* CC-BY-SA-3.0 *Contributors:* Transferred from en.wikipedia.
  Original artist: SVG version: K. Aainsqatsi at en.wikipedia

- **File:World_lines_and_world_sheet.svg** *Source:* https://upload.wikimedia.org/wikipedia/commons/2/25/World_lines_and_world_sheet.svg *License:* Public domain *Contributors:* Point&string.png *Original artist:* Kurochka, svg version by Actam

47.12.3 Content license

- Creative Commons Attribution-Share Alike 3.0

www.ingramcontent.com/pod-product-compliance
Lightning Source LLC
Chambersburg PA
CBHW080651190526
45169CB00006B/2075